NEUROMETHODS ☐ 30

Regulatory Protein Modification

NEUROMETHODS

Program Editors: Alan A. Boulton and Glen B. Baker

NEUROMETHODS □ 30

Regulatory Protein Modification

Techniques and Protocols

Edited by

Hugh C. Hemmings, Jr., MD, PhD

Cornell University Medical College, New York, NY

Humana Press ✳ Totowa, New Jersey

For additional copies, pricing for bulk purchases, and/or information about other Humana titles, contact Humana at the above address or at any of the following numbers: Tel.: 201-256-1699; Fax: 201-256-8341; E-mail: humana@interramp.com

Cover design by Patricia F. Cleary.

ISBN 0-89603-415-1
ISSN 0893-2336
Printed in the United States of America. 10 9 8 7 6 5 4 3 2 1

Preface

Molecular signal transduction has become an exciting and productive area of research in all aspects of biology, especially in neuroscience. This volume of *Neuromethods* focuses on the posttranslational modification of proteins in neurons, an important element in the regulation of neuronal function. It emphasizes protein phosphorylation, the most prominent molecular mechanism involved in the regulation of protein function by extracellular and intracellular signaling pathways. Other emerging mechanisms of regulatory protein modification are also covered to provide a concise overview of the major mechanisms currently known to play important roles in cellular regulation. The techniques used to analyze various forms of posttranslational protein modification are described, along with current protocols, discussion of the methodological limitations, and relevant examples from recent publications. This collection should be of use to investigators of protein modification who work in both neuronal and nonneuronal systems, and is meant to provide comprehensive coverage in specific areas for experienced investigators as well as a starting point for those new to this burgeoning field.

The nervous system has proved to be a fertile area for elucidation of the intricate protein regulatory mechanisms used by cells. This is particularly evident in the field of protein phosphorylation, in which fundamental discoveries have been made following the pioneering work of Paul Greengard. The first six chapters address various techniques relevant to the study of protein phosphorylation in the nervous system. The essential methods required to analyze protein phosphorylation in intact cells, cell extracts, and synaptosomes are covered in the first three chapters. Chapters 1 and 2 summarize general techniques used to demonstrate the physiological regulation of protein phosphorylation and dephosphorylation and the specific enzymes involved. The more specialized techniques applicable to isolated nerve terminals (synaptosomes) as well as common electrophoretic and immunochemical methods, are covered in Chapter 3. A detailed review of protein kinase and phosphatase inhibitors is provided in Chapter 4. Since this subject has not been covered previously in such depth, this chapter should provide a valuable review for scientists in any field who employ these reagents. The recent application of phosphorylation state-specific antibodies to the analysis of protein phosphorylation, a powerful technique that has significantly advanced the field, is described in Chapter 5. Protein–tyrosine phosphorylation has been recognized only recently as an important regulatory mechanism in the nervous system. The techniques used to analyze this form of protein modification are reviewed in Chapter 6.

Site-directed mutagenesis is a technique of molecular analysis that has recently been applied to the identification and characterization of protein modification sites. This powerful technique, which is described in detail in Chapter 7, is applicable to both protein phosphorylation studies as well as to studies of other types of protein modification. The following four chapters summarize techniques for the analysis of neuronal protein modification by mechanisms other than phosphorylation. The technical approaches relevant to the study of protein methylation are described in Chapter 8. Methods for the analysis of protein S-palmitoylation, a recently recognized component of neuronal signal transduction, are reviewed in Chapter 9. The recent identification of neuronal ADP-ribosylation and relevant analytical techniques are discussed in Chapter 10. Finally, Chapter 11 covers the techniques used to study protein glycosylation and glycophosphatidylinositol anchoring, a mechanism of covalent modification that is currently receiving considerable attention in the regulation of plasma membrane proteins.

This volume should provide a useful reference and manual for molecular neuroscience researchers interested in cell signaling, cell biology, and neurochemistry. We anticipate that the juxtaposition of critical technical reviews and experimental protocols will prove to be a valuable resource both at the desk and at the bench.

Hugh C. Hemmings, Jr., MD, PhD

Contents

**Protein Phosphorylation and Dephosphorylation
in Isolated Nerve Terminals (Synaptosomes)**
Talvinder S. Sirha

Protein Kinase and Phosphatase Inhibitors:
Applications in Neuroscience
Hugh C. Hemmings, Jr.

Phosphorylation State-Specific Antibodies
Andrew J. Czernik, Jeffrey Mathers, and Sheenah M. Mische

Protein Tyrosine Phosphorylation
Pascal Derkinderen and Jean-Antoine Girault

Identification of Posttranslational Modification Sites by Site-Directed Mutagenesis
James A. Bibb and Edgar F. da Cruz e Silva

Protein Methylation in the Nervous System
Darin J. Weber and Philip N. McFadden

Long-Chain Fatty Acylation of Proteins
Sean I. Patterson and J. H. Pate Skene

Protein ADP-Ribosylation
Keith D. Philibert and Henk Zwiers

Glycosylation and Glycosylphosphatidylinositol Membrane Anchors
Anthony J. Turner, Edward T. Parkin, and Nigel M. Hooper

Contributors

JAMES A. BIBB • *Laboratory of Molecular and Cellular Neuroscience, The Rockefeller University, New York, NY*

ANDREW J. CZERNIK • *Laboratory of Molecular and Cellular Neuroscience, The Rockefeller University, New York, NY*

EDGAR F. DA CRUZ E SILVA • *Instituto de Biologia Experimental e Tecnológica, Oeiras, Portugal*

PASCAL DERKINDEREN • *Chaire de Neuropharmacologie, INSERM, Collège de France, Paris, France*

JEAN-ANTOINE GIRAULT • *Chaire de Neuropharmacologie, INSERM, Collège de France, Paris, France*

SHELLEY HALPAIN • *Department of Cell Biology, The Scripps Research Institute, La Jolla, CA*

HUGH C. HEMMINGS, JR. • *Departments of Anesthesiology and Pharmacology, Cornell University Medical College, New York, NY*

NIGEL M. HOOPER • *Department of Biochemistry and Molecular Biology, University of Leeds, UK*

JEFFREY MATHERS • *The Protein/DNA Technology Center, The Rockefeller University, New York, NY*

PHILIP N. MCFADDEN • *Department of Biochemistry and Biophysics, Oregon State University, Corvallis, OR*

SHEENAH M. MISCHE • *The Protein/DNA Technology Center, The Rockefeller University, New York, NY*

ANNE CARINE ØSTVOLD • *Neurochemical Laboratory, University of Oslo, Norway*

EDWARD T. PARKIN • *Department of Biochemistry and Molecular Biology, University of Leeds, UK*

SEAN I. PATTERSON • *Cátedra de Fisiologia Normal, Facultad de Ciencias Médicinas, Universidad Nacuional de Cuyo, Mendoza, Argentina*

KEITH D. PHILIBERT • *Endocrine Research Group, Departments of Medical Physiology and Medical Biochemistry, Faculty of Medicine, The University of Calgary, Alberta, Canada*

TALVINDER S. SIHRA • *Department of Pharmacology, Royal Free Hospital School of Medicine, University of London, UK*

J. H. PATE SKENE • *Department of Neurobiology, Duke University Medical Center, Durham, NC*

ANTHONY J. TURNER • *Department of Biochemistry and Molecular Biology, University of Leeds, UK*

S. IVAR WALAAS • *Neurochemical Laboratory, University of Oslo, Norway*

DARIN J. WEBER • *Department of Biochemistry and Biophysics, Oregon State University, Corvallis, OR*

HENK ZWIERS • *Endocrine Research Group, Departments of Medical Physiology and Medical Biochemistry, Faculty of Medicine, The University of Calgary, Alberta, Canada*

Analysis of Protein Phosphorylation in Intact Cells and Extracts

S. Ivar Walaas and Anne Carine Østvold

1. Introduction

The role of protein phosphorylation in regulating and integrating the activity of the nervous system has become widely appreciated during the last few decades. Both extracellular signals (e.g., neurotransmitters, hormones, growth factors, and electrical activity of the nerve cell itself) and intracellular signals and effecters (e.g., cyclic nucleotides, Ca^{2+} ions and phospholipid derivatives) mediate many of their effects by regulating the activities of protein phosphorylation systems (Walaas and Greengard, 1991). A variety of methods have been developed to investigate these systems (Rudolph and Krueger, 1979; Palfrey and Mobley, 1987), many of which have been recently described in detail (Hunter and Sefton, 1991a,b). Here we will describe a number of well-established methods that have been used to investigate protein phosphorylation systems in intact cells and extracts from neural tissue, with emphasis placed on methods suitable for studies on vertebrate preparations. Complete coverage of the literature is not intended, and only a limited number of methods that have proven useful in our laboratory will be presented.

Cellular protein phosphorylation systems consist minimally of three components, all of which deserve analysis. These include the substrate phosphoproteins themselves, which may change their biological properties following

From: *Neuromethods, Vol. 30: Regulatory Protein Modification: Techniques and Protocols*
Ed: H. C. Hemmings, Jr. Humana Press Inc.

changes in their state of phosphorylation. Second, a large number of protein kinases catalyze the incorporation of phosphate from the terminal phosphate of ATP into serine, threonine, tyrosine, and possibly histidine residues (Hunter, 1991). A large number of these enzymes exist in brain (Nairn et al., 1985). Finally, a distinct set of phosphoprotein phosphatases appear to be responsible for dephosphorylation of the various phosphorylated residues on substrate proteins (Nairn and Shenolikar, 1992; Cohen, 1994). Studies on protein phosphorylation systems in intact tissues and cells would ideally include an examination of the levels, the cellular and intracellular localizations, and the biochemical characteristics and state of activity of each of these components.

2. Investigating Protein Phosphorylation Systems

2.1. General Considerations

The classical approach to measurement of protein phosphorylation employs either intact tissue and cells or cell-free preparations. The former approach includes preincubation of intact cells with [^{32}P]orthophosphate to label intracellular ATP pools, followed by stimulation of the tissue and solubilization in a denaturing buffer. The latter approach includes incubation of cell-free preparations with [γ–^{32}P]ATP under conditions in which specific protein kinases are activated, in the absence or presence of added substrate proteins, followed by solubilization of proteins. The proteins are then analyzed by gel electrophoresis. These methods should be viewed as complementary and should be used in combination, since questions raised by one might be resolved by the other approach.

2.2. Experimental Pitfalls

Protein phosphorylation is a highly dynamic phenomenon, and investigators should obtain a clear understanding of possible pitfalls inherent in the different experimental approaches before selections between different preparations and experimental conditions are made. One should consider whether one wants to examine identified or unknown phosphoproteins, and whether one wants to examine distinct phosphorylation sites, phosphorylation stoichiometry, or phosphoamino acid composition of individual proteins. Separate approaches will have to be used for examination of protein kinase or protein phosphatase activities. In crude systems containing intact cells or organelles, typical problems include unintended disturbances in plasma membrane ion and phosphate transport, changes in mitochondrial function, and unknown changes in the equilibration of various ATP pools inside cells (Rudolph et al., 1978; Hopkirk and Denton, 1986). Moreover, one must be aware of the potential for unexpected (e.g., Ca^{2+}-induced) changes in protein synthesis/degradation, which may occur during incubation.

3. Phosphorylation of Proteins in Intact Preparations

3.1. General Considerations

Because of the extreme complexity and specificity of the anatomical arrangements of neural systems, preparations containing individual neurons *in situ* would appear attractive for physiological and functional studies on neuronal protein phosphorylation. However, since subsequent biochemical analysis often depends on tissue disruption, this complexity is also a potential problem, since it might be difficult to ascertain in which type of cell the observed change took place.

The experimental preparations most frequently employed present considerable differences in levels of complexity, ease of manipulations, and types of questions that can be answered. This section briefly reviews basic properties and protocols for using some of these systems.

3.2. Intact Brain

3.2.1. General Considerations

Intact brain has only infrequently been used for investigations of protein phosphorylation. Although such studies might give some indications into how these systems are working in the living animal, the widespread electrical activity, multitude of signal molecules being released, interactions with other cells taking place simultaneously or sequentially, and receptor isoforms giving rise to distinct cellular responses despite identical primary signals, combine to make interpretations of such experiments difficult. However, some reports have clearly demonstrated the feasibility of studying in vivo regulation of specific protein phosphorylation systems in intact rodent brain (Strombom et al., 1979; Lewis et al., 1990; Haycock and Haycock, 1991).

3.2.2. Labeling and Stimulation of Intact Brain

Labeling of proteins in intact brain has been achieved by direct infusion of [^{32}P]orthophosphate into the brain. After a suitable period, which allows for ATP synthesis and incorporation of ^{32}P into proteins, experimental manipulations are carried out. This has been followed by denaturation and solubilization of proteins in SDS, one-dimensional or two-dimensional electrophoretic separations of proteins, and autoradiographic analysis of phosphoproteins (Mitrius et al., 1981; Agrawal et al., 1982; Rodnight et al., 1985; Haycock and Haycock, 1991).

Certain methods appear suitable for stimulation of intact brain. A wide variety of physiological or pharmacological treatments can be achieved by systemic (e.g., ip, iv) or local (intraventricular or stereotaxic, for example, by microdialysis) application of drugs. Electrical stimulation of identified fiber tracts, e.g., the perforant path from the entorhinal cortex to the dentate gyrus of the hippocampus (Akers et al., 1986) or the medial forebrain bundle carrying dopamine fibers to the caudate-putamen (Haycock and Haycock, 1991), has also been used.

The method used to terminate and inactivate protein phosphorylation in intact brain should be selected from the different demands of the relevant components. Rapid decapitation followed by cooling and dissection of the brain region of interest may be used on systems that do not change rapidly postmortem. Usually more demanding methods will have to be employed (McIlwain, 1975), such as rapid cooling in liquid nitrogen followed by brain dissection of frozen sections (Palkovits, 1973). Alternatively, direct application of liquid nitrogen inside the skull may sometimes be used (Haycock and Haycock, 1991). Irreversible protein inactivation can be achieved by focused microwave irradiation directed toward the head (Guidotti et al., 1974). This method heat-inactivates all proteins within seconds and allows brain dissection (O'Callaghan et al., 1983). However, in our experience, some proteins may resist solubilization from such heat-treated brain and this method might be more suitable for analysis of the phosphorylation state of heat-stable phosphoproteins (Lewis et al., 1990).

Protocol 1—Labeling of proteins with [^{32}P]orthophosphate in intact rat brain:

1. Anesthetize and place rats in stereotaxic frame and introduce guide cannulae at desired locations.
2. Connect infusion cannula to the pump, and set for infusion rate of 0.1–0.5 μL/min.
3. Load infusion cannula with predetermined volumes of a dye solution, air bubble, cold Tris-saline (TBS), and [^{32}P]orthophosphate (approx 10 Ci/mL, carrier-free in water), which has been dried and redissolved in TBS, to give approx 0.1 mCi/μL. Total isotope volume should usually be in the range of 3–5 μL (Rodnight et al., 1985).
4. Introduce tip of infusion cannula into desired brain location and infuse [^{32}P]orthophosphate for 20–60 min (e.g., infuse 2 μL at 0.1 μL/min). If desired, treat animal with drugs or electrical stimulation.
5. Terminate reaction by decapitation and rapid freezing of head or by direct introduction of liquid N_2 into skull (Haycock and Haycock, 1991). If necessary, store at –20°C until dissection.
6. Keep brain below –10°C and dissect brain areas of interest (Palkovits, 1973; Shahar et al., 1989). Solubilize tissue by tissue grinders; or sonicate in a buffer suitable for phosphoprotein analysis.

3.3. Brain Slices

3.3.1. General Considerations

Brain slices of different stages of anatomical intactness, as well as other dissected neuronal preparations (e.g., the superior cervical ganglion, the pituitary stalk-posterior pituitary preparation), have been examined extensively. Such preparations retain many important anatomical relations, but still allow more specific questions to be addressed than are possible by the study of intact brains. However, problems with dissected preparations, particularly with slices, include their less than complete structural preservation; their sensitivities to hypoxia, and their limited survival times (Garthwaite et al., 1979; Misgeld and Frotscher, 1982; Lipton, 1985; Schurr et al., 1987). Moreover, these preparations usually need to be preincubated for prolonged periods of time to "reset" physiological activities, such as receptor-regulated second messenger levels, ATP/ADP ratios, and phosphocreatine levels (Whittingham et al., 1984; Lipton, 1985; Whittingham, 1987), as well as the state of phosphorylation of various phosphoproteins (Forn and Greengard, 1978). Since these phenomena may be regulated differently in distinct brain regions, preliminary investigations into such problems are clearly important when embarking on new experimental systems. The following is a brief list of possible improvements that might be tried when working with a difficult preparation.

To avoid effects caused by chemical postmortem destruction (e.g., acidosis), transcardial perfusion with cold, well-oxygenated Krebs-Ringer buffer performed under light anesthesia before brain removal and dissection might be tried. Successful perfusion will give a white, cold, and firm brain that is easily handled and gives excellent slices. Improvement in cutting techniques may be obtained by avoiding mechanical "chopping," and instead using handcutting (McIlwain, 1975; Garthwaite et al., 1979) or vibrating microtomes. The composition of the artificial cerebrospinal fluid used during preincubation and incubation may also be varied, and inclusions of vitamins, amino acids, and extracellular proteins have been suggested (Lipton, 1985; Ames and Nesbett, 1981). Variations in Ca^{2+} levels may also be helpful, with improvements in slice characteristics sometimes following low concentrations of Ca^{2+} during extended parts of the preincubation. Since nonfunctional, leaky cells will be a source of intracellular proteolytic enzymes, it may also be worthwhile to examine the effects of various protease inhibitors.

In addition to these precautions, the selection of a suitable brain region from which to prepare the slices (Shahar et al., 1989) remains an important decision. One should carefully evaluate the anatomical arrangements and physiology of cells and fibers present, as well as the experimental modulations to be used in the system.

3.3.2. Preparation and Stimulation of Brain Slices

Preparations from brain or the peripheral nervous system may be stimulated by both electrical and pharmacological methods. Examples of preparations for which electrical activation of incoming fibers have been used to mimic normal physiological activity include the superior cervical ganglion (Nestler and Greengard, 1982) and the pituitary stalk-posterior pituitary preparation (Tsou and Greengard, 1982). Although such methods are potentially powerful, it is important to activate a sufficient number of fibers in these preparations in order to allow biochemical signals to become detectable above background. Rapid inactivation, employing freezing, boiling, and/or mechanical disruption in the presence of inactivating agents like the strong ionic detergent sodium dodecyl sulfate (SDS), is necessary when terminating these reactions.

Protocol 2—Protein phosphorylation in brain slices:

1. Dissect regions of interest from rat brains and place in 10–20 mL vials containing 5–10 mL ice-cold Krebs-Ringer bicarbonate medium (KRB) containing (final conc.): 125 mM NaCl, 3.5 mM KCl, 1.0 mM $CaCl_2$, 1.5 mM KH_2PO_4, 1.5 mM $MgSO_4$, 25 mM $NaHCO_3$, 10 mM glucose, pH 7.4, previously gassed with 95% O_2/5% CO_2.

2. Slice tissue at 4°C (preferably in a cold room) with a tissue slicer set at 0.3–0.4 mm. Return slices to KRB and swirl loosely to separate individual slices. Carefully aspirate buffer, add 10 mL fresh, Ca^{2+}-free KRB, and preincubate slices at 30°C for 30 min with continuous oxygenation in a shaking water bath.

3. After 30 min, aspirate medium, add Ca^{2+}-containing KRB, and continue preincubation for 30–90 min. When proteins are to be labeled with ^{32}P, phosphate-free KRB containing carrier-free [^{32}P]orthophosphate (0.5–1.0 Ci/mL) is used, each vial is purged with O_2:CO_2 (95:5, v/v), capped and incubated for 90 min with gentle shaking.

4. Initiate incubation by addition of relevant drug or medium.

5. Terminate reaction by rapid aspiration of the medium, and freeze slices in the vials by adding liquid N_2. Processing of tissue then depends on the specific variable to be investigated.

3.4. Isolated Cells and Nerve Terminals

3.4.1. General Considerations

Preparations containing isolated cells have become powerful tools in the analysis of protein phosphorylation systems at the single-cell level (Garrison, 1983; Saneto and De Vellis, 1987; Shahar et al., 1989; Banker and Goslin, 1991). Such preparations avoid the cellular and morphological complexity of intact neuronal tissue and allow one to examine homogenous populations of cells, all presumably with identical properties and responses to physiological stimuli. One should also note recent developments in molecular biological methods that allow targeted manipulations of gene expression in isolated cells and therefore allow examination of aspects of protein function regulated by phosphorylation with a high degree of precision. An example is the regulation of transmitter receptor phosphorylation following expression of these receptor genes in nonneuronal cells (Moss et al., 1993).

The major criteria used for selection of cell type include viability of the cell, responsiveness to relevant extracellular physiological stimuli and ability of the cells to perform the process under investigation. However, these cells frequently are neoplastic and transformed, and often are far from normal in terms of neuritis and synaptic organization. This might make the interpretations of results and their significance difficult.

Given the spatial separation of axonal terminals from nerve cell bodies, it is not surprising that nerve terminals, despite their lack of protein synthesis ability, in many ways are self-supporting, independent entities. Isolated nerve terminals (synaptosomes) are useful for the study of many presynaptic events in vitro (Whittaker, 1969; Bradford, 1975; Gordon-Weeks, 1987). Synaptosomes can take up ^{32}P-labeled inorganic phosphate, incorporate this into ATP, and label physiological phosphoproteins. The advantages and disadvantages of synaptosomal studies are further described by Sihra in Chapter 3.

Isolated cells or nerve terminals usually are stimulated by chemical methods. Depolarization of plasma membranes may be achieved by use of depolarizing agents like high concentrations of KCl (Forn and Greengard, 1978) and K^+-channel blockers like 4-aminopyridine (Sihra et al., 1992). These agents will act on all plasma membranes in the preparation, and synaptic specificity is therefore lost. However, they mimic certain aspects of both action potentials and excitatory postsynaptic potentials (Nicholls, 1994), and have allowed interesting neuronal mechanisms to be studied (Coffey et al., 1993).

Receptor-mediated activation or inhibition of physiological processes by pharmacological means are useful for identification of physiological messengers and receptor types which are involved in regulating intracellular protein phosphorylation systems. Some examples include identification of adrenoceptive and dopaminoceptive receptors regulating phosphorylation of the synaptic vesicle protein synapsin I in brain regions and the posterior pituitary (Mobley and Greengard, 1985; Treiman and Greengard, 1985; Walaas et al., 1989a).

Modulating levels of intracellular messengers may be achieved by compounds activating synthesis or decreasing degradation of these messengers. Examples include forskolin, which activates the catalytic domain of adenylyl cyclase, and phosphodiesterase inhibitors, which inhibit the breakdown of cyclic AMP (cAMP), such as isobutylmethylxanthine (Forn and Greengard, 1978). Alternatively, analogs that directly modulate the phosphorylation systems to be examined can be used. Examples include thiosubstituted analogs of cyclic nucleotides, which directly activate or inhibit the relevant cyclic nucleotide-dependent protein kinases (Gjertsen et al., 1995), and tumor-promoting phorbol esters, which mimic the action of diacylglycerols on the protein kinase C (PKC) family (Castagna et al., 1982). Finally, modulation of protein phosphatases may be achieved by using membrane-permeant phosphatase inhibitors (e.g., okadaic acid, calyculin A, cyclosporin A) (*see* Chapters 2 and 4).

3.4.2. Labeling and Stimulation of Isolated Cells and Nerve Terminals

Examination of specific radioactivity in cellular proteins or ATP has previously been recommended in order to determine suitable preincubation times and facilitate interpretation (Rudolph et al., 1978; Palfrey and Mobley, 1987). However, given the multitude of intracellular pools and different phosphate turnover rates on proteins in distinct cellular compartments, these studies would seem unnecessary.

Protocol 3—Protein phosphorylation in isolated cells and nerve terminals:

1. Grow cells or isolate nerve terminals by standard methods (Gordon-Weeks, 1987; Banker and Goslin, 1991).
2. Remove media by aspiration or centrifugation, add phosphate-free medium containing [^{32}P]orthophosphate (usually 0.5–2 Ci/mL) and incubate at 25–37°C for 30–120 min to allow the intracellular ATP pool to become labeled.
3. Remove medium, add fresh nonradioactive medium, and divide samples into aliquots, each receiving differ-

ent treatments. Incubate at optimal temperature for various time periods.

4. Terminate incubation either by precipitation (e.g., using 10% [w/v] trichloroacetic acid [TCA] or 50–90% [v/v] acetone), or by addition of a denaturing buffer suitable for protein separations and isolation.

4. Phosphorylation of Proteins in Cell-Free Preparations

4.1. General Considerations

Most protein phosphorylation systems in the central nervous system (CNS) were originally discovered in cell-free preparations, in which incubations can be performed under defined conditions, and the absence of diffusion barriers allows enzymes, substrates, and cofactors to interact (Wiegant et al., 1978; Walaas et al., 1983). In particular, use of radiolabeled ATP (usually employing the strong β-emitter ^{32}P) makes this approach suitable for partial characterization of previously unknown protein phosphorylation systems. (In this context, one should note that phosphotyrosine, which usually constitutes a minute fraction of total protein-bound phosphoamino acid, requires distinct methods; *see* Chapter 6). Such experiments will demonstrate the capacity of the preparation to phosphorylate and/or dephosphorylate unknown or identified proteins; conversely, the ability of tissue proteins to become phosphorylated can be demonstrated. Preliminary characterization of unknown ^{32}P-labeled phosphoproteins may include species and tissue expression as well as cellular and subcellular distributions, combined with biochemical characterization of phosphorylation sites and phosphoamino acid determination.

The limitations inherent in these preparations should be kept in mind. The cellular architecture is lost, the biochemical composition of the medium surrounding the protein phosphorylation systems becomes unphysiological, and the spatial and functional relations among enzymes, substrates, and regulators become distorted. Such artifacts make the physiological interpretation of the results difficult. In particular, physiological time-courses and proofs of phosphorylation of specific proteins in intact cells are not possible to obtain from these experiments (*see also* Palfrey and Mobley, 1987).

4.2. Labeling and Stimulation of Cell-Free Preparations

Investigation of cell-free protein phosphorylation systems necessitates selection of suitable and appropriate tissue fractions as well as incubation conditions (buffers, pH, ionic compositions, activating agents such as Ca^{2+} or cyclic nucleotides).

Protocol 4—Phosphorylation of proteins in vitro:

1. Dissect tissues (e.g., brain regions, peripheral nerves) and homogenize in 10 vol of a suitable medium, e.g., cold 10 mM Tris-HCl, pH 7.4, 2 mM EDTA, 1 mM dithiothreitol (DTT), and protease inhibitors like aprotinin (50 U/mL) and 0.1 mM phenylmethylsulfonyl fluoride (with leupeptin, antipain, pepstatin A, and chymostatin also added as deemed necessary).
2. Separate the homogenate into subcellular fractions by centrifugation (Ueda et al., 1979; Walaas et al., 1983). Resuspend particulate fractions by rehomogenization in original buffer.
3. Prepare aliquots (10–100 µg protein) for in vitro phosphorylation by mixing samples with ice-cold medium containing (final conc.) 20 mM HEPES, pH 7.4, 10 mM MgCl$_2$, 1 mM EDTA, 1 mM EGTA, 1 mM DTT, and store on ice.
4. Examine specific protein phosphorylation systems by addition of specific effecters. For example, cAMP-dependent protein kinase may be activated by addition of 1 µM 8-bromo cAMP; Ca^{2+}-calmodulin-dependent protein kinases may be activated by addition of 1.5 mM CaCl$_2$ and 1.5 µM calmodulin; and Ca^{2+}-diacylglycerol-phospholipid-dependent protein kinase (PKC) may be activated by addition of 1.5 mM CaCl$_2$, 50 µg/mL phos-

phatidylserine, and 10 nM phorbol-12-myristate 13-acetate. In the latter case, a stock solution of phosphatidylserine is prepared by dissolving the lipid in a small volume of chloroform, drying it under nitrogen, resuspending this to the proper stock concentrations in 20 mM Tris-HCl, pH 7.5, and preparing micelles by bath sonication for 5 min under nitrogen at 0°C. This stock solution should be stored in the dark at –20°C.

5. Preincubate mixture for 30–90 s at 30°C. Initiate reaction by addition of [γ-^{32}P]ATP (final conc. 1–10 µM, approx 1 µCi/assay; final vol 50–100 µL). Terminate reaction after 10–120 s by addition of stop solution, which can inactivate protein kinases, phosphatases, and proteases.

5. Analysis of Phosphoproteins

5.1. General Considerations

Phosphoproteins derived from the different preparations described above must be separated, characterized, and, if possible, identified. A variety of distinct methods are available for these purposes, detailed descriptions of which are outside the scope of this chapter. Rather, we will review some standard methods that have been found to be of general utility. For further information, *see,* e.g., Walker (1984), Darbre (1986), Harris and Angal (1989), Hames and Rickwood (1990), Hunter and Sefton (1991a,b).

5.1.1. Phosphoprotein Quantitation

In order to compare samples containing ^{32}P-labeled proteins reliably, aliquots of samples that contain equal amounts of the protein under study should be employed. Use of methods that allow specific quantitation of the protein (e.g., immunoblotting) therefore represents the preferred approach. However, since such methods often are not available, the total amount of protein present or the total amount of TCA- and alcohol-insoluble ^{32}P in the sample (most of which is incorporated into protein) has frequently been used as the reference value.

Protocol 5—Determination of ^{32}P-incorporation in intact cells and tissue:

1. Following labeling with ^{32}P and freezing, add 0.5 mL hot 1% (w/v) SDS and disrupt tissue by sonication. If necessary, transfer contents to glass-glass homogenizers and rehomogenize.
2. Determine protein content using the bicinchoninic acid method (Smith et al., 1985).
3. Measure TCA-/organic solvent-insoluble radioactivity by spotting 5–10 μL aliquots of sample onto GF-C filter paper that has been divided into 1 × 1 cm squares to form a grid.
4. Soak paper in 10% (w/v) TCA for 15 min. Repeat three times, each for 5 min. Rinse paper in 95% (v/v) ethanol for 5 min, dry paper, cut out squares, and measure ^{32}P content by Cerenkov or liquid scintillation counting.
5. Adjust samples to contain equal concentration of insoluble ^{32}P radioactivity and prepare aliquots for electrophoretic separations (Hames and Rickwood, 1990), immunoprecipitations, or other methods.

5.1.2. Phosphoprotein Separation

Electrophoresis of proteins in polyacrylamide gels remains unsurpassed as a general method for protein separation in terms of simplicity, resolution and reliability (Hames and Rickwood, 1990). Performing electrophoresis in the presence of the strong ionic detergent SDS has the added advantage that subunits of polymeric proteins are dissociated and proteins become denatured and inactivated. Although many electrophoresis buffers have been suggested, buffers containing SDS-Tris-chloride-glycine (Laemmli, 1970) appear most suitable for separation of proteins in the range of 15–300 kDa molecular mass, whereas buffers containing SDS-Tris-Tricine (Schägger and von Jagow, 1987) appear preferable for separation of proteins and peptides in the range of 2–50 kDa molecular mass. For further details about these methods, the reader is directed to general descriptions of biochemical separation methods (Walker 1984; Hames and Rickwood, 1990) and to Chapter 3.

The number of proteins that can be separated by one-dimensional electrophoresis remains limited. Introduction of two-dimensional electrophoresis methods allowed several hundred individual proteins to be detected on a single slab gel, particularly when using the methods employing either isoelectric focusing (IEF) or nonequilibrium pH-gradient gel electrophoresis (NEPHGE) (O'Farrell, 1975; O'Farrell et al., 1977) in the first dimension. Both methods rely heavily on urea to solubilize proteins prior to separation in the first dimension, and several particulate proteins remain difficult to handle (Kelly et al., 1985). An alternative method developed for membrane proteins, which includes SDS in the buffers used for both dimensions (Imada and Sueoka, 1980), represents an efficient way of solubilizing and separating several difficult neuronal phosphoproteins (Wang et al., 1988). However, this method does not possess the resolving power of the methods based on electrofocusing. Two-dimensional separation methods remain extremely time- and labor-consuming, and should, therefore, in our opinion, only be used for special purposes. For further information, the reader is directed to special publications dealing with these problems (Celis and Bravo, 1984; Hochstrasser et al., 1988; Dunbar et al., 1990) and to Chapter 3.

5.1.3. Protein Isolation by Immunomethods

The complexity of tissue preparations, which makes electrophoresis insufficient for study of individual proteins, can often be overcome by partial purification of the protein(s) in question. When a suitable antibody is available, a variety of immunoadsorption methods can be used to isolate the protein from highly complex mixtures (Harlow and Lane, 1988). For these purposes, proteins are solubilized in suitable lysis buffers, with compositions depending on the nature of tissue. For soluble proteins, a simple hypotonic medium will suffice, whereas preparations derived from cytoskeletal or membrane components may necessitate mixtures of chaotropes, strong detergents, and salts. Some of these components may then have to be removed in order to allow anti-body–antigen interactions to occur. A well-tested approach utilizes solubilization of proteins in media containing hot 1% (w/v) SDS, followed by addition of a fivefold excess of non-ionic detergents (e.g., Triton X-100) before the immunoreaction takes place (Goelz et al., 1981). For phosphoproteins, boiling in SDS has the added advantage of rapidly inactivating protein kinases, phosphatases, and proteases. For further information about these methods, the reader is referred to specialist literature (Harlow and Lane, 1988) and to Chapters 3 and 6.

5.1.4. Analysis of Phosphorylation Sites

Many proteins can be phosphorylated on multiple amino acid residues by distinct protein kinases. In the case of ^{32}P-labeled proteins, a classical approach for distinguishing between these phosphorylation sites has been to use phosphopeptide mapping. Reproducible cleavage of proteins in polyacrylamide gel pieces can be obtained with proteases or chemical treatments, e.g., trypsin, thermolysin, or the *Staphylococcus aureus* V8 protease (*see*, e.g., Mahboub et al., 1986; Wilkinson, 1986), or chemical treatment with cyanogen bromide, which cleaves proteins at methionine residues (Fontana and Gross, 1986; Jahnen et al., 1990). Similar approaches can be used on proteins adsorbed to polyvinylidene diflouride (PVDF) membranes (Scott et al., 1988).

To emphasize the importance of these approaches, we present one-dimensional and two-dimensional phosphopeptide mappings of proteins that are phosphorylated on multiple sites. A comparison between the neuronal protein synapsin I and the IP$_3$-receptor from rat cerebellum, both phosphorylated by cAMP-dependent protein kinase in vitro, is shown in Fig. 1. In this experiment, a method for one-dimensional peptide mapping based on incomplete proteolysis (Cleveland et al., 1977) as described by Huttner and Greengard (1979) was used (SAP-MAP; *see* Chapter 3). Gel pieces were cut out from the original gels, reswollen, inserted into the wells of a second SDS-PAGE gel, and overlayed

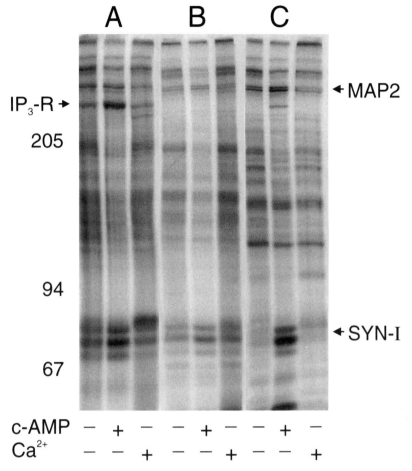

Fig. 1. Autoradiogram showing in vitro phosphorylation of particulate proteins in rat brain (*see* Protocol 4). Crude particulate preparations from rat cerebellum **(A)**, pons-medulla **(B)**, and neostriatum **(C)** were incubated for 10 s at 30°C with [γ-^{32}P]ATP in the absence or presence of 2 μM cAMP plus 1 mM isobutylmethylxanthine (cAMP), or 1.5 mM CaCl$_2$ and 1 mM EGTA (Ca^{2+}) (Walaas et al., 1983). The reactions were terminated by addition of SDS-containing "sample buffer," proteins were separated by SDS-PAGE on gels containing 7.5% polyacrylamide, and phosphoproteins were visualized by autoradiography. Identified proteins include: IP$_3$-R, inositol 1,4,5-trisphosphate receptor; MAP2, microtubule-associated protein-2; SYN-I, synapsin I. Apparent molecular mass is indicated on the left in kilodaltons. Only the part of the gel containing proteins above 55 kDa is shown (Walaas, S. I., unpublished).

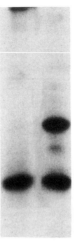

A   B

Fig. 2. Autoradiogram showing one-dimensional phosphopeptide mapping of synapsin I from neostriatum **(A)** and the IP_3 receptor from cerebellum **(B)** following phosphorylation with cAMP-dependent protein kinase. Pieces from the gel shown in Fig. 1 containing the proteins were cut out, reswollen, and subjected to incomplete proteolytic digestion with the *S. aureus* V8 protease during electrophoresis (Huttner and Greengard, 1979; Walaas, S. I., unpublished).

with a solution containing 5 µg/lane of the V8 protease from *S. aureus* (Sigma, St. Louis, MO). Proteolysis occurring during electrophoresis through the stacking gel at low voltage (e.g., 50–60 V) generated phosphopeptides, that could be separated through normal electrophoresis in the separating gel. As shown in Fig. 2, synapsin I generated one, whereas the neuronal IP_3-receptor generated two phosphopeptides. This accurately reflects the presence of one phosphorylation site for cAMP-dependent protein kinase in synapsin I and two in the IP_3-receptor, respectively (Huttner and Greengard, 1979; Weeks et al., 1988; Ferris et al., 1991).

One-dimensional mapping of peptides derived from incomplete proteolysis is frequently not sufficient to detect all phosphorylation sites in a protein, and complete digestion of phosphoproteins must be performed. Figure 3 demonstrates this approach (*see also* Chapter 3), using as an example the nuclear phosphoprotein P1, a protein that can be phosphorylated by a variety of protein kinases in vitro (Østvold et al., 1985; Walaas et al., 1989b; Meijer et al., 1991). Following prelabeling of HeLa cells or hepatocytes (Fig. 3) with [^{32}P]orthophosphate (Protocol 3), the protein was subjected to complete digestion with thermolysin and phosphopeptides separated by two-dimensional thin-layer electrophoresis and chromatography as described (Walaas and Nairn, 1989). Note the unusually large number of distinct phosphorylation sites found in proliferating cells.

Finally, the importance of phosphoamino acid analysis, which sometimes may suffice for identification of the regulated phosphorylation site (Halpain et al., 1990), should also be stressed. Such analysis may be performed on phosphoproteins in gels, on proteolytic phosphopeptides eluted from gels, or on phosphoproteins blotted onto PVDF membranes (Hildebrandt and Fried, 1989; Kamps, 1991).

5.1.5. Analysis of the State of Phosphorylation

Understanding the mechanisms involved in physiological regulation of phosphorylation systems requires that phosphoproteins not only are characterized biochemically, but that the state of phosphorylation of the different phosphoamino acid residues in the protein in intact cells also is determined. This is not a trivial problem (*see,* e.g., discussion in Rudolph et al., 1978). Chemical quantification of phosphate content in identified proteins might be used, but this approach requires a purified protein and is rather insensitive. More sophisticated nonradioactive methods, based on mass spectrometry, are also available, but employ highly specialized equipment not usually available to standard laboratories (Cohen et al., 1991). Most investigators will therefore rely on the use of labeled ATP for these experiments.

A rather specialized method termed "back-phosphorylation" has proven to be useful under certain circumstances (Forn and Greengard, 1978; Walaas et al., 1989a). This proce-

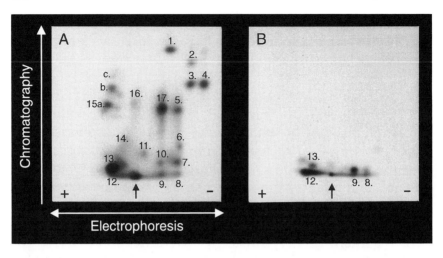

Fig. 3. Autoradiogram showing two-dimensional phosphopeptide mapping of the nuclear phosphoprotein P1, following labeling of HeLa cells **(A)** or hepatocytes **(B)** with [^{32}P]orthophosphate. Cells were labeled as described in Protocol 3. P1 was isolated by acid extraction (Østvold et al., 1985) and SDS-PAGE, the gel pieces were subjected to complete digestion with thermolysin, and the resulting phosphopeptides were separated by thin-layer electrophoresis at pH 3.5 and ascending chromatography as described (Walaas et al., 1989b). Arrow indicates application point, and numbers indicate distinct phosphopeptides derived from the protein. Note the major difference in phosphorylation pattern between proliferating (A) and quiescent (B) cells (Østvold, A. C., unpublished).

dure involves stimulation of unlabeled tissue and termination of the reaction under conditions in which further changes in phosphorylation state do not take place. Using the assumption that physiological phosphorylation sites not occupied by "cold" phosphate *in situ* will be accessible during in vitro incubation, the protein is extracted and phosphorylated with a protein kinase in the presence of [γ-^{32}P]ATP, which will label the specific phosphorylation sites in the protein. Hence, a high level of ^{32}P-labeled protein indicates that the protein was phosphorylated to a low extent *in situ*, whereas a low level of ^{32}P-labeled protein indicates that the protein was phosphorylated to a high extent *in situ*.

Protocol 6—Analysis of the state of phosphorylation by back-phosphorylation: regulation of synapsins I and II in cerebrocortical slices:

1. Prepare and incubate cortical slices as described in Protocol 1.
2. Terminate reaction by aspiration of KRB, add 5 mL of 5 mM Zn(OAc)$_2$, and rapidly homogenize the tissue in cold tissue grinders to precipitate proteins and inactivate kinase and phosphatase activities.
3. Centrifuge the suspensions (3000g for 10 min) and aspirate the supernatants. Extract the pellets in 1–2 mL of cold 10 mM citric acid, pH 2.9, containing 0.1% (v/v) Triton X-100 by repeated vigorous swirling and mixing. This treatment solubilizes both synapsins I and II (Forn and Greengard, 1978; Walaas et al., 1988).
4. Centrifuge the suspensions (12,000g for 10 min) and collect the extracts. Add Na$_2$HPO$_4$ (1/10 vol from a 0.5M stock solution) to neutralize the extracts, to pH 6.0–7.0, recentrifuge samples (12,000g for 10 min), and collect the supernatants.
5. Assay for protein content (Smith et al., 1985), and adjust volumes of supernatants to equal protein concentrations.
6. Analyze amount of dephospho-synapsins, using a reaction mixture containing 10 μM [γ-^{32}P]ATP, 50 mM HEPES, pH 7.4, 10 mM MgCl$_2$, 1 mM EDTA, and 10 nM of the purified catalytic subunit of cAMP-dependent protein kinase (Sigma) together with approx 10–30 μg extracted protein.
7. Incubate at 30°C for 30 min, terminate reaction by addition of SDS-containing "sample buffer" (Laemmli, 1970), and perform conventional SDS-PAGE on gels containing 7.5% acrylamide.
8. Excise synapsins I and II from dried gels and count radioactivity.

Following this protocol, the ^{32}P label incorporated into synapsins I and II has been shown to represent synapsin molecules, that were not phosphorylated at the cAMP-dependent phosphorylation site *in situ*. In order to compare this to the total number of sites available, the latter can be determined by incubation of the extract with phosphatases, which will completely dephosphorylate the protein before the phosphorylation assay (Forn and Greengard, 1978; Walaas et al., 1989a).

Finally, a more recent method relies on the use of phosphospecific antibodies directed against distinct phosphorylation sites in identified proteins. Under optimal conditions, the preparation and use of such antibodies allow quantitative determination of the state of phosphorylation of specific sites in proteins (Czernik et al., 1991; Snyder et al., 1992). Further description is given by Czernik et al. (Chapter 5).

6. Analysis of Protein Kinases

6.1. General Considerations

Although the physiologically relevant result of protein phosphorylation in intact cells is a change in the state of phosphorylation of a protein, estimation of protein kinase and phosphatase activities may also be of interest in these investigations. Measurement of enzyme activities is also necessary during purification of these enzymes, as well as during biochemical characterization of the phosphorylation system. Since methods for studying protein kinases may differ somewhat from those used in phosphoprotein analysis, a brief discussion is presented here. We will restrict ourselves to enzymes specific for phosphoserine and phosphothreonine residues, and the reader is directed to Chapter 6 for information on tyrosine-specific protein kinases.

6.2. Analysis of Protein Kinase Activity In Vitro

6.2.1. General Considerations

The basic approach to protein kinase analysis consists of preparation of an extract containing the activity of interest, and of measuring the ability of this extract to catalyze phosphorylation of a suitable exogenous substrate under "initial rate" conditions. The extracts must therefore be prepared in a buffer that ensures the ability to control pH, with suitable amounts of reducing agents and activating factors being added, and with means of preventing protein aggregation (by salts and/or detergents) and proteolysis. Addition of phosphatase inhibitors may be necessary to prevent these enzymes from overpowering the protein kinases in crude extracts. Finally, unspecific inhibitory compounds may be removed by dilutions or by simple chromatographic methods.

Protein kinase activities can also be examined in "solid-phase" assays, where proteins of interest have been adsorbed onto filter papers, immunoadsorbents, or other kinds of binding matrices, which then are used as a source for kinase activities (Glover and Allis, 1991). Crude extracts spotted on filter paper may contain numerous contaminants, whereas affinity-bound proteins, which usually have been washed extensively, may represent highly purified enzymes virtually devoid of contaminating factors.

The substrates to be added for the phosphorylation reaction include a phosphate donor and a phosphate acceptor. The most frequently employed methods use radioisotopic techniques, which involve transfer of ^{32}P from [γ-^{32}P]ATP to the substrate, separation of the phosphorylated substrate from unreacted [γ-^{32}P]ATP, and quantitation of labeled product. This approach is generally similar to the broken cell protein phosphorylation discussed in Section 4., but is performed with added exogenous substrates and under initial rate conditions, i.e., with activity linear with amount of enzyme added and incubation time, with substrate concentrations in excess of apparent K_m values, and with not more than 5–10% of substrate being consumed (Segel, 1975).

All protein kinases can employ Mg-ATP as phosphate donor, which therefore is the preferred labeled substrate for analysis of most of these enzymes. The enzymes display apparent K_m values for ATP below 100 μM (usually around 10 μM), and ATP levels near saturating concentrations (5- to 10-fold over K_m) will usually give the best results. The amount of radioactive tracer added can vary, and specific activities of 100–400 dpm/pmol are usually adequate.

Several types of protein substrates can be used as phosphate acceptors. Ideally, the protein should represent a physiological, preferentially monospecific substrate. Such substrates are, however, rare, and multikinase substrates like histones and casein are more frequently employed. To distinguish between enzymes, it then becomes necessary to exploit other enzymatic properties, most frequently by comparing activity in the absence and presence of specific activators or inhibitors (e.g., cyclic nucleotides, Ca^{2+}-calmodulin, phospholipid derivatives, specific inhibitor peptides, and so forth; *see* Chapter 4). Alternatively, synthetic peptides with sequences derived from the known phosphorylation sites of physiological protein substrates have become popular. Since such "consensus sequences" represent major specificity determinants for many protein kinases (Pearson and Kemp, 1991), these synthetic peptides can tentatively be used as "monospecific" substrates. They also have the added advantages that they can be designed to suit specific needs, for example, by addition of charged residues to simplify product purification (Casnellie, 1991; Kemp and Pearson, 1991). Finally, small nonpeptide substrates or synthetic random amino acid polymers may also be used with advantage under certain conditions (*see* Racker, 1991; Racker and Sen, 1991).

Isolation and quantitation of reaction product is usually achieved by separation of ^{32}P-labeled product from [γ-^{32}P]ATP, using protocols that exploit: differences between these components in solubilities, size, and charge, employ precipitation with TCA, binding of positively charged peptides/proteins to phosphocellulose paper, binding of ATP and acidic components to anion exchange resins, or size separation by SDS-PAGE. These methods all have their advantages and disadvantages.

Most, although not all, proteins will be denatured and precipitated in the presence of 5–25% (w/v) TCA, whereas free phosphate and ATP remain in solution. This method can therefore be used for a large variety of substrates. A carrier protein, bovine serum albumin (BSA), is usually added to ensure quantitative recovery of substrate. In contrast, phosphocellulose paper will preferentially bind peptides or proteins with predominantly basic residues (e.g., arginine) (Witt and Roskoski, 1975; Casnellie, 1991). Hence, this method is suitable for protein kinases in which substrates with basic residues predominate. In cases of acidic substrates, synthetic peptides can be designed with the necessary basic residues added. Because of the possibility of running a large number of samples on the same phosphocellulose paper, this is a highly efficient method for screening of multiple samples.

Passing the reaction mixture through anionic exchangers will allow basic peptide substrates to elute in the flow-through, whereas ATP is quantitatively retained. This method is recommended for an initial purification of peptide substrates when further characterization of product is contemplated (e.g., HPLC, determination of phosphoamino acid, amino acid sequencing).

Separation with SDS-PAGE generally gives the lowest background, and is also suitable when more than one protein is present in the substrate preparation. However, this time-consuming method is not suitable when small peptides are employed (Schägger and von Jagow, 1987). We present a general method based on phosphocellulose paper (Witt and Roskoski, 1975) that works well with most of the enzymes and substrates we have examined.

Protocol 7—Analysis of protein kinase activity by substrate binding to phosphocellulose:

1. Mark phosphocellulose paper (P81, Whatman, Kent, UK) with a pencil into 2 × 2 cm squares to form a grid.
2. Mix media containing suitable buffers, salts, activating factors, and/or inhibitors with a suitable phosphate acceptor substrate (usually positively charged Arg- or Lys-containing peptide).
3. Prepare extracts containing the enzyme, store on ice. Highly dilute preparations might be protected from spontaneous inactivation by addition of glycerol (e.g.,

10–30%) or carrier proteins like BSA (e.g., 0.1 mg/mL).
4. Mix media and enzyme and pre-equilibrate at 30°C for 1 min. Initiate reaction by addition of [γ-^{32}P]ATP (final conc. 0.05–0.1 mM). Preliminary experiments must be performed to determine incubation time and amount of protein, which will give "initial rate" conditions (Segel, 1975).
5. Terminate reaction by directly spotting aliquots onto phosphocellulose paper. Alternatively, glacial acetic acid can be added (final conc. 10% [v/v]) before aliquots are spotted on paper.
6. Soak paper in 10% (w/v) TCA for 15 min. Repeat three times, each for 5 min. Rinse paper in water, dry, cut out squares, and measure ^{32}P content by Cerenkov or scintillation counting.

6.3. Analysis of Specific Protein Kinases In Vitro

6.3.1. cAMP-Dependent Protein Kinase

This enzyme consists of two regulatory, inhibitory subunits which bind two catalytic subunits. Added cAMP binds to the regulatory subunits and releases the catalytic units, which are now free to phosphorylate suitable substrates. Even under "basal" conditions, many tissues contain some free catalytic subunit, the activity of which gives a certain "background." The assay should ideally be performed both in the presence of a cAMP analog (this will give total activity of the enzyme) and in the presence of the heat-stable protein kinase inhibitor peptide (PKI), which binds strongly to the catalytic domain of the enzyme, and completely prevents any substrate binding and catalysis from taking place (this will give background activity). In addition, the assay should be performed without any addition, which will give background activity plus that caused by a free catalytic subunit.

Preferred medium consists of Tris-HCl or HEPES, pH 7.4, 10 mM MgCl$_2$, 1 mM EGTA, and 1 mM DTT. Histones are frequently used substrates, with histone H2B being particularly useful. Since extensive studies on primary structure determinants have shown that Arg-X-X-Ser(Thr) constitutes the preferred motif for this kinase (Pearson and Kemp, 1991), synthetic peptides containing this sequence may also be used. The most popular is the octapeptide termed kemptide (commercially available; e.g., Sigma, St. Louis, MO), which is derived from the phosphorylation site of pyruvate kinase (Pearson and Kemp, 1991).

Protocol 8—Analysis of cAMP-dependent protein kinase:

1. Prepare ice-cold medium containing (final conc.) 20 mM HEPES, pH 7.4, 10 mM MgCl$_2$, 1 mM, EDTA, 1 mM EGTA, and 1 mM DDT, and store on ice. Add kemptide peptide substrate (0.1–1 mM).
2. Prepare tissue extracts on ice. Add replicate samples (0.1–10 µg protein) to tubes containing medium in the absence or presence of 1–5 µM 8-bromo cAMP or 1–5 µM PKI (commercially available).
3. Mix medium and extract. Preincubate mixture for 30–90 s at 30°C. Initiate reaction by addition of [γ-^{32}-P]ATP (final conc. 0.1 mM).
4. Terminate reaction after 10–120 s, and analyze phosphorylation by spotting aliquots on phosphocellulose paper as described in Protocol 7.

6.3.2. Ca^{2+}-Calmodulin-Dependent Protein Kinase II (CaMKII)

This enzyme, which is particularly enriched in neural tissues, is composed of several subunits with binding sites for Ca^{2+}-calmodulin and several activating and inhibitory autophosphorylation sites (Hanson and Schulman, 1992). Addition of Ca^{2+}-calmodulin initiates activation of the enzyme, which then preferentially autophosphorylates a site (Thr-286 in the α-subunit) which creates an autonomous enzyme no longer dependent on Ca^{2+}-calmodulin (Lai et al., 1986; Miller and Kennedy, 1986; Colbran and Soderling, 1990; Hanson and Schulman, 1992). It should be noted that addition of detergents may interfere in the binding of the hydrophobic Ca^{2+}-calmodulin complex to the enzyme, and therefore could be detrimental to enzyme assays in vitro. CaMKII has been characterized using exogenous protein substrates like synapsin I or microtubule-associated protein-2 (Bennett et al., 1983; Schulman, 1984), but histone H3, glycogen synthase, or casein may also be used (Nairn et al., 1985). However, a minimal consensus sequence has been determined (Pearson et al., 1985) and the enzyme is more conveniently assayed using synthetic peptizes, e.g., "syntide-2," derived from glycogen synthase, or "autocamtide-2" (Hanson et al., 1989) (both commercially available; e.g., Calbiochem Novabiochem Intl., La Jolla, CA), in the presence of 1.5 mM $CaCl_2$–1.5 µM calmodulin or 1 mM EGTA (for basal independent activity), respectively.

Protocol 9—Analysis of CaMKII:

1. Prepare ice-cold medium containing (final conc.) 20 mM HEPES, pH 7.4, 10 mM $MgCl_2$, 1 mM DTT, 0.05–0.1 mM peptide substrate, and store on ice.
2. Prepare tissue extracts on ice. Add replicate samples (0.1–10 µg protein) to tubes containing medium in the absence or presence of either 1 mM EGTA, or 1 mM EGTA, 1.5 mM $CaCl_2$, and 1.5 µM calmodulin. Depending on contaminating kinases in the tissue, peptide inhibitors of cAMP-dependent protein kinase (PKI, 2 µM) and PKC [19–36] (5 µM) may be included (Fukunaga et al., 1992).
3. Mix medium and extract, and preincubate mixture for 30–90 s at 30°C. Initiate reaction by addition of [γ-^{32}P]ATP (final conc. 0.1 mM).
4. Terminate reaction after 10–60 s, and analyze phosphorylation by spotting aliquots on phosphocellulose paper, washing, and counting as described in Protocol 7. Total activity is defined as that obtained following addition of 1 mM EGTA, 1.5 mM $CaCl_2$, plus 1.5 µM calmodulin, whereas autonomous activity is that obtained following addition of 1 mM EGTA alone.

6.3.3. PKC (Diacylglycerol-Activated, Ca^{2+}-Phospholipid-Dependent Protein Kinase)

This complicated enzyme family is made up of monomeric enzymes, most of which have regulatory domains with binding sites for unsaturated diacylglycerols (tumor-promoting phorbol esters act as artificial ligands here), and distinct binding domains for Ca^{2+} and acidic phospholipids (Nishizuka, 1995). It should be noted that addition of detergents may interfere in the binding of both diacylglycerol and phospholipids to the protein, and therefore could be detrimental to enzyme assays in vitro. The enzyme has a pseudo-substrate domain (residues 19–36) that binds tightly to the catalytic domain, and a synthetic peptide (PKC[19–36]) can be used as a specific inhibitor (Pearson and Kemp, 1991).

The enzyme activity is conveniently assayed using histone H1 (Type IIIS from Sigma) followed by TCA precipitations or the phosphocellulose paper method to isolate the product. Alternatively, synthetic peptides containing the motif Arg-X-X-<u>Ser(Thr)</u>-X-Arg, several of which are commercially available (e.g., myelin basic protein [4–14] available from Sigma and Calbiochem-Novabiochem) can be used.

Protocol 10—Analysis of PKC:

1. Prepare ice-cold medium containing (final conc.) 20 mM Tris-HCl, pH 7.4, 10 mM MgCl$_2$, 1 mM DTT, 10 μM leupeptin. Add 0.05–0.1 mM peptide substrate (e.g., myelin basic protein [4–14]) or histone IIIS (100–500 μg/mL) and store on ice.

2. Prepare stock solutions of activating factors, i.e., separate 10X stocks of 15 mM CaCl$_2$, and 500 μg/mL phosphatidylserine, and 10 μg/mL diolein (or alternatively phorbol esters like 1 μM phorbol 12,13-dibutyrate). Stock solutions of phosphatidylserine plus diolein are prepared as described in Protocol 4.

3. Prepare tissue extracts in buffer containing 10 μM leupeptin and 0.1 mM EGTA, and store on ice. Add replicate samples (0.1–10 μg protein) to tubes containing medium in the absence or presence of either 1 mM EGTA or 1 mM EGTA, 1.5 mM CaCl$_2$, and phosphatidylserine (50 μg/mL) plus diolein (1 μg/mL). Diolein may be substituted by tumor-promoting phorbol esters, which are metabolically more stable, but must be handled with caution because of their actions as tumor promoters.

Depending on contaminating kinases in the tissue, peptide inhibitors of cAMP-dependent protein kinase (PKI, 2 μ*M*) and CaMKII (5 μ*M*) may also be included.

4. Mix medium and extract, and preincubate mixture for 30–90 s at 30°C. Initiate reaction by addition of [γ-^{32}P]ATP (final conc. 0.05–0.1 m*M*).

5. Terminate reaction after 10–120 s, and analyze phosphorylation by spotting aliquots on phosphocellulose paper, washing, and counting as described in Protocol 7. Background activity is estimated by substituting 1 m*M* EGTA for Ca^{2+} and leaving out phospholipids and diacylglycerols. Alternatively, 1–10 μ*M* PKC(19–36) can be included.

6.3.4. Casein Kinase 2

This multifunctional and widely distributed enzyme is characterized by its ability to use GTP as well as ATP as substrate, its inhibition by low amounts of heparin, and its activation by relatively high salt concentrations (Tuazon and Traugh, 1991). Casein kinase 2, which appears to phosphorylate a variety of proteins *in situ*, preferentially phosphorylates peptides with acidic residues carboxy-terminal to the phosphorylated serine/threonine, e.g., containing the motif Ser-X-X-Glu. For such peptides, basic residues like arginine must be added at the amino-terminal before the substrate will bind to phosphocellulose paper, e.g., Arg-Arg-Arg-Asp-Asp-Asp-Ser-Asp-Asp-Asp (commercially available from Bachem Bioscience, King of Prussia, PA).

Protocol 11—Analysis of casein kinase 2:

1. Prepare ice-cold medium containing (final conc.) 50 m*M* Tris-HCl, pH 7.6, 10 m*M* MgCl$_2$, with NaCl or KCl added (depending on the amount present in the enzyme sample) to a final salt concentration of 200 m*M*. Add 0.5 m*M* peptide substrate and store on ice.

2. Prepare tissue extracts and add aliquots (0.1–10 μg protein) to replicate tubes. Store on ice.

3. Mix medium and extract. Preincubate mixture for 30–90 s at 30°C. Initiate reaction by addition of [γ-^{32}P]ATP (final conc. 0.1 m*M*). Terminate reaction after 1–5 min by spot-

ting aliquots on phosphocellulose paper. Immediately immerse paper into beaker with 75 mM H_3PO_4. Wash five times, each for 5 min with gentle stirring, and decant radioactive solution. Dry phosphocellulose paper, cut out, and count individual squares as described in Protocol 7.

6.4. Analysis of Protein Kinase Activity in Intact Cells

The state of activity of a given protein kinase in intact tissue is of considerable interest to the investigator. At the present time, methods for determination of *in situ* activity of cAMP-dependent protein kinase and Ca^{2+}-calmodulin-dependent protein kinase II appear valuable. Activation of protein kinases *in situ* might also be examined by studying the state of phosphorylation of specific substrates in the intact cell. Examples include analyzing the activity of PKC through MARCKS phosphorylation (Robinson et al., 1993) or the activity of neuronal CaMKII through phosphorylation of sites 2/3 on synapsin I (Czernik et al., 1991).

6.4.1. cAMP-Dependent Protein Kinase

The fraction of free, dissociated catalytic subunit present in the tissue indicates the state of activity of cAMP-dependent protein kinase *in situ* (Walaas et al., 1973; Corbin et al., 1975; Palmer et al., 1980). This fraction can be estimated by subtracting the cAMP-induced increase in activity from the total activity of cAMP-dependent protein kinase. This approach is valid only if the enzyme can be extracted in a medium that prevents further dissociation and reassociation of the holoenzyme from occurring, using either high levels of NaCl (0.5M) and/or phosphodiesterase inhibitors. Addition of exogenous enzyme as an internal standard has also been suggested (Palmer et al., 1980). Here we present a method used in neuronal slices and isolated cells that relies on the ability of Zn^{2+} to aggregate proteins and inhibit protein kinase and phosphatase activities (Forn and Greengard, 1978; Chneiweiss et al., 1991) (*see also* Protocol 6).

Protocol 12—Analysis of cAMP-dependent protein kinase *in situ*:

1. Incubate cells or tissue in suitable media, wash in cold saline, and homogenize in 5 mM Zn(OAc)$_2$ containing 0.1% (v/v) Triton X-100. These preparations can be kept frozen (–70°C) until assay.
2. Prepare extracts from thawed suspensions by centrifugation (10,000g for 10 min), remove supernatants, and resuspend the pellets in ice-cold medium containing 5 mM Zn(OAc)$_2$, 5 mM EDTA, and 0.1% (v/v) Triton X-100.
3. Determine enzyme activity as described in Protocol 7 in the absence or presence of optimal concentrations of cAMP analogs (e.g., 1–5 µM 8-bromo cAMP). Background is determined in the presence of the protein kinase inhibitor peptide (PKI, 2 µM).
4. Activity owing to the free catalytic subunit is determined by subtracting the increase in activity induced by addition of 8-bromo-cAMP from total activity. When expressed as a fraction of total activity, this value indicates the "activity ratio," i.e., the state of activation of the enzyme *in situ*.

6.4.2. Ca^{2+}-Calmodulin-Dependent Protein Kinase II

Autophosphorylation is a particularly important mechanism for activation of protein kinases. A well-studied example of this mechanism is represented by CaMKII (Colbran and Soderling, 1990; Hanson and Schulman 1992). As described in Section 6.3.2., following Ca^{2+}-calmodulin-induced activation and autophosphorylation, this enzyme becomes independent of Ca^{2+}-calmodulin (Lai et al, 1986; Miller and Kennedy, 1986). Comparison of CaMKII activity in the absence and presence of Ca^{2+}-calmodulin can therefore be used to estimate the amount of "autonomous" activated enzyme present (Gorelick et al., 1988; Fukunaga et al., 1992; MacNicol and Schulman, 1992). It should be noted that this activation mechanism can be reversed by phosphatases, and lysis buffers should therefore contain broad-spectrum phosphatase inhibitors, such as pyrophosphate. We describe a method used to analyze regulation of CaMKII in isolated nerve terminals (Gorelick et al., 1988).

   As alternative methods, it is possible to follow the state of phosphorylation of Thr-286 in the α-subunit, either by labeling *in situ* with ^{32}P and phosphoamino acid analysis or phosphopeptide fingerprinting (Gorelick et al., 1988), or by using phosphospecific antibodies specifically recognizing this epitope (Suzuki et al., 1992).

   Protocol 13—Analysis of CaMKII activity *in situ*:

1. Incubate isolated cells, nerve terminals, or tissue preparations. Add 10 vol of ice-cold lysis buffer containing, e.g., 10 mM HEPES, pH 7.4, 1 mM EDTA, 1 mM EGTA, 10 mM sodium pyrophosphate, 1 mM DTT, and protease inhibitors, like leupeptin, pepstatin A, chymostatin, aprotinin, and phenylmethylsulfonyl fluoride, at appropriate concentrations (in cell cultures, medium may be aspirated and a smaller volume of lysis buffer may be used). Lyse tissue by sonication, taking care to keep the sample cold. If necessary, detergents like Triton X-100 or Chaps (0.1–0.5% [v/v]) may also be added.

2. Assay CaMKII by following phosphorylation of peptides like syntide-2 or autocamtide-2 as described in Protocol 9. Total activity is defined as that obtained following addition of 1.5 mM $CaCl_2$ plus 1.5 μM calmodulin, whereas autonomous activity is that obtained following addition of 1 mM EGTA alone.

6.5. Analysis of Protein Kinase Activity by Renaturation In Vitro

6.5.1. General Considerations

   Alternative analytical methods, based on rapid denaturation and inactivation followed by subsequent renaturation, may be helpful in order to examine a variety of protein kinases (Ferrell and Martin, 1989; Celenza and Carlson, 1991). In this approach, the tissue samples are inactivated, and proteins are rapidly solubilized in denaturating buffer containing, e.g., SDS to inactivate the enzymes completely. This is usually followed by electrophoretic separation before the proteins are transferred to membranes, subjected to careful renaturation, and incubated with Mg^{2+}-[γ-^{32}P]ATP. This approach allows demonstration of those kinase activities that are able to survive de-/renaturation, and that can either autophosphorylate or phosphorylate proteins that comigrate during electrophoresis.

A major advantage of this method is the ability to discover new, unknown kinases, such as the demonstration of kinase activity of a yeast gene product (Celenza and Carlson, 1991), and 14 previously unknown protein kinases, which were found in platelets, several of which responded to Ca^{2+} influx, thrombin, or phorbol esters (Ferrell and Martin, 1989). The method clearly has limitations, particularly since the enzymes must express activity following treatment with SDS. Moreover, since there appear to be clear differences in the specific activities of cAMP-dependent protein kinase, CaMKII, PKC, and tyrosine kinases using this assay when compared to assays in solution (Ferrell and Martin, 1991), there may be significant differences among different enzymes in the ability to survive denaturation and renaturation.

Certain proteins devoid of protein kinase activity appear to be able to bind ATP directly. To ensure that the phosphorylated protein bands contain phosphoryl groups bound in phosphoester bonds, control experiments should be performed with $[\alpha\text{-}^{32}P]ATP$. Additionally, the presence of phosphoamino acids in the labeled bands may be analyzed; this may be carried out directly on the PVDF membrane (Hildebrandt and Fried, 1989; Kamps, 1991).

6.5.2. Renaturation of Protein Kinases

The method described here is based on that reported by Ferrell and Martin (1989, 1991), which we have found suitable for neural tissue (unpublished observations).

Protocol 14—Determination of protein kinase activity by renaturation in vitro:

1. Following experimental treatments, inactivate tissue by rapid sonication/homogenization in hot 1% (w/v) SDS and protease inhibitors (optional).
2. Separate proteins by SDS-PAGE, usually running aliquots containing approx 50 µg protein in a minigel

apparatus. Electrotransfer proteins to PVDF membranes (e.g., Immobilon, Millipore) in the cold room, using 192 mM glycine, 25 mM Tris (Towbin et al., 1979), or 10 mM CAPS, pH 11.0 (Matsudaira, 1987).

3. Wash the filters briefly in TBS and subject them to protein renaturation (4°C, overnight) in 140 mM NaCl, 10 mM Tris-HCl, pH 7.4, 2 mM EDTA, 1% (w/v) BSA, 0.1% (v/v) Triton X-100, and 2% (v/v) 2-mercaptoethanol. Time for renaturation should not exceed 18 h.

4. Block filters for 30–60 min at room temperature by addition of 2–5% (w/v) nonfat dry milk or BSA (optional).

5. Assay activity of the renatured kinases by adding the filters to media containing 25 mM Tris-HCl, pH 7.4, 10 mM MgCl$_2$, 2 mM MnCl$_2$, and [γ-^{32}P]ATP (3000 Ci/mmol, final conc. usually 0.03 μM, containing 0.1 mCi/mL). Incubate with rocking behind Plexiglas shield at room temperature for 30 min. Terminate reaction by transferring filters to washing container.

6. Wash filters until washing solution is devoid of radioactivity (check with Geiger counter), using, for example, TBS, TBS/Triton X-100 (0.1%, v/v), and (in the case of PVDF membranes) brief treatment with 1M KOH followed by 10% (v/v) acetic acid. Final washes can then be performed in distilled water. Dry the blots and subject to autoradiography or other methods for detection of ^{32}P-labeling.

References

Agrawal, H. C., Martenson, R. E., and Agrawal, D. (1982) In vivo incorporation of [^{32}P]orthophosphate into myelin basic protein of developing rabbit brain: its location in components 3 and 5 in a new protein tentatively identified as basic protein component 7. *J. Neurochem.* **39,** 1755–1758.

Akers, R. F., Lovinger, D. M., Colley, P. A., Linden, D. J., and Routtenberg, A. (1986) Translocation of protein kinase C activity may mediate hippocampal long-term potentiation. *Science* **231,** 587–589.

Ames, A., III and Nesbett, F. B. (1981) *In vitro* retina as an experimental model of the central nervous system. *J. Neurochem.* **37,** 867–877.

Banker, G. and Goslin, K. (1991) *Culturing Nerve Cells.* MIT, Cambridge, MA, p. 453.

Bennett, M. K., Erondu, N. E., and Kennedy, M. B. (1983) Purification and characterization of a calmodulin-dependent protein kinase that is highly concentrated in brain. *J. Biol. Chem.* **258,** 12,735–12,744.

Bradford, H. F. (1975) Isolated nerve terminals as an in vitro preparation for the study of dynamic aspects of transmitter metabolism and release, in *Handbook of Psychopharmacology, vol. I, Biochemical Principles and Techniques in Neuropharmacology* (Iversen, L. L., Iversen, S. D., and Snyder, S. H., eds.), Plenum, pp. 191–252.

Casnellie, J. E. (1991) Assay of protein kinases using peptides with basic residues for phospho-cellulose binding. *Methods Enzymol.* **200,** 115–120.

Castagna, M., Takai, Y., Kaibuchi, K., Sano, K., Kikkawa, U., and Nishizuka, Y. (1982) Direct activation of calcium-activated, phospholipid-dependent protein kinase by tumor-promoting phorbol esters. *J. Biol. Chem.* **257,** 7847–7851.

Celenza, J. L. and Carlson, M. (1991) Renaturation of protein kinase activity on protein blots. *Methods Enzymol.* **200,** 423–430.

Celis, J. E. and Bravo, R (1984) *Two-Dimensional Gel Electrophoresis of Proteins. Methods and Applications.* Academic, Orlando, FL.

Chneiweiss, H., Cordier, J., and Glowinski, J. (1991) Cyclic AMP accumulation induces a rapid desensitization of the cAMP-dependent protein kinase in mouse striatal neurons. *J. Neurochem.* **57,** 1708–1715.

Cleveland, D. W., Fischer, S. G., Kirschner, M. W., and Laemmli, U. K. (1977) Peptide mapping by limited proteolysis in sodium dodecyl sulfate and analysis by gel electrophoresis. *J. Biol. Chem.* **252,** 1102–1106.

Coffey, E. T., Sihra, T. S., and Nicholls, D. G. (1993) Protein kinase C and the regulation of glutamate exocytosis from cerebrocortical synaptosomes. *J. Biol. Chem.* **268,** 21,060–21,065.

Cohen, P. (1994) The discovery of protein phosphatases: from chaos and confusion to an understanding of their role in cell regulation and human disease. *Bioessays* **16,** 583–588.

Cohen, P., Gibson, B. W., and Holmes, C. F. (1991) Analysis of the *in vivo* phosphorylation states of proteins by fast atom bombardment mass spectrometry and other techniques. *Methods Enzymol.* **201,** 153–168.

Colbran, R. J. and Soderling, T. R. (1990) Calcium/calmodulin-dependent protein kinase II. *Curr. Top. Cell. Regul.* **31,** 181–221.

Corbin, J. D., Keely, S. L., and Park, C. R. (1975) The distribution and dissociation of cyclic adenosine 3',5'-monophosphate-dependent protein kinases in adipose, cardiac, and other tissues. *J. Biol. Chem.* **250,** 218–225.

Czernik, A. J., Girault, J.-A., Nairn, A. C., Chen, J., Snyder, G., Kebabian, J., and Greengard, P. (1991) Production of phosphorylation state-specific antibodies, in *Methods in Enzymology,* vol. 201 (Hunter, T. and Sefton, B. M., eds.), Academic, San Diego, CA, pp. 264–283.

Darbre, A. (1986) *Practical Protein Chemistry—A Handbook.* John Wiley, New York.

Dunbar, B. S., Kimura, H., and Timmons, T. M. (1990) Protein analysis using high-resolution two-dimensional polyacrylamide gel electrophoresis. *Methods Enzymol.* **182,** 441–459.

Ferrell, J. E., Jr. and Martin, G. S. (1989) Thrombin stimulates the activities of multiple previously unidentified protein kinases in platelets. *J. Biol. Chem.* **264,** 20,723–20,729.

Ferrell, J. E., Jr. and Martin, G. S. (1991) Assessing activities of blotted protein kinases. *Methods Enzymol.* **200,** 430–435.

Ferris, C. D., Cameron, A. M., Bredt, D. S., Huganir, R. L., and Snyder, S. H. (1991) Inositol 1,4,5-trisphosphate receptor is phosphorylated by cAMP-dependent protein kinase at serines 1755 and 1589. *Biochem. Biophys. Res. Commun.* **175,** 192–198.

Fontana, A. and Gross, E. (1986) Fragmentation of polypeptides by chemical methods in *Practical Protein Chemistry—A Handbook* (Darbre, A., ed.), John Wiley, New York, pp. 67–120.

Forn, J. and Greengard, P. (1978) Depolarizing agents and cyclic nucleotides regulate the phosphorylation of specific neuronal proteins in rat cerebral cortex slices. *Proc. Natl. Acad. Sci. USA* **75,** 5195–5199.

Fukunaga, K., Soderling, T. R., and Miyamoto, E. (1992) Activation of Ca^{2+}/calmodulin-dependent protein kinase II and protein kinase C by glutamate in cultured rat hippocampal neurons. *J. Biol. Chem.* **267,** 22,527–22,533.

Garrison, J. C. (1983) Measurement of hormone-stimulated protein phosphorylation in intact cells, in *Methods Enzymol.,* vol. 99 (Corbin, J. D. and Hardman, J. G., eds.) Academic, New York, pp. 20–36.

Garthwaite, J., Woodhams, P. L., Collins, M. J., and Balazs, R. (1979) On the preparation of brain slices: morphology and cyclic nucleotides. *Brain Res.* **173,** 373–377.

Gjertsen, B. T., Mellgren, G., Otten, A., Maronde, E., Genieser, H.-G., Jastorff, B., Vintermyr, O. K., McKnight, G. S., and Døskeland, O. (1995) Novel (Rp)-cAMPS analogs as tools for inhibition of cAMP-kinase in cell culture. *J. Biol. Chem.* **270,** 1–9.

Glover, C. V. C. and Allis, C. D. (1991) Enzyme activity dot blots for assaying protein kinases. *Methods Enzymol.* **200,** 85–90.

Goelz, S. E., Nestler, E. J., Chehrazi, B., and Greengard, P. (1981) Distribution of protein I in mammalian brain as determined by a detergent-based radioimmunoassay. *Proc. Natl. Acad. Sci. USA* **78,** 2130–2134.

Gordon-Weeks, P. R. (1987) Isolation of synaptosomes, growth cones and their subcellular components, in *Neurochemistry—A Practical Approach* (Turner, A. J. and Bachelard, H. S., eds.), IRL, Oxford, UK, pp. 1–26.

Gorelick, F. S., Wang, J. K T., Lai, Y., Nairn, A. C., and Greengard, P. (1988) Autophosphorylation and activation of Ca^{2+}/calmodulin-dependent protein kinase II in intact nerve terminals. *J. Biol. Chem.* **263,** 17,209–17,212.

Guidotti, A., Cheney, D. L., Trabucchi, M., Doteuchi, M., Wang, C., and Hawkins, R. A. (1974) Focussed microwave radiation: a technique to minimize post mortem changes of cyclic nucleotides, dopa and choline and to preserve brain morphology. *Neuropharmacology* **13**, 1115–1122.

Halpain, S., Girault, J. A., and Greengard, P. (1990) Activation of NMDA receptors induces dephosphorylation of DARPP-32 in rat striatal slices. *Nature* **343**, 369–372.

Hames, B. D. and Rickwood, D. (eds.) (1990) *Gel Electrophoresis of Proteins—A Practical Approach,* IRL, Oxford, UK, p. 290.

Hanson, P. I. and Schulman, H. (1992) Neuronal Ca^{2+}/calmodulin-dependent protein kinases. *Ann. Rev. Biochem.* **61**, 559–601.

Hanson, P. I., Kapiloff, M. S., Lou, L. L., Rosenfeld, M. G., and Schulman, H. (1989) Expression of a multifunctional Ca^{2+}/calmodulin-dependent protein kinase and multifunctional analysis of its autoregulation. *Neuron* **3**, 59–70.

Harlow, E. and Lane, D. (1988) *Antibodies. A Laboratory Manual,* Cold Spring Harbor Laboratory, Cold Spring Harbor, New York, p. 726.

Harris, E. L. V. and Angal, S. (eds.) (1989) *Protein Purification Methods—A Practical Approach,* IRL, Oxford University Press, Oxford, UK, p. 317.

Haycock, J. W. and Haycock, D. A. (1991) Tyrosine hydroxylase in rat brain dopaminergic nerve terminals. Multiple-site phosphorylation *in vivo* and in synaptosomes. *J. Biol. Chem.* **266**, 5650–5657.

Hildebrandt, E. and Fried, V. A. (1989) Phosphoamino acid analysis of protein immobilized on polyvinylidene diflouride membrane. *Anal. Biochem.* **177**, 407–412.

Hochstrasser, D. F., Harrington, M. G., Hochstrasser, A.-C., Miller, M. J., and Merril, C. R. (1988) Methods for increasing the resolution of two-dimensional protein electrophoresis. *Anal. Biochem.* **173**, 424–435.

Hopkirk, T. J. and Denton, R M. (1986) Studies on the specific activity of $[\gamma^{32}P]ATP$ in adipose and other tissue preparations incubated with medium containing $[^{32}P]$phosphate. *Biochim. Biophys. Acta* **885**, 195–205.

Hunter, T. (1991) Protein kinase classification. *Methods Enzymol.* **200**, 3–37.

Hunter, T. and Sefton, B. M. (eds.) (1991a) *Methods in Enzymology,* vol. **200**, *Protein Phosphorylation*, Part A, *Protein Kinases: Assays, Purification, Antibodies, Functional Analysis, Cloning and Expression.* Academic, San Diego, CA, p. 763.

Hunter, T. and Sefton, B. M. (eds.) (1991b) *Methods in Enzymology,* vol. **201**, *Protein Phosphorylation*, Part B, *Analysis of Protein Phosphorylation, Protein Kinase Inhibitors, and Protein Phosphatases.* Academic, San Diego, CA, p. 547.

Huttner, W. B. and Greengard, P. (1979) Multiple phosphorylation sites in protein I and their differential regulation by cAMP and calcium. *Proc. Natl. Acad. Sci. USA* **76**, 5402–5406.

Imada, M. and Sueoka, N. (1980) A two-dimensional polyacrylamide gel electrophoresis system for the analysis of mammalian cell surface proteins. *Biochim. Biophys. Acta* **625,** 179–192.

Jahnen, W., Ward, L. D., Reid, G. E., Moritz, R. L., and Simpson, R. J. (1990) Internal amino acid sequencing of proteins by *in situ* cyanogen bromide cleavage in polyacrylamide gels. *Biochem. Biophys. Res. Commun.* **166,** 139–145.

Kamps, M. P. (1991) Determination of phosphoamino acid composition by acid hydrolysis of protein blotted to immobilon, in *Methods in Enzymology,* vol. 201 (Hunter, T. and Sefton, B. M., eds.), Academic, San Diego, CA, pp. 21–27.

Kelly, P. T., Yip, R. K., Shields, S. M., and Hay, M. (1985) Calmodulin-dependent protein phosphorylation in synaptic junctions. *J. Neurochem.* **45,** 1620–1634.

Kemp, B. E. and Pearson, R B. (1991) Design and use of peptide substrates for protein kinases. *Methods Enzymol.* **200,** 121–134.

Laemmli, U. K. (1970) Cleavage of structural proteins during the assembly of the head of bacteriophage T4. *Nature* **227,** 680–685.

Lai, Y., Nairn, A. C., and Greengard, P. (1986) Autophosphorylation reversibly regulates the Ca^{2+}/calmodulin-dependence of Ca^{2+}/calmodulin-dependent protein kinase II. *Proc. Natl. Acad. Sci. USA* **83,** 4253–4257.

Lewis, R. M., Levari, I., Ihrig, B., and Zigmond, M. J. (1990) *In vivo* stimulation of D1 receptors increases the phosphorylation of proteins in the striatum. *J. Neurochem.* **55,** 1071–1074.

Lipton, P. (1985) Brain slices. Uses and abuses, in *Neuromethods 1. General Neurochemical/Techniques* (Boulton, A. A. and Baker, G. B., eds.) Humana, Clifton, NJ, pp. 69–115.

MacNicol, M. and Schulman, H. (1992) Cross-talk between protein kinase C and multifunctional Ca^{2+}/calmodulin-dependent protein kinase. *J. Biol. Chem.* **267,** 12,197–12,201.

Mahboub, S., Richard, C., Delacourte, A., and Han, K.-K. (1986) Applications of chemical cleavage procedures to the peptide mapping of neurofilament triplet protein bands in sodium dodecyl sulfate-polyacrylamide gel electrophoresis. *Anal. Biochem.* **154,** 171–182.

Matsudaira, P. (1987) Sequence from picomole quantities of proteins electroblotted onto polyvinylidene difluoride membranes. *J. Biol. Chem.* **262,** 10,035–10,038.

McIlwain, H. (ed.) (1975) *Practical Neurochemistry,* 2nd ed., Churchill Livingstone, Edinburgh, UK, p. 337.

Meijer, L., Østvold, A. C., Walaas, S. I., Lund, T., and Laland, S. G. (1991) High-mobility-group proteins P1, I and Y as substrates of the M-phase-specific p34[cdc2]/cyclin[cdc13] kinase. *Eur. J. Biochem.* **196,** 557–567.

Miller, S. G. and Kennedy, M. B. (1986) Regulation of brain type II Ca^{2+}/calmodulin-dependent protein kinase by autophosphorylation: a Ca^{2+}-triggered molecular switch. *Cell* **44,** 861–870.

Misgeld, U. and Frotscher, M. (1982) Dependence of the viability of neurons in hippocampal slices on oxygen supply. *Brain Res. Bull.* **8,** 95–100.

Mitrius, J. C., Morgan, D. G., and Routtenberg, A. (1981) *In vivo* phosphorylation following [^{32}P]orthophosphate injection into neostriatum or hippocampus: selective and rapid labeling of electrophoretically separated brain proteins. *Brain Res.* **212,** 67–81.

Mobley, P. and Greengard, P. (1985) Evidence for widespread effects of noradrenaline on axon terminals in the rat frontal cortex. *Proc. Natl. Acad. Sci. USA* **82,** 945–947.

Moss, S. J., Blackstone, C. D., and Huganir, R. L. (1993) Phosphorylation of recombinant non-NMDA glutamate receptors on serine and tyrosine residues. *Neurochem. Res.* **18,** 105–110.

Nairn, A. C., Hemmings, H. C., Jr., and Greengard, P. (1985) Protein kinases in the brain. *Ann. Rev. Biochem.* **54,** 931–976.

Nairn, A. C. and Shenolikar, S. (1992) The role of protein phosphatases in synaptic transmission, plasticity and neuronal development. *Curr. Opinion Neurobiol.* **2,** 296–301.

Nestler, E. J. and Greengard, P. (1982) Nerve impulses increase the phosphorylation state of protein I in rabbit superior cervical ganglion. *Nature* **296,** 452–454.

Nicholls, D. G. (1994) *Proteins, Transmitters and Synapses,* Blackwell Scientific, Oxford, UK, p. 253.

Nishizuka, Y. (1995) Protein kinase C and lipid signaling for sustained cellular responses. *FASEB J.* **9,** 484–496.

O'Callaghan, J. P., Lavin, K. L., Chess, Q., and Clouet, D. H. (1983) A method for dissection of discrete regions of rat brain following microwave irradiation. *Brain Res. Bull.* **11,** 31–42.

O'Farrell, P. H. (1975) High resolution two-dimensional electrophoresis of proteins. *J. Biol. Chem.* **250,** 4007–4021.

O'Farrell, P. Z., Goodman, H. M., and O'Farrell, P. H. (1971) High resolution two-dimensional electrophoresis of basic as well as acidic proteins. *Cell* **12,** 1133–1142.

Østvold, A. C., Holtlund, J., and Laland, S. G. (1985) A novel, highly phosphorylated protein, of the high-mobility group type, present in a variety of proliferating and non-proliferating mammalian cells. *Eur. J. Biochem.* **153,** 469–475.

Palfrey, H. C. and Mobley, P. (1987) Second messengers and protein phosphorylation in the nervous system, in *Neurochemistry—A Practical Approach* (Turner, A. J. and Bachelard, H. S., eds.), IRL, Oxford, UK, pp. 161–191.

Palkovits, M. (1973) Isolated removal of hypothalamic or other brain nuclei of the rat. *Brain Res.* **59,** 449,450.

Palmer, W. K., McPherson, J. M., and Walsh, D. A. (1980) Critical controls in the evaluation of cAMP-dependent protein kinase activity ratios as indices of hormonal action. *J. Biol. Chem.* **255,** 2663–2666.

Pearson, R. B. and Kemp, B. E. (1991) Protein kinase phosphorylation site sequences and consensus specificity motifs: tabulations. *Methods Enzymol.* **200,** 63–81.

Pearson, R. B., Woodgett, J. R., Cohen, P., and Kemp, B. E. (1985) Substrate specificity of a multifunctional calmodulin-dependent protein kinase. *J. Biol. Chem.* **260,** 14,471–14,476.

Racker, E. (1991) Use of synthetic amino acid polymers for assay of protein-tyrosine and protein serine kinases. *Methods Enzymol.* **200,** 107–111.

Racker, E. and Sen, P. C. (1991) Assay of phosphorylation of small substrates and of synthetic random polymers that interact chemically with adenosine 5'-triphosphate. *Methods Enzymol.* **200,** 112–114.

Robinson, P. J., Liu, J.-P., Chen, W., and Wenzel, T. (1993) Activation of protein kinase C *in vitro* and in intact cells or synaptosomes determined by acetic acid extraction of MARCKS. *Anal. Biochem.* **210,** 172–178.

Rodnight, R., Trotta, E. E., and Perrett, C. (1985) A simple and economical method for studying protein phosphorylation *in vivo* in the rat brain. *J. Neurosci. Methods* **13,** 87–95.

Rudolph, S. A. and Krueger, B. K. (1979) Endogenous protein phosphorylation and dephosphorylation, in *Advances in Cyclic Nucleotide Research,* vol. 10 (Brooker, G., Greengard, P., and Robison, G. A., eds.) Raven, New York, pp. 107–133.

Rudolph, S. A., Beam, K. G., and Greengard, P. (1978) Studies of protein phosphorylation in relation to hormonal control of ion transport in intact cells, in *Membrane Transport Processes,* vol. 1 (Hoffman, J. F., ed.), Raven, New York, pp. 107–123.

Saneto, R P. and De Vellis, J. (1987) Neuronal and glial cells: cell culture of the central nervous system, in *Neurochemistry—A Practical Approach* (Turner, A. J. and Bachelard, H. S., eds.), IRL, Oxford, UK, pp. 27–63.

Schägger, H. and von Jagow, G. (1987) Tricine-sodium dodecyl sulfate-polyacrylamide gel electrophoresis for the separation of proteins in the range from 1 to 100 kDa. *Anal. Biochem.* **166,** 368–379.

Schulman, H. (1984) Phosphorylation of microtubule-associated proteins by a Ca^{2+}/calmodulin-dependent protein kinase. *J. Cell. Biol.* **99,** 11–19.

Schurr, A., Teyler, T. J., and Tseng, M. T. (eds.) (1987) *Brain Slices: Fundamentals, Applications and Implications* (Conf., Louisville, KY, 1986) Karger, Basel, p. 193.

Scott, M. G., Crimmins, D. L., McCourt, D. W., Tarrand, J. J., Eyerman, M. C., and Nahm, M. H. (1988) A simple *in situ* cyanogen bromide cleavage method to obtain internal amino acid sequence of proteins electroblotted to polyvinyldifluoride membranes. *Biochem. Biophys. Res. Commun.* **155,** 1353–1359.

Segel, I. H. (1975) *Enzyme Kinetics: Behavior and Analysis of Rapid Equilibrium and Steady-State Enzyme Systems.* Wiley, New York, p. 957.

Shahar, A., De Vellis, J., Vernadakis, A., and Haber, B. (eds.) (1989) *A Dissection and Tissue Culture Manual of the Nervous System.* Liss, New York, p. 371.

Sihra, T. S., Bogonez, E., and Nicholls, D. G. (1992) Localized Ca^{2+} entry preferentially effects protein dephosphorylation, phosphorylation, and glutamate release. *J. Biol. Chem.* **267**, 1983–1989.

Smith, P. K., Krohn, R. I., Hermanson, G. T., Mallia, A. K., Gartner, F. H., Provenzano, M. D., Fujimoto, E. K., Goeke, N. M., Olson, B. J., and Klenk, D. C. (1985) Measurement of protein using bicinchoninic acid. *Anal. Biochem.* **150**, 76–85.

Snyder, G. L., Girault, J.-A., Chen, J. Y. C., Czernik, A. J., Kebabian, J. W., Nathanson, J. A., and Greengard, P. (1992) Phosphorylation of DARPP-32 and protein phosphatase inhibitor-1 in rat choroid plexus: regulation by factors other than dopamine. *J. Neurosci.* **12**, 3071–3083.

Strombom, U., Forn, J., Dolphin, A. C., and Greengard, P. (1979) Regulation of the state of phosphorylation of specific neuronal proteins in mouse brain by *in vivo* administration of anesthetic and convulsant agents. *Proc. Natl. Acad. Sci. USA* **76**, 4687-4690.

Suzuki, T., Okumura-Noji, K., Ogura, A., Kudo, Y., and Tanaka, R. (1992) Antibody specific for the Thr-286-autophosphorylated alpha subunit of Ca^{2+}/calmodulin-dependent protein kinase II. *Proc. Natl. Acad. Sci. USA* **89**, 109–113.

Towbin, H., Staehelin, T., and Gordon, J. (1979) Electrophoretic transfer of proteins from polyacrylamide gels to nitrocellulose sheets: procedure and some applications. *Proc. Natl. Acad. Sci. USA* **76**, 4350–4354.

Treiman, M. and Greengard, P. (1985) D-1 and D-2 dopaminergic receptors regulate protein phosphorylation in the rat neurohypophysis. *Neuroscience* **15**, 713–722.

Tsou, K. and Greengard, P. (1982) Regulation of phosphorylation of proteins 1, IIIa, and IIIb in rat neurohypophysis in vitro by electrical stimulation and by neuroactive agents. *Proc. Natl. Acad. Sci. USA* **79**, 6075–6079.

Tuazon, P. T. and Traugh, J. A. (1991) Casein kinase I and II—multipotential serine protein kinases: structure, function, and regulation. *Adv. Second Messenger Phosphoprotein Res.* **23**, 123–164.

Ueda, T., Greengard, P., Berzins, K., Cohen, R. S., Blomberg, F., Grab, D. J., and Siekevik, P. (1979) Subcellular distribution in cerebral cortex of two proteins phosphorylated by a cAMP-dependent protein kinase. *J. Cell Biology* **83**, 308–319.

Walaas, S. I., Browning, M. D., and Greengard, P. (1988) Synapsin Ia, synapsin Ib, protein IIIa and protein IIIb, four related synaptic vesicle-associated phosphoproteins, share regional and cellular localization in rat brain. *J. Neurochem.* **51**, 1214–1220.

Walaas, S. I. and Greengard, P. (1991) Protein phosphorylation and neuronal function. *Pharmacol. Rev.* **43**, 299–349.

Walaas, S. I. and Nairn, A. C. (1989) Multisite phosphorylation of micro-tubule-associated protein 2 (MAP-2) in rat brain: peptide mapping distinguishes between cAMP-, calcium/calmodulin-, and calcium/phospholipid-regulated phosphorylation mechanisms. *J. Mol. Neurosci.* **1,** 117–127.

Walaas, S. I., Nairn, A. C., and Greengard, P. (1983) Regional distribu-tion of calcium- and cyclic adenosine 3',5'-monophosphate-regu-lated protein phosphorylation systems in mammalian brain. I. Particulate systems. *J. Neurosci.* **3,** 291–301.

Walaas, S. I., Sedvall, G., and Greengard, P. (1989a) Dopamine-regu-lated phosphorylation of synaptic vesicle-associated proteins in rat neostriatum and substantia nigra. *Neuroscience* **29,** 9–19.

Walaas, S. I., Østvold, A. C., and Laland, S. G. (1989b) Phosphorylation of P1, a high mobility-like protein, catalyzed by casein kinase II, protein kinase C, cAMP-dependent protein kinase and calcium/calmodulin-dependent protein kinase II. *FEBS Lett.* **258,** 106–108.

Walaas, O., Walaas, E., and Gronnerod, O. (1973) Hormonal regulation of cyclic-AMP-dependent protein kinase of rat diaphragm by epi-nephrine and insulin. *Eur. J. Biochem.* **40,** 465–477.

Walker, J. M. (ed.) (1984) *Methods Molecular Biology, vol. 1, Proteins,* Humana, Clifton, NJ, p. 365.

Wang, J. K., Walaas, S. I., and Greengard, P. (1988) Protein I phosphorylation in nerve terminals: comparison of calcium/calmodulin-dependent and calcium/diacylglycerol-dependent systems. *J. Neurosci.* **8,** 281–288.

Weeks, G., Picciotto, M., Nairn, A. C., Walaas, S. I., and Greengard, P. (1988) Purification and characterization of PCPP-260: a Purkinje cell-enriched cAMP-regulated membrane phosphoprotein of M_r 260,000. *Synapse* **2,** 89–96.

Whittaker, V. P. (1969) The synaptosome, in *Handbook of Neurochemistry,* vol. II (Lajtha, A., ed.), Plenum, New York, pp. 327–364.

Whittingham, T. S. (1987) Metabolic studies in the hippocampal slice preparation, in *Brain Slices: Fundamentals, Applications and Implica-tions* (Conf., Louisville, KY 1986) (Schurr, A., Teyler, T. J., and Tseng, M. S., eds.), Karger, Basel, pp. 59–69.

Whittingham, T. S., Lust, W. D., Christakis, D. A., and Passonneau, J. V. (1984) Metabolic stability of hippocampal slice preparations dur-ing prolonged incubation. *J. Neurochem.* **43,** 689–696.

Wiegant, V. M., Zwiers, H., Schotman, P., and Gispen, W. H. (1978) Endoge-nous phosphorylation of rat brain synaptosomal plasma membranes in vitro: some methodological aspects. *Neurochem. Res.* **3,** 443–453.

Wilkinson, J. M. (1986) Fragmentation of polypeptides by enzyme meth-ods, in *Practical Protein Chemistry—A Handbook* (Darbre, A., ed.), Wiley, New York, pp. 121–148.

Witt, J. J. and Roskoski, R., Jr. (1975) Rapid protein kinase assay using phosphocellulose-paper absorption. *Anal. Biochem.* **66,** 253–258.

Analysis
of Protein Dephosphorylation
in Intact Cells and Extracts

Shelley Halpain

1. Introduction

Protein phosphorylation/dephosphorylation systems have evolved as a primary means through which cellular processes are regulated posttranslationally. Indeed, most transmembrane signals exert their biological effects, at least in part, by altering the balance of protein kinase and phosphatase activity within cells. The nervous system is particularly enriched both in protein kinases and protein phosphatases, and these mediate a spectrum of transient to long-term signaling events. Protein kinases are enzymes that transfer a phosphoryl group from ATP to serine, threonine, or tyrosine residues. Protein phosphatases reverse this action by hydrolyzing phosphoryl groups from these residues.

However, a protein phosphatase reaction is not simply the reverse of a kinase reaction, because the substrate protein targeted by the phosphatase is different from that targeted by the kinase. The addition of a negatively charged phosphate moiety alters the protein's conformation and thereby alters its functional state. Thus, protein kinases and their opposing phosphatases must recognize structurally different targets. Substrate specificity for protein kinases is encoded in their ability to recognize short stretches of specific amino acids within the primary sequence. In contrast, protein phos-

From: *Neuromethods, Vol. 30: Regulatory Protein Modification: Techniques and Protocols*
Ed: H. C. Hemmings, Jr. Humana Press Inc.

phatase specificity appears to be dictated not by primary sequence, but rather by secondary and tertiary structural features that are still poorly characterized (Kennedy and Krebs, 1991).

As a result of these different structural requirements, it is not possible to match a particular protein phosphatase to a particular kinase, nor is it possible at this time to identify consensus sequences for protein phosphatases. In practical terms, this means that two substrate proteins, or even two sites within a single protein, that are phosphorylated by a given kinase might be dephosphorylated by two distinct protein phosphatases (Fig. 1). A well-known example of this is provided by the two sites phosphorylated by cyclic AMP (cAMP)-dependent protein kinase on phosphorylase kinase, a substrate first used in identifying multiple protein phosphatases (*see* Section 1.1.).

1.1. Serine/Threonine Protein Phosphatases

Protein phosphatases fall generally into two categories: those that target exclusively tyrosine (Tyr) residues or those that target exclusively serine (Ser) and threonine (Thr) residues, although recent exceptions have been demonstrated (Walton and Dixon, 1993). Aspects of protein tyrosine phosphorylation systems are addressed elsewhere in this volume (*see* Chapter 6). This chapter will focus on phosphatases that selectively dephosphorylate proteins at Ser/Thr residues. Knowledge about Ser/Thr protein phosphatases has evolved rapidly in the past several years, and has been the subject of several excellent and detailed reviews (e.g., Shenolikar and Nairn, 1991; Mumby and Walter, 1993; Brautigan, 1994; Depaoli-Roach et al., 1994). Basic features of these enzyme systems will be discussed here briefly.

Biochemical analyses led to a classification scheme for Ser/Thr phosphatases that is still used. It separates the phosphatases into two broad categories. Type 1 phosphatase was originally defined as an activity that dephosphorylated the β subunit of phosphorylase kinase (Ingebritsen et al., 1980). Type 2 phosphatase activity dephosphorylated the α subunit

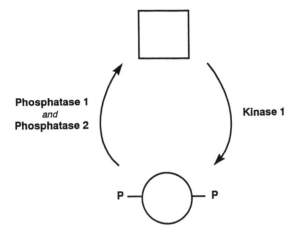

Fig. 1. Schematic diagram to illustrate two points: (1) Protein kinases and phosphatases recognize different structural forms of a given substrate protein, and (2) sites that can be phosphorylated by a single protein kinase might each be dephosphorylated by a distinct protein phosphatase.

of phosphorylase kinase. Later, the type 2 phosphatase activities were further subdivided into three categories on the basis of biochemical properties and cation dependence, termed protein phosphatase 1 (PP1), protein phosphatase 2A (PP2A), protein phosphatase 2B (PP2B), and protein phosphatase 2C (PP2C) (Ingebritsen and Cohen, 1983). Some distinguishing features of the Ser/Thr protein phosphatases are given in Table 1.

The catalytic subunits of PP1, PP2A, and PP2B belong to a single gene family and their amino acid sequences are highly conserved across organisms (Shenolikar and Nairn, 1991). PP2C belongs to a different gene family and is less well characterized, in terms of both substrates and regulation. It can be distinguished from the other Ser/Thr phosphatases by its Mg^{2+}-dependence. In vitro PP1 and PP2A exhibit broad substrate specificity and their activities are similar against many substrates. However, PP1 is readily distinguished by its sensitivity to either of two small protein inhibitors, termed inhibitor-1 and inhibitor-2 (Huang and Glinsmann, 1976), which are endogenous regulators of PP1 activity (*see* Chapter 4). Small protein inhibitors specific for PP2A have recently been

Table 1
Properties of Ser/Thr Protein Phosphatases

	Prefers α or β subunit of phosphorylase kinase	Inhibited by inhibitor-1 and inhibitor-2	Stimulated by Ca^{2+} calmodulin	Requires Mg^{2+}	IC$_{50}$ for okadaic acid in vitro	Catalytic subunits	Regulatory subunits	Regulatory subunit M$_r$
PP1	β	Yes	No	No	100 nM	C1 α, γ1, γ2, δ		
ATP-Mg^{2+}-dependent							R (inhibitor-2)	22,900
Glycogen/SR-associated							R$_{GL}$	124,000
Myrofibril-associated							R$_{MY}$	130/20,000
Nuclear							NIPP-1	18,000
PP2A	α	No	No	No	1 nM	C2 α, β		
PP2A$_2$							A α, β	60,000–65,000
PP2A$_1$							B α, β, γ	52,000–54,000
PP2A$_0$							B' α, β, γ, δ	53,000–55,000
PPPCS$_M$							B''	72,000–130,000
PP2B (calcineurin)	α	No	Yes	No	5 μM	A α, β, γ	B Calmodulin	18,000 17,000
PP2C	α	No	No	Yes	>>100 μM	C α, β	None	

48

described (Li et al., 1995) that may also become useful tools in protein phosphatase research.

Among the Ser/Thr protein phosphatases, only PP2B is activated directly by binding of a second messenger. PP2B, also known as calcineurin, is stimulated by binding of Ca^{2+} to its B subunit, which is structurally similar to parvalbumin, and by binding of Ca^{2+}/calmodulin to its A subunit, which contains the catalytic domain. PP2B was initially purified as a calmodulin binding protein highly enriched in vertebrate nervous system (Klee et al., 1988). In comparison to the other Ser/Thr protein phosphatases it exhibits a highly restricted substrate specificity in vitro.

1.2. Regulation of Phosphatase Activity

For over three decades, it has been clear that protein kinase activity is regulated, for example, by second messengers like cAMP and Ca^{2+}. However, until recently it was thought that protein phosphatases simply served a housekeeping function, constitutively reversing the action of protein kinases. This traditional view emerged, in part, because protein phosphatase activities could be purified as monomeric catalytic units. More recent evidence contradicts this traditional view. The protein phosphatases, too, are highly regulated enzymes, both via posttranslational modification (Johansen and Ingebritsen, 1986; Chen et al., 1992; Guo and Damuni, 1993) and by regulatory subunits that control activity, substrate specificity, and subcellular targeting (Shenolikar and Nairn, 1991; Depaoli-Roach et al., 1994).

Studies by several laboratories have demonstrated the importance of regulatory subunits in controlling the activity and specificity of PP1, and more recently PP2A, by targeting the catalytic subunit (C) to specific subcellular domains. As mentioned above, the C subunits of PP1 and PP2A can function in the absence of other subunits and exhibit only moderate specificity among a variety of substrates in vitro. In vivo, however, the situation is different, and these catalytic subunits are restricted in their substrate specificity by associating with various regulatory subunits.

Cell-specific functions of PP1 and PP2A arise through expression of regulatory subunits that differ widely in structure. At least four distinct regulatory subunits (R) target PP1 activity to sarcoplasmic reticulum, smooth muscle myosin, the nucleus, or the cytosol (Depaoli-Roach et al., 1994). Native PP2A appears to exist as either a heterodimer (designated AC) or heterotrimer (designated ABC). Two forms of the A regulatory subunit have been identified, and at least eight B subunits of PP2A have been cloned (Mumby and Walter, 1993; Brautigan, 1994; Depaoli-Roach et al., 1994; Wera and Hemmings, 1995). The functions and cellular expression of these PP2A subunits are only beginning to be elucidated. For example, a brain-specific isoform of Bβ has been reported (Mayer et al., 1991), and a Bα subunit has been suggested to target PP2A activity to microtubules in cultured cells (Sontag et al., 1995). PP2A activity is also associated with neurofilaments (Saito et al., 1995), although specific targeting subunits have yet to be identified.

Regulation via subcellular targeting is probably also important for the other protein phosphatases. Relatively little is known about PP2C, and only two isoforms of C subunit and no regulatory subunits have so far been identified (Depaoli-Roach et al., 1994). In cultured primary neurons PP2B is selectively enriched in growth cones (Ferreira et al., 1993), suggesting that mechanisms exist for targeting PP2B to specific subcellular domains. Indeed, PP2B binding was demonstrated to the same anchoring protein that targets cAMP-dependent protein kinase to specific sites within cells (Coghlan et al., 1995). The next several years promise to reveal additional examples of subcellular localization of signaling enzymes and to provide increasing insight into the subtleties of protein phosphatase regulation in vivo.

2. Measurement of Phosphatase Activity in Neural Tissue

Several different approaches have been developed to study the role of protein phosphatases in neural mechanisms. It is of interest to evaluate protein phosphatase levels and activity, to delineate signal transduction mechanisms that regulate such phosphatase activity, and to identify endogenous substrates of specific phosphatases. Many of the techniques discussed in detail elsewhere in this volume can be applied to evaluate the contribution of protein phosphatases in the nervous system. For example, the development of specific inhibitors of protein phosphatases has greatly facilitated identification of phosphatase activities in vitro as well as in living cells (*see* Chapter 4). The use of ^{32}P metabolic labeling, back phosphorylation assays (*see* Chapter 1), and phospho-epitope-specific antibodies (*see* Chapter 5) are equally useful for evaluating protein phosphatase pathways as well as protein kinase pathways.

One simple way to evaluate protein phosphatases is to measure the levels of their catalytic subunits via quantitative immunoblot analysis (*see* Chapter 5). Antibodies to various protein phosphatase catalytic subunits have been described and some are available commercially (e.g., Transduction Laboratories, Lexington, KY; Upstate Biotechnology, Lake Placid, NY; and others). An alternative, and perhaps more meaningful, approach is to assay protein phosphatase activity in extracts (*see* Section 3.1.). Both approaches have been used to measure developmental expression and tissue distribution of protein phosphatases and changes associated with pathological states (Polli et al., 1991; Gong et al., 1993; Dudek and Johnson, 1995).

One problem with such an approach, however, is that, as discussed above, protein phosphatase activity is highly regulated by a variety of factors. The relative amount of a catalytic subunit may not be completely indicative of its *in situ* activity and such activity may be poorly preserved during the process of tissue extraction. Homogenization of tissue in some cases disrupts the binding or stability of phosphatase subunits, and since the C subunit is active in monomer form it is possible that preparations of extracts alter the activity of protein phosphatases, thereby obscuring stimulation-induced changes in their activity (Brautigan and Shriner, 1988). For example, an enzyme that in vivo exhibits tightly controlled activity by virtue of subcellular targeting close to membrane-associated signal transduction systems may become either activated or inactivated on dissociation from its natural intracellular environment. Therefore, in vitro assays of phosphatase activity can provide only a snapshot of enzymatic "potential" under given conditions. They are limited in their ability to quantify the physiologically relevant amount of a given protein phosphatase's activity. Thus, the challenge is to extract protein phosphatases and/or their endogenous substrates in a manner that preserves their *in situ* functional state.

3. Protocols

3.1. Measurement of Protein Phosphatase Activity in Extracts

3.1.1. Sample Preparation

As discussed above it can be difficult to measure reliably in vitro the activity of highly regulated enzymes in a way that accurately represents their state of activation in intact cells. This is true for many protein kinases and phosphatases, because activation steps (i.e., binding of a second messenger or regulatory subunit) are highly reversible. For assaying phosphatase activity, tissue must be homogenized under conditions that neither inactivate nor stimulate the activity that was present when the tissue was collected. In practice this is difficult, if not impossible, to achieve for certain enzymes, especially in the case of the Ca^{2+}-dependent protein phosphatase PP2B. Standard homogenization buffer for phosphatase assay contains EGTA, which serves to chelate Ca^{2+}, thereby inactivating any PP2B that was active *in situ*. On the other hand, if homogenization is carried out in the absence of chelators, the release of Ca^{2+} from internal stores during cell disruption would stimulate PP2B activity above its *in situ* state. Despite difficulties in assessing PP2B activation *in situ*, it is nevertheless possible to assay total phosphatase activity in tissue derived from CNS slices, and changes in Ca^{2+}-independent activity can be detected.

Starting material for phosphatase activity assays can be fresh or frozen brain tissue, brain slices, or cultured cells. Tissue is homogenized at 4°C with a Potter (glass/Teflon) apparatus in a hypotonic buffer, in the presence of protease inhibitors: 10 m*M* Tris-HCl, pH 7.5, 2 m*M* EGTA, 1 m*M* EDTA, 10 m*M* 2-mercaptoethanol, 20 µg/mL leupeptin, 2 µg/mL pepstatin A, 20 µg/mL Trasylol (aprotinin), and 0.5 m*M* phenylmethylsulfonyl fluoride (PMSF).

3.1.2. Phosphatase Activity Assay

Endogenous phosphatase activity is assayed as the release of ^{32}P from an appropriate purified substrate phosphorylated in vitro. For example, phosphorylase kinase phosphorylated by cAMP-dependent protein kinase is appropriate for all four classes of Ser/Thr phosphatase (Ingebritsen and Cohen, 1983). Alternatively, glycogen phosphorylase *a* phosphorylated by phosphorylase kinase can be used as a substrate for PP1 and PP2A (Brautigan and Shriner, 1988). Several convenient substrates for assaying PP2B have been described, including the RII regulatory subunit of cAMP-dependent protein kinase (Parsons et al., 1994), myosin light chains phosphorylated by their associated kinase (Klee et al., 1988), or inhibitor-1 or DARPP-32 phosphorylated by cAMP-dependent protein kinase (King et al., 1984). The purified protein is first phosphorylated using purified kinase in the presence of $[\gamma\text{-}^{32}P]$-ATP (*see* Chapter 1). It is then separated from $[\gamma\text{-}^{32}P]$-ATP by chromatography on a Dowex column followed by repurification and concentration of the labeled protein (e.g., via reverse-phase-HPLC or ammonium sulfate precipitation and dialysis). After purification, more than 98% of the radioactivity must be precipitable by trichloroacetic acid (TCA).

The phosphatase assay is run in a final volume of 40 µL containing: 10 µL 4X phosphatase assay buffer, 10 µL tissue extract, 10 µL water or pharmacological agents to activate or inhibit activity (*see* Section 3.1.3.), and 10 µL phosphorylated substrate (1000–5000 cpm). The recipe for 4X phosphatase assay buffer is: 200 m*M* Tris-HCl, pH 7.0, 4 mg/mL bovine serum albumin (BSA), and 20 m*M* 2-mercaptoethanol.

Typically the homogenate is preincubated for 5 min at 30°C and the labeled substrate is added last to start the reaction. After a 10 min incubation at 30°C, the reaction is stopped by the addition of 100 µL of ice-cold 20% (w/v) TCA and 100 µL of 6 mg/mL BSA as a carrier protein. After allowing the tubes to sit on ice for 5 min, they are centrifuged at 10,000*g* and 150 µL of the supernatant is removed and subjected to Cerenkov or liquid scintillation counting. The optimal concentration of protein in the homogenate added to the assay should be determined in a preliminary experiment in order to ascertain that the assay is run under initial rate conditions. The concentration (usually around 0.1 mg/mL) should be such that the radioactivity released is <10% of the total and is linear with respect to time and protein concentration.

An alternative to the use of radiolabeled substrates has been provided by the development of colorimetric assays for protein phosphatase activity (e.g., Serine/Threonine Phosphatase Assay System, Promega, Madison, WI). Although fairly expensive, such kits can be advantageous because they provide the phosphatase substrate, thereby eliminating the need to select, purify, and phosphorylate substrate. One drawback of this kit is that the peptide substrate provided does not work well for Type 1 phosphatase activity. However, the substrate can be dephosphorylated by all the known Type 2 Ser/Thr protein phosphatases and thus various activities must be distinguished by the use of inhibitors and activators added to the assay, as described in Section 3.1.3. In addition, several compounds (for example, reducing agents) have the potential to interfere with the assay, so the manufacturer's recommended protocol should be carefully evaluated for compatibility with the desired tissue preparation and buffer choice.

3.1.3. Distinguishing Different Protein Phosphatases

Specific phosphatase activities can be distinguished with the help of activators and inhibitors (Table 2; *see* Chapter 4). Okadaic acid, a selective inhibitor of PP1 and PP2A, can be added to the assay to distinguish these phosphatase activities from those of PP2B and PP2C. (Microcystin-LR is a less expen-

Table 2
Assay Conditions for Distinguishing Phosphatase Activities

Addition (final concentration)	PP1	PP2A	PP2B	PP2C
Inhibitor-2[a] (50 ng/μL)	–	+	+	+
Okadaic acid (0.5 μM)	–	–	+	+
5 mM CaCl$_2$	–	–	+	–
Calmodulin[b] (60 μg/mL)	–	–	+	–
10 mM MgCl$_2$	–	–	–	+

[a]Available from Promega (Madison, WI).
[b]Available from Sigma (St. Louis, MO).

sive alternative to okadaic acid for selectively inhibiting PP1 and PP2A activity in extracts. However, unlike okadaic acid, it is not membrane permeable, so it cannot be used to block phosphatase activity in cells). Specific inhibition of PP1 is achieved by addition of inhibitor-2 or phospho-inhibitor-1 to the assay. PP2B is stimulated by the addition of Ca^{2+}/calmodulin and PP2C is assayed in the presence of 10 mM MgCl$_2$.

3.2. Assaying the Phosphorylation State of Phosphoprotein Substrates in Tissue

Three methods for evaluating the state of phosphorylation of specific proteins from neural tissue will be discussed.

3.2.1. Metabolic Labeling with [^{32}P] Orthophosphate

Activation of protein phosphatases by transmembrane signals can be studied in slices or cultures metabolically prelabeled to allow endogenous protein kinases to phosphorylate proteins in the presence of [^{32}P]ATP. Since ATP does not readily cross the plasma membrane, it is necessary to incubate tissue with ^{32}P-labeled inorganic phosphate (e.g., [^{32}P] orthophosphoric acid in water, 25 mCi/mL; New England Nuclear, Boston, MA), which is then transported across the plasma membrane. A portion of this labeled phosphate is incorporated into the γ phosphate of ATP produced by the ongoing metabolic activity of the cells. The ^{32}P found in such labeled tissue is not only in the form of ATP and protein, but is also incorporated into phospholipids, sugars, and various nucleotides. Since incorporation into the particular protein of interest has a relatively low efficiency, it is necessary to use a large amount of ^{32}P (0.2–2 mCi/mL is recommended; in our laboratory we typically use 0.5 mCi/mL). Using an isotope with a high specific activity (25–100 mCi/mL) facilitates this. In addition, the incubation buffer must be free of unlabeled phosphate, which would reduce the specific activity of the isotope. Typically we preincubate tissue for 45–60 min in phosphate-free medium prior to incubation in phosphate-free medium containing [^{32}P] orthophosphate. An optimal period of labeling must be established for the proteins of interest. Using rat hippocampal slices, ^{32}P incorporation into the proteins MAP2 and DARPP-32 reaches a steady state within 90 min of incubation in the presence of 0.5–2 mCi/mL [^{32}P] orthophosphate. Using such methods, MAP2 and DARPP-32 were identified as targets of glutamate receptor-stimulated dephosphorylation in rat brain slices (Halpain et al., 1990; Halpain and Greengard, 1990). Similar methods work well for cultured hippocampal neurons (S. Halpain, unpublished observation).

It is essential to use appropriate safety precautions when handling such large amounts of radioisotope. These include the use of eye protection, double gloves, sleeve protectors, lab coats, and Plexiglas shielding at least 1 in. thick (we typically use 35-mm Plexiglas shields with an additional 1/8-in. lead sheet placed between the Plexiglas and the experimenter to contain secondary Bremsstrahlung radiation). Solid waste (pipet tips, tubes, contaminated gloves) is collected behind the shield. At the end of the prelabeling period, drugs are added as appropriate to each incubation condition for the specified time. Then the ^{32}P-containing buffer is carefully removed and blotted onto paper towels placed in a plastic bag behind the shield. The ^{32}P-containing buffer is thus disposed of as solid waste. For brain slices we typically stop the incubation by immersing the lower half of the tube containing the slice in liquid nitrogen to rapidly freeze the tissue. For cultured cells, we add boiling hot 1% (w/v) sodium dodecyl sulfate (SDS) to the dish, scrape loose cells, then sonicate immediately, as described below.

In order to maintain proteins in their *in situ* phosphorylation state, it is necessary to homogenize tissue in a way that
completely inactivates all endogenous enzymatic activity,
including kinases, phosphatases, and proteases. The best way
to achieve this is to rapidly solubilize tissue under denaturing conditions. Tubes containing frozen slices are placed on
powdered dry ice adjacent to a probe-type microsonicator
(MS-50, Heat Systems Ultrasonics, Farmingdale, NY) and a
boiling water bath containing a tube of 1% (w/v) SDS. Each
slice is solubilized by brief bursts of the sonicator after adding 0.2 mL of the boiling hot 1% (w/v) SDS. The time between
removing the slice from dry ice and sonication should be minimized. Brain slices (e.g., 0.5–1 mm slices of adult rat hippocampus) can be solubilized within 1–2 s of detergent addition.

Protein concentrations are determined using the bicinchoninic acid assay (Pierce). Equal concentrations of homogenate are placed in SDS sample buffer and separated by one- or
two-dimensional gel electrophoresis or are subjected to
immunoprecipitation using specific antibodies (*see* Chapter 3).
For immunoprecipitation, homogenates are diluted 1:1 with
a buffer consisting of: 50 mM Tris-HCl, pH 7.4, 50 mM NaCl,
5 mM EDTA, 50 mM NaF, 1 mM PMSF, and 6% (v/v) Nonidet
P-40 (NP40).

The nonionic detergent NP40 serves to "quench" the
denaturing detergent SDS, thereby enabling antibody–antigen binding to occur. Samples are incubated with antibody
at 4°C for 1 h, followed by incubation for an additional 30 min
after the addition of formaldehyde-fixed protein A-bearing
Staphylococcus aureus cells (Pansorbin; Calbiochem, La Jolla,
CA) or another immunoaffinity resin, such as Protein A
Sepharose. Samples are centrifuged lightly (3400g at 4°C) to
pellet the immune complexes. Pellets are resuspended in
the above buffer by vigorous vortexing, then repelleted as
before. They are washed twice more in the same buffer as
above, minus NP40, then immunoprecipitates are separated
by SDS-polyacrylamide gel electrophoresis (PAGE). Gels are
subjected to autoradiography or to phosphorimager analysis to visualize and quantify ^{32}P incorporation into proteins.

3.2.2. Back-Phosphorylation Assay

Back-phosphorylation was a method originally designed for the study of synapsin phosphorylation (*see* Chapter 1 for details). The principle of this approach is to extract the protein(s) of interest using conditions that inactivate kinases, phosphatases, and proteases without denaturing the protein(s). Proteins are then phosphorylated with [γ-^{32}P]ATP in the presence of an appropriate purified protein kinase. The enzyme that has been used most successfully in this approach is the catalytic subunit of cAMP-dependent protein kinase. With this method, the greater the occupation by endogenous nonradioactive phosphate of relevant sites for the kinase *in situ*, the less the ^{32}P-radioactivity incorporated by the protein in vitro. In other words, the signal one obtains in a back phosphorylation assay is an estimation of the available in unphosphorylated sites of phosphorylation. Such an approach has several drawbacks for assays of protein phosphatase activity toward a substrate in vivo. First, one must know *a priori* the identity of a kinase to use in phosphorylating the substrate of interest. Second, one needs to have a sufficient supply of purified kinase with high activity. Third, the protein of interest must be extractable under conditions that inactivate endogenous enzymes without denaturing the protein itself. Finally, the method is not sensitive for proteins that are phosphorylated to a low stoichiometry in intact cells (*see* Halpain and Girault, 1995, for a discussion). Nevertheless, this technique may provide a useful alternative for evaluating changes in endogenous protein dephosphorylation, especially when metabolic labeling approaches are impractical.

3.2.3. Phosphoepitope-Specific Antibody

A very powerful approach that can be used to study protein dephosphorylation in intact cells is the use of phosphorylation state-specific antibodies (*see* Chapter 5). Such antibodies can be raised against a phosphorylated synthetic peptide whose sequence encompasses the phosphorylated residue of interest. Selected peptides must be short and can be phosphorylated enzymatically or synthesized with a phosphorylated residue. In several instances antibodies selective for the phosphorylated form of the native protein have thus been obtained (Czernik et al., 1991; Ginty et al., 1993). Using unphosphorylated peptides as immunogens, antibodies that react only with the dephosphorylated form of a protein have been produced in the case of synapsin 1 (Czernik et al., 1991) and MAP2 (Sanchez et al., 1995). Phosphoepitope-specific antibodies can be used to estimate the amount of the phosphorylated protein using immunoblot analysis. An important advantage offered by this approach over the above two methods is that it eliminates the need to metabolically label tissue with ^{32}P, and the tissue of interest can be rapidly isolated under conditions that preserve the endogenous phosphorylation state of substrates (e.g., by solubilizing tissue in boiling 1% (w/v) SDS or directly in SDS-PAGE sample buffer). In addition, with the help of other antibodies that bind to the protein in a phosphoepitope-independent manner, and by using known quantities of the phosphoprotein to produce a "standard curve," it is possible to measure the actual stoichiometry of phosphorylation of the protein in intact cells.

Occasionally, production of phosphorylation state-specific antibodies occurs "accidentally" when a protein of undefined phosphorylation state is used as immunogen. Examples include monoclonal antibody AP18, which recognizes phosphoserine-136 of the microtubule-associated protein MAP2 (Berling et al., 1994); monoclonal antibody AT8, which is specific for a phosphoepitope in the microtubule-associated protein tau (Mercken et al., 1992); monoclonal antibody tau1, which recognizes a dephosphorylated phosphoepitope in tau (Szendrei et al., 1993); and several examples of anti-neurofilament antibodies that selectively detect neurofilament proteins in either phosphorylated or dephosphorylated states (Sternberger and Sternberger, 1983). AT8, AP18, tau1, and other phosphorylation state-specific antibodies have been successfully used to examine the regulation of protein dephosphorylation in intact cells (Garver et al., 1994; Mawal-Dewan et al., 1994; Burak and Halpain, 1996).

3.3. Pharmacological Tools for Identifying
   Phosphatases that Target Proteins in Intact Cells

As discussed elsewhere in this volume (Chapter 4), pharmacological agents that specifically block various protein phosphatases in cell-free systems have been described. Unfortunately, only a few of these inhibitors are useful in studies of intact cells, because of their limited membrane permeability. Okadaic acid and calyculin A are compounds that readily cross plasma membranes and are effective at low concentrations. In vitro these compounds inhibit dilute PP1 and PP2A activity completely in the low nanomolar concentration range, however higher concentrations (0.1–2 μM) are necessary to achieve inhibition *in situ* (Cohen et al., 1990; Hardie et al., 1991). Calyculin A is equally potent in inhibiting both PP1 and PP2A in vitro (IC_{50} = 1–2 nM; Ishihara et al., 1989), whereas okadaic acid exhibits >50-fold selectively toward PP2A (IC_{50} = 0.1–1 nM toward PP2A, vs 2–300 nM toward PP1; Bialojan and Takai, 1988; Cohen et al., 1990). Therefore, differential sensitivity to calyculin A and okadaic acid can be used to distinguish PP1 and PP2A activities in extracts. Nevertheless, in practice it is difficult to determine with absolute confidence whether a dephosphorylation event *in situ* is mediated by PP1 vs PP2A using this criterion, since it is difficult to ascertain the intracellular concentrations of such compounds when applied to intact cells. A potent, lipid-permeable inhibitor specific for either PP1 or PP2A is still lacking. On the other hand, okadaic acid inhibits PP2B in vitro with an IC_{50} that is 500-fold higher than for either PP1 or PP2A. Inhibition of an event by high concentrations (but not low concentrations) of okadaic acid is suggestive of a PP2B-mediated event, although some caution on interpretation must be exercised. Two additional types of cell-permeant inhibitors of PP2B have been described: the immunosuppressant compounds cyclosporin A and FK-506 (Kunz and Hall, 1993), and the type II pyrethroids, including deltamethrin and cypermethrin (Enan and Matsumura, 1993). Both of these PP2B-specific inhibitors have been used to inhibit PP2B activity in intact nervous tissue. For example, differential sensitivity to okadaic acid and deltamethrin was used to demonstrate that, although PP1 and/or PP2A regulate dephosphorylation of MAP2 under basal conditions, PP2B is primarily responsible for dephosphorylation of MAP2 stimulated by glutamate receptor activity (Quinlan and Halpain, 1996).

3.4. Dephosphorylation vs Proteolysis: A Caveat

A difficulty in studying the regulation of protein phosphatases, as opposed to protein kinases, in living cells is that a decrease in ^{32}P incorporation into metabolically labeled protein or a decrease in binding of a phosphorylation state-specific antibody can be potentially misinterpreted as dephosphorylation rather than proteolysis. Indeed, proteolysis is of particular concern when studying signal transduction events that raise intracellular Ca^{2+} levels, which could activate Ca^{2+}-dependent proteases and lead to selective degradation of substrates. For this reason, it is important to rule out proteolysis by verifying that the total amount of the protein of interest is not decreased by the stimulus. This is readily achieved by performing quantitative immunoblot analysis on samples treated identically to those examined in the dephosphorylation assay. However, if only a minor proportion of a substrate is labeled during ^{32}P incubation (i.e., <<10%), then it may be difficult to detect proteolytic changes affecting only that specific population via immunoblot analysis of the total pool.

Acknowledgments

The author wishes to thank David Brautigan for helpful discussions and Diana Crosswhite for assistance in preparation of the manuscript. Work in the author's laboratory is supported by the National Institutes of Health.

References

Berling, B., Wille, H., Roll, B., Mandelkow, E.-M., Garner, C., and Mandelkow, E. (1994) Phosphorylation of microtubule-associated proteins MAP2a,b and MAP2c at Ser 136 by proline-directed kinases in vivo and in vitro. *Eur. J. Cell Biol.* **64,** 120–130.

Bialojan, C. and Takai, A. (1988) Inhibitory effect of a marine-sponge toxin, okadaic acid, on protein phosphatases. *Biochem. J.* **256,** 283–290.

Brautigan, D. (1994) Protein phosphatases. *Rec. Prog. Hormone Res.* **49,** 197–214.

Brautigan, D. L. and Shriner, C. L. (1988) Methods to distinguish various types of protein phosphatase activity. *Methods Enzymol.* **159,** 339–346.

Burak, M. A. and Halpain, S. (1996) Site-specific regulation of Alzheimer-like tau phosphorylation in living neurons. *Neuroscience* **72,** 167–184.

Chen, J., Martin, B. L., and Brautigan, D. L. (1992) Regulation of protein serin-threonine phosphatase type-2A by tyrosine phosphorylation. *Science* **257,** 1261–1264.

Coghlan, V. M., Perrino, B. A., Howard, M., Langeberg, L. K., Hicks, J. B., Gallatin, W. M., and Scott, J. D. (1995) Association of protein kinase A and protein phosphatase 2B with a common anchoring protein. *Science* **267,** 108–111.

Cohen, P., Holmes, C. F. B., and Tsukitani, Y. (1990) Okadaic acid: a new probe for the study of cellular regulation. *Trends Biochem. Sci.* **15,** 98–102.

Czernik, A. J., Girault, J. A., Nairn, A. C., Chen, J., Snyder, G., Kebabian, J. W., and Greengard, P. (1991) Production of phosphorylation state-specific antibodies. *Methods Enzymol.* **201,** 264–283.

Depaoli-Roach, A. A., Park, I.-K., Cerovsky, V., Csortos, C., Durbin, S. D., Kuntz, M. J., Sitikov, A., Tang, P. M., Verin, A., and Zolnierowicz, S. (1994) Serine/threonine protein phosphatases in the control of cell function. *Adv. Enzyme Reg.* **34,** 199–224.

Dudek, S. M. and Johnson, G. V. W. (1995) Postnatal changes in serine/threonine protein phosphatases and their association with microtubules. *Dev. Brain Res.* **90,** 54–61.

Enan, E. and Matsumura, F. (1993) Activation of phosphoinositide/protein kinase c pathway ion rat brain tissue by pyrethroids. *Biochem. Pharmacol.* **45,** 703–710.

Ferreira, A., Kincaid, R., and Kosik, K. S. (1993) Calcineurin is associated with the cytoskeleton of cultured neurons and has a role in the acquisition of polarity. *Mol. Biol. Cell* **4,** 1225–1238.

Garver, T. D., Harris, K. A., Lehman, R. A. W., Lee, V. M-.Y., Trojanowski, J. Q., and Billingsley, M. L. (1994) τ phosphorylation in human, primate, and rat brain: evidence that a pool of τ is highly phosphorylated in vivo and is rapidly dephosphorylated in vitro. *J. Neurochem.* **63,** 2279–2287.

Ginty, D. D., Kornhouser, J. M., Thompson, M. A., Bading, H., Mayo, K. E., Takahashi, J. S., and Greenberg M. E. (1993) Regulation of CREB phosphorylation in the suprachiasmatic nucleus by light and a circadian clock. *Science* **260,** 238–241.

Gong, C.-X., Singh, T. J., Grundke-Iqbal, I., and Iqbal, K. (1993) Phosphoprotein phosphatase activities in Alzheimer disease brain. *J. Neurochem.* **61,** 921–927.

Guo, H. and Damuni, Z. (1993) Autophosphorylation-activated protein kinase phosphorylates and inactivates protein phosphatase 2A. *Proc. Natl. Acad. Sci. USA* **90,** 2500–2504.

Halpain, S., Girault, J. A., and Greengard, P. (1990) Activation of NMDA receptors induces dephosphorylation of DARPP-32 in rat striatal slices. *Nature* **343,** 369–372.

Halpain, S. and Girault, J.-A. (1995) The use of brain slices to study protein phosphatase regulation and function. *Neuroprotocols* **6,** 46–55.

Halpain, S. and Greengard, P. (1990) Activation of NMDA receptors induces rapid dephosphorylation of the cytoskeletal protein MAP2. *Neuron* **5,** 237–246.

Hardie, D. G., Haystead, T. A. J., and Sim, A. T. R. (1991) Use of okadaic acid to inhibit protein phosphatases in intact cells. *Methods Enzymol.* **201,** 469–476.

Huang, F. L. and Glinsmann, W. H. (1976) Separation and characterization of two phosphorylase phosphatase inhibitors from rabbit skeletal muscle. *Eur. J. Biochem.* **70,** 419–426.

Ingebritsen, T. S., Foulkes, J. G., and Cohen, P. (1980) The broad specificity protein phosphatase from mammalian liver separation of the M_r 35,000 catalytic subunit into two distinct enzymes. *FEBS Lett.* **119,** 9–15.

Ingebritsen, T. S. and Cohen, P. (1983) The protein phosphatases involved in cellular regulation. 1. Classification and substrate specificities. *Eur J. Biochem* **132,** 255–261.

Ishihara, H., Martin, B. L., Brautigan, D. L., Karaki, H., Ozaki, H., Kato, Y., Fusetani, N., Watabe, S., Hashimoto, K., Uemura, D., and Hartshorne, D. J. (1989) Calyoulin A and okadaic acid: inhibitors of protein phosphatase activity. *Biochem. Biophys. Res. Commun.* **159,** 871–877.

Johansen, J. W. and Ingebritsen, T. S. (1986) Phosphorylation and inactivation of protein phosphatase 1 by pp60[v-src]. *Proc. Natl. Acad. Sci. USA* **83,** 207–211.

Kennelly, P. J. and Krebs, E. G. (1991) Consensus sequences as substrate specificity determinants for protein kinases and protein phosphatases. *J. Biol. Chem.* **266,** 15,555–15,558.

King, M. M., Huang, C. Y., Chock, P. B., Nairn, A. C., Hemmings, H. C., Jr., Chan, K.-F. J., and Greengard, P. (1984) Mammalian brain phosphoproteins as substrates for calcineurin. *J. Biol. Chem.* **259,** 8080–8083.

Klee, C. B., Draetta, G. F., and Hubbard, M. J. (1988) Calcineurin. *Adv. Enzymol.* **61,** 149–200.

Kunz, J. and Hall, M. N. (1993) Cyclosporin A, FK506 and rapamycin, more than just immunosuppression. *Trends Biochem. Sci.* **18,** 334–338.

Li, M., Guo, H., and Damuni, Z. (1995) Purification and characterization of two potent heat stable protein inhibitors of protein phosphatase 2A from bovine kidney. *Biochemistry* **34,** 1988–1996.

Mawal-Dewan, M., Henley, J., Van de Voorde, A., Trojanowski, J. Q., and Lee, V. M.-Y. (1994) The phosphorylation state of tau in the developing rat brain is regulated by phosphoprotein phosphatases. *J. Biol. Chem.* **269,** 30,981–30,987.

Mayer, R. E., Hendrix, P., Cron, P., Matthies, R., Stone, S. R., Goris, J., Merlevede, W., Hofsteenge, J., and Hemmings, B. A. (1991) Structure of the 55-kDa regulatory subunit of protein phosphatase 2A: evidence for a neuronal-specific isoform. *Biochemistry* **268,** 3589–3597.

Mercken, M., Vandermeeren, M., Lübke, U., Six, J., Boons, J., Van der Voorde, A., Martin, J.-J., and Gheuens, J. (1992) Monoclonal antibodies with selective specificity for Alzheimer tau are directed against phosphatase-sensitive epitopes. *Acta Neuropathol.* **84,** 265–272.

Mumby, M. C. and Walter, G. (1993) Protein serine/threonine phosphatases: structure, regulation, and functions in cell growth. *Physiol. Rev.* **73,** 673–699.

Parsons, J. N., Wiederrech, G. J., Salowe, S., Burbaum, J. J., Rokosz, L. L., Kincaid, R., and O'Keefe, S. J. (1994) Regulation of calcineurin phosphatase activity and interaction with the FK-506-FK-506 binding protein complex. *J. Biol. Chem.* **269,** 19,610–19,616.

Polli, J. W., Billingsley, M. L., and Kincaid, R. L. (1991) Expression of the calmodulin-dependent protein phosphatase, calcineurin, in rat brain: developmental patterns and the role of nigrostriatal innervation. *Dev. Brain Res.* **63,** 105–119.

Quinlan, E. M. and Halpain, S. (1996) Postsynaptic mechanisms for bidirectional control of MAP2 phosphorylation by glutamate receptors. *Neuron* **16,** 357–368.

Saito, T., Shima, H., Osawa, Y., Nagao, M., Hemmings, B. A., Kishimoto, T., and Hisanaga, S.-I. (1995) Neurofilament-associated protein phosphatase 2A: its possible role in preserving neurofilaments in filamentous states. *Biochemistry* **34,** 7376–7384.

Sanchez, C., Diaz-Nido, J., and Avila, J. (1995) Variations in *in vivo* phosphorylation at the proline-rich domain of the microtubule-associated protein 2 (MAP2) during rat brain development. *Biochem. J.* **306,** 481–487.

Shenolikar, S. and Nairn, A. C. (1991) Protein phosphatases: recent progress. *Adv. Second Messengers Phosphoprotein Res.* **23,** 1–121.

Sontag, E., Nunbhakdi-Craig, V., Bloom, G. S., and Mumby, M. C. (1995) A novel pool of protein phosphatase 2A is associated with microtubules and is regulated during the cell cycle. *J. Cell Biol.* **128,** 1131–1144.

Sternberger, L. A. and Sternberger, N. H. (1983) Monoclonal antibodies distinguish phosphorylated and nonphosphorylated forms of neurofilaments *in situ. Proc. Natl. Acad. Sci. USA* **80,** 6126–6130.

Szendrei, G. I., Lee, V. M.-Y., and Otvos, L., Jr. (1993) Recognition of the minimal epitope of monoclonal antibody Tau-1 depends on the presence of a phosphate group but not its location. *J. Neurosci. Res.* **34,** 243–249.

Walton, K. M. and Dixon, J. E. (1993) Protein tyrosine phosphatases. *Annu. Rev. Biochem.* **62,** 101–120.

Wera, S. and Hemmings, B. A. (1995) Serine/threonine protein phosphatases. *Biochem. J.* **311,** 17–29.

Protein Phosphorylation and Dephosphorylation in Isolated Nerve Terminals (Synaptosomes)

Talvinder S. Sihra

1. Introduction

Nerve terminal depolarization leads to Ca^{2+} influx and the exocytotic release of neurotransmitters (Jessell and Kandel, 1993; Kelly, 1993). Elucidation of the regulatory influences on this primary function of the nerve terminal is of fundamental importance in understanding synaptic function. Posttranslational modification of proteins by phosphorylation/dephosphorylation is a means of altering the characteristics of a protein and thus represents a key molecular mechanism through which neuronal function can be regulated (Greengard, 1987; Walaas and Greengard, 1991). The aim of this chapter is to provide a working knowledge of the isolated nerve terminal preparation (synaptosomes) as a model for studying protein phosphorylation changes that are specific to the presynaptic digit of the synapse. Each section will briefly review the literature to provide a historical overview of how present working paradigms have been derived. Experimental details will be provided, with appropriate examples, to enable the investigator to analyze a putative regulatory presynaptic phosphoprotein to increasing molecular detail.

From: *Neuromethods, Vol. 30: Regulatory Protein Modification: Techniques and Protocols*
Ed: H. C. Hemmings, Jr. Humana Press Inc.

2. Isolated Nerve Terminals (Synaptosomes)

2.1. Background (Historical)

Early investigations aimed at obtaining a purified synaptic vesicle preparation led to observations by DeRobertis et al. (1961) and Gray and Whittaker (1962) that paved the way for the development of isolated nerve terminal preparations. These workers noted that, under specific preparative conditions, brain homogenates contained anucleate, membrane-bound sacs (0.7–1 μm in diameter) that contained numerous small synaptic vesicles (50 nm in diameter), and in some cases mitochondria. These structures were proposed to be isolated nerve endings sheared from axons and subsequently resealed into intact nerve terminals or synaptosomes (Whittaker et al., 1964). Subsequent work refined the conditions for the reproducible preparation and partial purification of synaptosomes through differential centrifugation steps and sucrose density gradients (Whittaker, 1964).

Preparation of synaptosomes requires conditions of moderate shear and iso-osmolality. The shear force is provided by glass (container)-Teflon (pestle)-based (Potter-Elvehjem) tissue homogenizers. In order to obtain reproducible shear force, it is critical to control the spacing between glass container and Teflon pestle, and the speed of pestle rotation. Iso-osmotic conditions are important during homogenization; specifically, sucrose-based media are optimal for obtaining purer preparations. Ionic media are unsuitable because of a marked aggregating effect (coacervation) of most ions on synaptosomes (Whittaker, 1969). This is a particular problem with divalent cations, which must be rigorously eliminated during the early stages of synaptosome preparation.

The homogenate obtained after initial disruption of brain tissue is a mix of isolated nerve terminals, nuclear debris, membrane fragments, free synaptic vesicles, free mitochondria, microsomal membranes, ribosomes, and myelin. Differential centrifugation allows the elimination of nuclear debris (P_1), free synaptic vesicles (S_2), microsomal membranes (S_2), and ribosomes (S_2) from the homogenate to yield a crude

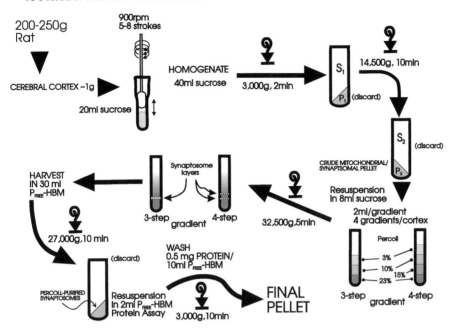

Fig. 1. General scheme of synaptosome preparation by differential centrifugation and Percoll gradient purification.

mitochondrial/synaptosomal fraction (P_2) (Fig. 1). The smooth cerebral cortices of rodent brains appear to be particularly amenable to the isolation of synaptosomes. The ability to remove a significant proportion of the white matter during dissection of the brain is probably a key factor in determining whether a good synaptosomal preparation can be obtained from any given brain area. Apart from the cerebral cortex, synaptosomes have been prepared from a number of other rodent brain regions, including the striatum (Rapier et al., 1990), cerebellum (Terrian et al., 1988), and hippocampus (Thorne et al., 1991), as well as the neurohypophysis (Nordmann et al., 1982). Additionally, the basic protocols have proven successful in the preparation of synaptosomes from neuronal tissue obtained from a number of species, including teleost fish (Whittaker and Greengard, 1971), squid (Dowdall and Whittaker, 1973), chick (Babitch et al., 1976), *Torpedo* (Israel et al., 1985), *Aplysia* (Chin et al., 1989), and crustacea (Santiapillai et al., 1989). This chapter discussion will be limited to the preparation and use of mammalian cerebrocortical synaptosomes.

The crude mitochondrial/synaptosomal pellet (P_2) (Fig. 1) is highly contaminated with free mitochondria, membrane fragments, and myelin. In order to enrich nerve terminals, further purification of this crude synaptosomal fraction on the basis of relative densities is necessary. The principal method used in early studies was density gradient centrifugation using discontinuous sucrose gradients (Gray and Whittaker et al., 1962), which work by a combination of rate-zonal and isopycnic effects. Sucrose gradients with steps of 0.2, 0.4, 0.8, and $1.2M$ are used to separate synaptosomes which sediment at a buoyant density between 0.8 and $1.2M$ sucrose. Gradients have to be run close to equilibrium using swing-out rotors; thus, resolution of membrane fractions requires 3–4 h of centrifugation time. Continuous sucrose gradients provide a quicker method (1.5 h) (Whittaker, 1968). However, separation by either discontinuous or continuous sucrose gradients does expose synaptosomes to very hypertonic conditions. Indeed, this hypertonicity is probably one of the major factors in facilitating the separation of synaptosomes from contaminating fractions by causing differential dehydration of membrane structures (Whittaker, 1968). Thus, although sucrose gradients expedite efficient separation of synaptosomes from mitochondria, light membranes, and myelin, the physiological competence of the final preparation is compromised by the exposure of the synaptosomes to hypertonic conditions for lengthy periods (Whittaker, 1968).

The advent of polymers of sucrose (with epichlorohydrin), such as Ficoll 400 (Pharmacia) (Abdel-Latif, 1966; Autilio et al., 1968), facilitated the use of less hyperosmotic gradients. Ficoll requires shorter centrifugation times for gradient resolution, and Ficoll-gradient-purified synaptosomes appear to be metabolically more competent (Verity, 1972) than those separated on sucrose gradients (Whittaker, 1969). Thus, Ficoll-gradient-purified synaptosomes, with their improved metabolic competency (Nicholls, 1978; Kauppinen and Nicholls, 1986), have been extensively used in transmitter release studies (reviewed in McMahon and Nicholls, 1991). Arguably, however, these synaptosomes are more heterogenous than those purified by sucrose gradients (Kyriazi and Basford, 1986). This may be owing, in part, to the lack of differential osmotic dehydration that is instrumental in separation of synaptosomes in sucrose gradients and to problems introduced by the increased viscosity of Ficoll-polymer at higher densities (Whittaker, 1988). Some improvements in synaptosome purification using Ficoll gradients are evident in protocols using flotation techniques (Booth and Clark, 1978). However, in protein phosphorylation studies, these synaptosomes produce a higher degree of background signal on sodium dodecyl sulfate-polyacrylamide gel electrophoresis (SDS-PAGE) (Sihra, unpublished), perhaps because of the tendency of the polymer to adhere to biological membranes. Alternative density media that have been used for the purification of synaptosomes include sodium diatrizoate/sucrose (Tamir et al., 1974) and dextran/polyethylene glycol/sorbitol (flotation method) (Enriquez et al., 1990). Both produce relatively pure preparations, the latter being of particular use when a large amount of material is required (cf Percoll technique below), however, these preparations have not been extensively characterized with respect to their utility in protein phosphorylation experiments.

A method of purification that has gained popularity since its introduction by Nagy and Delgado-Escueta (1984) is based on the use of Percoll gradients (polyvinylpyrrolidone-coated colloidal silica [15–30 nm]) (Pharmacia), which have several advantages. First, the gradients are iso-osmotic and nontoxic, causing no dehydration effects. Second, the Percoll beads are easily washed away from biological membranes. Third, because Percoll has a low viscosity and the gradients effect separation on the basis of nonequilibrium/rate principles, resolution times are substantially shorter than for other density gradient systems. By careful utilization of angle rotors, the speed of resolution has been further improved (Dunkley et al., 1986). Finally, Percoll-gradient-purification yields the greatest enrichment of synaptosomes (Nagy and Delgado-

Escueta, 1984; Dunkley et al., 1986). Perhaps the only disadvantage of this method is the relatively small amount of synaptosomal protein (approx 1 mg) that can be resolved on each gradient. Nevertheless, this limitation is outweighed by the purity and high metabolic competency of Percoll-purified synaptosomes (Nagy and Delgado-Escueta, 1984; Dunkley et al., 1986; Harrison et al., 1988) as evident in their characteristics of Ca^{2+} homeostasis/influx and excellent properties of transmitter glutamate release (Sihra et al., 1992,1993). A number of modifications of the original technique (Nagy and Delgado-Escueta, 1984) have now been described and utilized in preparation of synaptosomes from several brain regions (Dunkley et al., 1986a, 1988; Wang et al., 1988; Thorne et al., 1991).

2.2. Preparation of Synaptosomes and Purification by Percoll Gradients

2.2.1. Reagents

2.2.1.1. SUCROSE MEDIUM

Make up a 5X stock (1.6M, 547.7 g/L). The purest possible grade of sucrose should be used. Divalent cations are a major problem in causing aggregation of synaptosomal protein and lead to a decrease in yield. Some protocols use EDTA (1 mM final) in the initial homogenization to obviate this possibility. However, this may prove deleterious in some types of experiments, since it is known that a finite proportion of the homogenization medium can be incorporated into the cytosol of the synaptosomes when they reseal. Store the stock at 4°C and dilute to 1X with ice-cold, ultrapure (double-distilled and deionized) water immediately before use.

2.2.1.2. HEPES-BUFFERED MEDIUM (HBM)

Make up a 10X stock (1.4M NaCl, 50 mM KCl, 200 mM HEPES, 50 mM $NaHCO_3$, 10 mM $MgCl_2$, 12 mM Na_2HPO_4, and 100 mM D-glucose, pH to 7.4 with NaOH). Store the stock at 4°C and dilute to 1X with ice-cold, ultrapure water immediately before use. The 1X HBM should be bubbled with 95% O_2/5% CO_2 before use. Although the buffering capacity of HEPES is sufficient in most instances, this is necessary to oxygenate the medium before incubation of synaptosomes in static incubations, which minimizes anaerobic conditions. Active stirring of synaptosomal suspensions during experimentation also obviates anaerobic conditions.

2.2.1.3. PHOSPHATE-FREE HEPES BUFFERED MEDIUM (P_{FREE}-HBM)

This is as above with the exclusion of Na_2HPO_4.

2.2.1.4. PERCOLL DISCONTINUOUS GRADIENTS

Percoll (Pharmacia [Uppsala, Sweden] #17-0891–01) solutions should be made up regularly (3–4 wk). If the original Percoll is aged, it may be necessary to filter it (Millipore [Milford, MA] #AP15 042 00) to remove aggregates. Make up four solutions containing 23, 15, 10, and 3% (v/v) Percoll in 320 mM sucrose, 1 mM EDTA, and 0.25 mM dithiothreitol (DTT). The pH of the solutions should be adjusted to pH 7.4 using very dilute HCl. Stock solutions should be prepared every 3–4 wk and kept at 4°C.

2.2.2. Equipment

2.2.2.1. CENTRIFUGE AND ROTORS

Use a high-speed centrifuge with angle rotors (Sorvall [Du Pont, Wilmington, DE] SS34 [8 × 50 mL] and SM24 [24 × 16 mL] or Beckman [Palo Alto, CA] JA-20 [8 × 50 mL] and JA-20.1 [32 × 15 mL] rotors). Precool centrifuge and rotors to 4°C.

2.2.2.2. CENTRIFUGE TUBES

Polycarbonate tubes are most appropriate (50 mL, 28.7 × 103 mm, Nalgene [Rochester, NY] #3117-0500 and 12 mL, 16 × 101 mm, Nalgene #3117-0120). New tubes should be "washed" with a 10 mg/mL solution of bovine serum albumin (BSA) to block protein binding. After use, rinse tubes with warm water followed by distilled water. Avoid the use of detergent.

2.2.2.3. HOMOGENIZERS

One type of homogenizer is the Potter-Elvehjem tissue grinder (30 mL). A Teflon pestle with 0.13–0.18 mm radial clearance is optimal (Thomas Scientific [Swedesboro, NJ]

Type BB #3431–E-18). A motor-drive (with torque control for safety) capable of 900 rpm is used to rotate the pestle. A Dounce glass-glass homogenizer (7 mL) is useful, but not essential, for resuspension of pellets (Wheaton [Millville, NJ] #357542).

2.2.2.4. PERISTALTIC PUMP

This is useful, but not essential when preparing a large number of gradients. Connect the outlet to a piece of stiff plastic tubing with a narrow internal diameter (0.5 mm).

2.2.3. Procedure (Fig. 1)

1. Prepare Percoll gradients: Pipet 2 mL of 23% Percoll into a 12 mL polycarbonate tubes and sequentially layer 2 mL of 15, 10, and 3% Percoll solutions down the side of the tube using a peristaltic pump with a narrow outlet tube. With four-step gradients, synaptosomes sediment in two layers. For most purposes, the 15% step can be excluded and synaptosomes sediment as a single layer at the initial 23/10% interface. Be careful not to agitate the gradients, and keep them at 4°C. The number of gradients required depends on the amount of P_2 resuspension (*see* step 6). In general, 1 g (wet wt) of brain tissue requires four Percoll gradients (g wet wt/brain 200–250 g rat; cerebral cortices ~1; cerebellum ~0.3; hippocampi ~0.2, and striatum 0.15). Subsequent details refer to preparation of synaptosomes from the cerebral cortices of one rat (Fig. 1).
2. Precool centrifuge rotors, centrifuge tubes, and homogenizers. Sacrifice a male rat (200–250 g) and remove the brain into 25–30 mL of ice-cold 320 mM sucrose. Rapid removal and cooling of the brain are critical in minimizing proteolysis and retaining greater functionality of the subsequent preparation. Rapidly dissect the cerebral cortices away from the hippocampi and striatum on a cooled glass plate placed on ice. Carefully remove as much white matter as possible from the cortices.
3. Place the two halves of the cerebral cortex into the Potter-Elvehjem tissue grinder containing 20 mL of 320 mM sucrose at 4°C. Homogenize the tissue with 5–8 up/

down strokes of the Teflon pestle, motor-driven at 900 rpm. Keep the tissue cold during the homogenization by carrying out manipulations in a cold room if possible.

4. Divide the homogenate between two cooled 50-mL tubes and make the volumes up to approx 40 mL with ice-cold, 320 mM sucrose. Centrifuge the homogenate at 3000g (5000 rpm in a SS34 rotor) for 2 min.

5. Decant the supernatants (S$_1$) into two fresh 50-mL tubes at 4°C and recentrifuge at 14,500g (11,000 rpm in a SS34 rotor). Decant and discard the supernatants (S$_2$). Save the pellets (P$_2$).

6. Resuspend the pellets in approx 2 mL of 320 mM sucrose with the aid of a Dounce homogenizer. Make the total volume up to 8 mL with sucrose in a 15-mL Falcon tube, and mix well by gentle inversion (the protein concentration in the P$_2$ resuspension loaded onto the gradient should be below ~4 mg/mL). Layer 2 mL of the suspension onto each of the Percoll gradients using a Pasteur pipet. Centrifuge at 32,500g (16,200 rpm in a SM24 rotor) for 5 min at 4°C. Because these are rate gradients, the timing is critical; centrifugation should be timed from final speed.

7. Myelin and light membranes are separated in the top (3% Percoll layer) of the gradient. Synaptosomes layer as a single band at the position of the original 23/10% Percoll interface in three-step gradients (or as two bands at the original position of the 15% Percoll layer in four-step gradients) (Fig. 1). The loose layer/pellet at the bottom of the gradient is composed of mitochondria. Carefully aspirate and discard the material above the synaptosome layer(s) and harvest the synaptosomes into a 50-mL tube containing 30 mL P$_{FREE}$-HBM at 4°C. Harvest four gradients into the same tube. This step dilutes and washes the synaptosomes free of most of the Percoll.

8. Centrifuge at 27,000g (15,000 rpm in SS34 rotor) for 10 min at 4°C. Discard the supernatants and gently resuspend the Percoll-purified synaptosome pellet with 1–2 mL P$_{FREE}$-HBM with the aid of a Dounce homogenizer.

9. Determine the protein concentration of the resuspension using 5- and 10-µL aliquots and a standard curve of 0, 10, 20, 30, and 40 µL of 0.5 mg/mL BSA. Add 1 mL of Bradford reagent (1:4 dilution of the Bio-Rad Bradford Reagent concentrate), and read the absorbance at 595 nm.

10. Pipet the desired amount of protein into cooled centrifuge tubes containing P_{FREE}-HBM (10 mL/0.5 mg protein) at 4°C. Centrifuge at 3000g (5000 rpm in SS34) for 10 min at 4°C to pellet synaptosomes. Decant supernatants completely, store the pellets on ice, and resuspend just before use.

3. Protein Phosphorylation/Dephosphorylation in Synaptosomes

3.1. Background

There are two major methods for studying protein phosphorylation in synaptosomes. The first is a *post hoc* paradigm, in which synaptosomal fractions are isolated after lysis and incubated in vitro with $[\gamma\text{-}^{32}P]ATP$ under conditions simulating the synaptosomal cytosol. The second method involves "prelabeling" intact synaptosomes with inorganic $[^{32}P]$orthophosphate $(^{32}P_i)$ to allow its incorporation into endogenous ATP. The major advantage of the first method is that a high specific activity of labeling of proteins can be obtained (O'Callaghan et al., 1980; Sieghart et al., 1980; Rauch and Roskoski, 1984; Gower et al., 1986). However, inappropriate substrate and protein kinase or phosphatase interactions in the lysate may yield misleading results. Indeed, the pattern of labeling using the *post hoc* method and prelabeling is very different (Sieghart and Greengard, 1979). The *post hoc* method does have some utility when the protein of interest can be conveniently isolated and phosphorylation/dephosphorylation reactions are carried out in vitro using purified kinases/phosphatases to effect "back-phosphorylation" of the pure substrate (Forn and Greengard, 1978). This approach requires significant characterization of the phosphoprotein of interest to effect its isolation, and is thus of limited use with novel phosphoproteins. This chapter will focus on the use of the "prelabeling" protocol for intact nerve terminals, which has several notable advantages:

1. Only intact and metabolically competent terminals take up $^{32}P_i$ and incorporate it into endogenous ATP to produce [^{32}P]ATP;
2. Phosphoproteins, kinases, and phosphatases are in their native compartments;
3. Endogenous kinases and phosphatases can be activated by physiologically relevant stimuli; and
4. Causal links between protein phosphorylation/dephosphorylation and nerve terminal functions, such as membrane excitability, Ca^{2+}-influx, and transmitter release, can be explored.

3.2. Prelabeling Synaptosomes with [^{32}P]Orthophosphate

3.2.1. Background

Depolarization-dependent changes in protein phosphorylation of some 20 proteins in a ^{32}P-labeled, crude P_2/synaptosomal fraction was first demonstrated by Krueger et al. (1977). Some proteins displayed an increase in phosphorylation in response to Ca^{2+} influx (P86–80, P50–60), representing substrates for Ca^{2+}-dependent kinases, whereas others decreased in ^{32}P-labeling (P96, P140), representing substrates for Ca^{2+}-dependent protein phosphatases. Although specific assignment of some of these proteins to the presynaptic digit could be made on the basis of an enrichment of the observed phosphorylation changes in Ficoll-purified synaptosomes, the latter still represents a heterogenous population with respect to the transmitter type of each terminal. The use of a source of neuronal tissue with only one transmitter type would *a priori* obviate this problem, but this is difficult using mammalian brain. When synaptosomes prepared from sources with homogenous populations of nerve terminal type, such as cholinergic terminals from *Torpedo* electric organ (Michaelson and Avissar, 1979) or glutamatergic terminals from optic lobe (Pont et al., 1979) are assessed, the patterns of phosphorylation differ significantly from those obtained with synaptosomes purified from mammalian sources. In examining phosphorylation changes in mammalian synaptosomes, the heterogeneity of the preparation with respect to nerve terminal origin remains an inherent problem and needs to be accounted for when assessing the data.

3.2.2. Reagents

1. [^{32}P]orthophosphate (10 mCi/mL);
2. P$_{FREE}$-HBM (*see* Section 2.2.1.);
3. HBM (*see* Section 2.2.1.);
4. 10 mM CaCl$_2$; and
5. 100 mM CaCl$_2$.

3.2.3. Equipment

1. Radiation screens: 2–3 cm thickness Perspex;
2. Magnetic stirrer: Water bath submersible; and
3. Radioactive waste disposal.

3.2.4. Procedure

1. Resuspend synaptosomes with P$_{FREE}$-HBM to obtain a protein concentration of 2 mg/mL. Add ^{32}P$_i$ to achieve an activity of 1 mCi/mL:0.5 mCi/mg and incubate at 37°C for 45 min. The preincubation time is based on the assumption that the efficacy of synaptosomal phosphate uptake (Salamin, 1981) is sufficient to allow ^{32}P$_i$ equilibration with ATP (Robinson and Dunkley, 1985). Although ^{32}P-labeling of synaptosomes continues to increase for up to 4 h (the longest time examined), at 45 min the incorporation is in its linear phase (Robinson and Dunkley, 1985). During prelabeling, the synaptosomal suspension should be agitated occasionally or, preferably, continually stirred using a magnetic stirrer.

2. Add CaCl$_2$ (0.1 mM final) 5 min after the start of prelabeling. A large number of different incubation conditions with respect to external [Ca^{2+}] have been used previously (Dunkley and Robinson, 1986). However, the use of low concentrations of Ca^{2+} during prelabeling

appears to be optimal in that it minimizes Ca^{2+}-dependent proteolysis (reflected in high background phosphorylation) during the extended prelabeling period and, on subsequent incubation with normal external $[Ca^{2+}]$ (1 mM), maximizes phosphorylation/dephosphorylation responses (Robinson and Dunkley, 1985,1987; Sihra et al., 1992). Using this protocol for incubation, transmitter glutamate release characteristics from mock-prelabeled synaptosomes are comparable to freshly resuspended synaptosomes (Sihra et al., 1992; Coffey et al., 1993).

3. At the end of the prelabeling period, remove the unincorporated extrasynaptosomal $^{32}P_i$ by centrifugation of the synaptosomes in a microfuge at 10,000g for 15 s. Completely aspirate the supernatant and discard it carefully by blotting it onto absorbent paper before appropriate radioactive disposal.

4. Resuspend the labeled synaptosome pellet in normal phosphate-containing HBM at a protein concentration of 1 mg/mL and incubate at 37°C with stirring. Add $CaCl_2$ (1 mM final) 3–5 min after initiating the incubation and allow equilibration for 5–10 min. The preincubation of 3–5 min before the addition of Ca^{2+} allows synaptosomes to repolarize and prevents premature Ca^{2+}-entry. Depolarization or addition of other effectors is achieved by the addition of aliquots (100 µg protein/100 µL) of the synaptosomal suspension to tubes containing 0.25 vol of HBM with 5X the final required concentration of effector.

5. Incubate the stimulated synaptosomes at 37°C with stirring/shaking for the required time. Terminate the incubations with the addition of a 5X concentration of the appropriate STOP solution according to the analysis to be used (Section 3.4.).

3.3. Modulators of Protein Kinases and Phosphatases

In order to obtain specific information about the phosphorylation of a synaptosomal phosphoprotein, effectors and inhibitors of proteins kinases and phosphatases are of great utility, bearing in mind that their absolute specificity cannot always be presumed (*see* Chapter 4 for detailed discussion of protein kinase and phosphatase inhibitors).

3.3.1. Ca^{2+}-Dependent Phosphorylation/Dephosphorylation

The major second messenger mobilized in synaptosomes is Ca^{2+}, the concentration of which increases in the cytosol on the activation of voltage-dependent Ca^{2+}-channels. *In situ,* Ca^{2+}-channels are stimulated by action potentials propagated down axons impinging on the nerve endings to effect depolarization of the plasma membrane potential. With isolated nerve terminals, in the absence of an axon, biochemical means of depolarization are required to simulate physiological excitation. Veratridine (100 μm) (Krueger et al., 1977; Robinson and Dunkley, 1983), high external [K^+] (30–60 mM) (Robinson and Dunkley, 1983,1985; Wang et al., 1988; Sihra et al., 1992) and 4-aminopyridine (0.05–3 mM) (Sihra et al., 1992; Coffey et al., 1994) all effect biochemical depolarization of synaptosomes, but by different mechanisms.

Veratridine depolarizes synaptosomes by preventing the inactivation of voltage-dependent Na^+-channels, such that they remain in the open state for seconds rather than milliseconds. The entry of Na^+ results in depolarization, but the extensive Na^+ influx elicited by veratidine is somewhat unphysiological and may affect synaptosomal bioenergetics, as well as cause a reversal in the plasma membrane Na^+/ Ca^{2+} exchanger (McMahon and Nicholls, 1991).

Since the K^+ conductance is high in resting nerve terminals, increased external [K^+] depolarizes and clamps the plasma membrane potential according to the external/internal [K^+] gradient across the plasma membrane. This method of depolarization, although extensively utilized, is arguably also rather unphysiological as evidenced by the lack of tetrodotoxin sensitivity, which implies that voltage-dependent Na^+ channels, normally operating during electrical stimulation, are inactivated. Nevertheless, high external [K^+]-induced depolarization remains a useful paradigm in that it produces simultaneous depolarization and clamping of the membrane potential of the whole population of synaptosomes.

The recent use of 4-aminopyridine in synaptosomal release and phosphorylation studies (Sihra et al., 1992,1993; for review, *see* Sihra and Nichols, 1993) is based on the observation that this K^+-channel blocker, by presumably destabilizing the resting plasma membrane potential, allows greater tetrodotoxin-sensitive activation of voltage-dependent Na^+-channels and consequent depolarization of the synaptosomal plasma membrane potential (Tibbs et al., 1989). This provides a closer approximation to physiological depolarization resulting from action potentials impinging on nerve terminals *in situ*. However, because the depolarization of individual synaptosomes in a population is presumably a random, stochastic event, the measured phosphorylation response will be an average of different states of nerve terminal activation.

Direct entry of Ca^{2+} can be mediated without the need for voltage-dependent Ca^{2+}-channel activation by the use of Ca^{2+} ionophores. This paradigm for effecting Ca^{2+}-dependent phosphorylation changes has been employed in a number of studies using either A23187 (40 μM) (Krueger et al., 1977; Robinson and Dunkley, 1987) or ionomycin (0.5–2 μM) (McMahon and Nicholls, 1991; Verhage et al., 1991; Sihra et al., 1992). Although Ca^{2+} ionophores are an effective means of bypassing Ca^{2+}-channels, their use should be viewed critically, since at higher concentrations (Krueger et al., 1977; Robinson and Dunkley, 1987) these ionophores can intercalate into mitochondria and release the Ca^{2+} therein (Sihra et al., 1984; Nicholls et al., 1987). Accordingly, it is advisable to titrate the amount of ionophore at a given synaptosomal concentration to ascertain the lowest concentration usable for the required Ca^{2+} influx. Finally, it is important to note that ionophores effect "delocalized" Ca^{2+} entry at random points on the plasma membrane. As such, this differs from the "localized" Ca^{2+} entry that occurs through voltage-dependent Ca^{2+} channels (McMahon and Nicholls, 1991; Verhage et al., 1991; Sihra et al., 1992). These different forms of Ca^{2+} entry can result in markedly different patterns of phosphorylation of phosphoproteins that are localized at the active zone (Robinson and Dunkley, 1987; Gomez-Puertas et al., 1991; Sihra et al., 1992).

The activation of Ca^{2+}/calmodulin-dependent protein kinases (Kennedy and Greengard, 1981), most notably Ca^{2+}/calmodulin kinase II (Kennedy et al., 1983) (CaMKII), resulting from Ca^{2+} entry, leads to the phosphorylation of a number of synaptosomal substrates (Krueger et al., 1977; Robinson and Dunkley, 1985). There are several ways to analyze the specific activation of Ca^{2+}-dependent enzymes and Ca^{2+}-dependent phosphorylation changes, each having particular merits. The most obvious is to compare responses in the presence of Ca^{2+} with those in which external Ca^{2+} is eliminated using a Ca^{2+} chelator, such as EGTA. It is usually not sufficient simply to omit $CaCl_2$ from incubation medium, since there may be significant divalent cation contamination, and even a small amount Ca^{2+} is sufficient to effect phosphorylation and considerable transmitter release (Sihra et al., 1993; and Sihra, unpublished observations). In the absence of added Ca^{2+}, addition of 0.2 mM EGTA produces conditions of basal phosphorylation, above which additional phosphorylation changes owing to activation of Ca^{2+}-dependent kinases or phosphatases can be examined. Intrasynaptosomal Ca^{2+} mobilization is an unlikely source of interference in synaptosomes since, other than the mitochondria, there is little evidence for mobilizable Ca^{2+}-stores (Sihra et al., 1984; Nicholls et al., 1987).

Although the short-term use of EGTA is appropriate for the examination of Ca^{2+} dependency, long-term elimination of Ca^{2+} is deleterious to synaptosomal bioenergetics (Kauppinen et al., 1986) and may adversely affect protein phosphorylation studies (Robinson and Dunkley, 1985; Dunkley and Robinson, 1986; and Sihra, unpublished observations). It is therefore pertinent in some instances to block the entry of external Ca^{2+} using Ca^{2+}-channel inhibitors. The simplest way this can be effected is by the use of 10 μM Cd^{2+} and 10 μM Co^{2+} (Sihra et al., 1993) in combination, in the absence of any added Ca^{2+}. The utilization of more specific Ca^{2+}-channel-blocking toxins, such as ω-AgaGI (Coffey et al., 1994) or ω-AgaIVA (Turner et al., 1992), is of some value, but limited because, even at high concentrations, the blockade of Ca^{2+}-influx into synaptosome is not complete.

Another alternative approach in dissecting the role of Ca^{2+}/calmodulin-dependent enzymes in synaptosomes is by comparison of responses obtained in the presence of Ba^{2+} and Ca^{2+}. Ba^{2+} does not bind calmodulin (Chao et al., 1984) and thus, fails to support Ca^{2+}/calmodulin-dependent protein phosphorylation in synaptosomes (Robinson and Dunkley, 1985; Sihra et al., 1993). More direct demonstration of Ca^{2+}/calmodulin-dependent events using calmodulin-inhibitors has been attempted (DeLorenzo, 1982; Robinson et al., 1984). Although calmodulin inhibitors decrease Ca^{2+}/calmodulin-dependent phosphorylation changes, the interpretation of these data is confounded by nonspecific effects of these compounds on mitochondrial function and synaptosomal bioenergetics (Snelling and Nicholls, 1984). The development of more specific inhibitors based on the ATP-binding domains of particular kinases has led to the introduction of KN-62 as a relatively specific inhibitor of CaMKII (Tokumitsu et al., 1990). In this instance, although KN-62 can be shown to be a potent Ca^{2+}/CaMKII inhibitor in vitro using purified kinase, its use with synaptosomes is severely limited because of its direct inhibitory effects on synaptosomal Ca^{2+} channels (Sihra and Pearson, 1995). KN-62 may nevertheless be quite useful in studies in which Ca^{2+} entry is mediated by Ca^{2+} ionophore.

3.3.2. Ca^{2+}/Phosphoplipid-Dependent Phosphorylation— Protein Kinase C (PKC)

Phosphorylation by the Ca^{2+}/phospholipid-dependent protein kinase (PKC) in synaptosomes has been addressed extensively (reviewed in Robinson, 1991a). Of particular utility are the phorbol esters (e.g., phorbol 12, myristate 13-acetate PMA; phorbol 12,13-dibutyrate, PDBU—0.1–1 mM), which are potent activators of PKC (Wang et al., 1988, 1989; Dekker et al., 1990; Robinson, 1992; Coffey et al., 1994). Phorbol esters remain indispensable tools in the study of PKC, but not all of the multiple PKC isoforms are activated by phorbol esters (Robinson, 1992). The situation with the use of inhibitors of PKC is no less complex, since differential inhibition of phosphorylation of PKC substrates is apparent (Robinson, 1992).

This is perhaps not surprising in view of the distinct substrate specificities for the different PKC isozymes depending on activation conditions (Robinson, 1992). In studies employing pharmacological manipulations of PKC activity, it is therefore important to ascertain the PKC isozyme phosphorylating a particular substrate before unequivocal conclusions can be reached.

3.3.3. cAMP-Dependent Phosphorylation (PKA)

In vitro studies using lysates from synaptosomal fractions clearly implicate cAMP-dependent protein kinase (PKA) in nerve terminal protein phosphorylation (Huttner and Greengard, 1979; Huttner et al., 1981; Rauch and Roskoski, 1984). However, early studies using exogenous, membrane-permeant cyclic-nucleotides (1 mM) designed to stimulate PKA in intact synaptosomes failed to confirm a role for cAMP-dependent phosphorylation (Krueger et al., 1977). Subsequent studies using higher concentrations of dibutyryl cAMP (10 mM) (Dunkley et al., 1986) or the adenylyl cyclase activator forskolin (50 µM) (Krueger et al., 1977; Musgrave et al., 1994) clearly demonstrate the presence and activation of synaptosomal PKA. Notably, levels of cAMP in synaptosomes are already high in control conditions (Krueger et al., 1977; Musgrave et al., 1994). This perhaps explains the relative ineffectiveness of another way to increase intrasynaptosomal cAMP, i.e., blocking its catabolism by inhibiting phosphodiesterases with 3-isobutyl-1-methylxanthine (IBMX) (1 mM) (Krueger et al., 1977) or theophylline 1,3-dimethylxanthine (1 mM) (Dunkley et al., 1986b).

3.3.4. Protein Phosphatases

Modulation of protein phosphatases in synaptosomes had received relatively little attention until the recent advent of inhibitors. Okadaic acid (100 nM), an inhibitor of protein phosphatase 1 and 2A, causes an increase in basal levels of phosphorylation of synaptosomal phosphoproteins, but has an inconsistent effect on the depolarization-dependent increases of phosphorylation (Sim et al., 1991; Sihra et al., 1992; Nichols et al., 1994). Immunosuppressants, such as cyclosporin A and FK506 (1 μM), are potent inhibitors of protein phosphatase 2B (calcineurin) (Liu et al., 1991). These inhibitors, unlike okadaic acid, cause no significant changes in total basal levels of phosphorylation, but are very effective in preventing the depolarization-dependent dephosphorylation of identified calcineurin substrates (Nichols et al., 1994).

3.4. Analysis of Synaptosomal Phosphoproteins

3.4.1. One-Dimensional Electrophoresis: Discontinuous Sodium Dodecyl Sulfate-Polyacrylamide Gel Electrophoresis (SDS-PAGE) (Laemmli, 1970)

3.4.1.1. REAGENTS

		mL/30 mL, %		
1. Resolving gel	Stock	7.5	10	15
Acrylamide/bis-acrylamide	30/0.8%	7.5	10	15
375 mM Tris/0.1% SDS, pH 8.8	1.5M Tris/ 0.4% SDS	7.5	7.5	7.5
H_2O		15	12.5	7.5
0.5 µL/mL TEMED[a]		15 µL	15 µL	15 µL
2.5 µL/mL APS[a]	10%	75 µL	75 µL	75 µL

[a]TEMED—tetra ethylmethylethylenediamine, APS—ammonium persulfate.

2. Stacking gel	Stock	mL/10 mL
3.75% Acryl/bis-acrylamide	30/0.8%	1.25
125 mM Tris/0.1 SDS, pH 6.8	1M Tris/0.8% SDS	1.25
H_2O		7.5
1 µL/mL TEMED		10 µL
5 µL/mL APS	10%	50 µL

3. Running buffer: 25 mM Tris (3.03 g/L), 192 mM glycine (14.3 g/L), 0.1% SDS (1 g/L). Adjust to pH 8.6 and dilute with H_2O as required for gel apparatus.

4. Mol-wt markers: A mixture of standards encompassing the appropriate mol-wt range with respect to the proteins of interest should be solubilized in SDS-STOP (29–205 kDa, Sigma #SDS-6H; 14–66 kDa, #SDS-7).

3.4.1.2. PROCEDURE

1. After prelabeling, resuspend synaptosomes in normal phosphate containing HBM and treat with effectors as described in Section 3.2.2. To terminate the incubations, add 0.25 vol of 5X concentrated SDS sample buffer (SDS-STOP) to achieve 1% (w/v) SDS, 6.25 mM Tris, pH 6.8, 5% 2-mercaptoethanol, 10% (v/v) glycerol, and 0.001% (w/v) bromophenol blue in the final sample. Boil the samples for 5 min. Centrifuge the cooled samples to retrieve condensed liquid from tube walls.

2. Any number of slab gel formats can be used (e.g., Hoefer [San Francisco, CA] or Bio-Rad [Hercules, CA]); longer formats provide better resolution. With the volume of sample obtained with most synaptosome experiments, a gel thickness of 1.5 mm is optimal, particularly since this allows loading of radioactive samples with normal pipet tips (50–200 μL). Prepare 7.5, 10, or 15% resolving gel mixes as detailed above. Degas the mixtures using a vacuum before effecting polymerization by addition of TEMED and APS. Pour the gels leaving 3 cm at the top for the stacking gel. Overlay the acrylamide with water-saturated isobutanol. Allow 30 min for polymerization.

3. For a discontinuous SDS-PAGE, prepare the stacking gel mix as detailed above. Degas the mixture using a vacuum before effecting polymerization by addition of TEMED and APS. With distilled water, thoroughly wash the top of the polymerized resolving gel free of any isobutanol, or it will inhibit the polymerization of the stacking gel. Pour the stacking gel and insert sample well combs. Any one of the standard 10, 15, or 20 sample well combs is appropriate, depending on the sample volume. The respective maximum volumes are ~330, ~250, and ~160 μL, but to avoid crosscontamination between lanes, it is best to use only two-thirds of the capacity. Allow 15–30 min for polymerization.

4. Remove combs and wash the sample wells thoroughly with water to remove any unpolymerized acrylamide

mix. Load samples with a standard 200-µL pipet. Wash each of the tips by using them to overlay the sample with running buffer.

5. Molecular weight markers should be used at either end of each gel (29–205 kDa for 7.5% gel and 14–66 kDa for 10 or 15% gels). Approx 10–30 µg of the standard mix should be diluted in a "pseudo-sample solution" containing the same proportions of HBM and SDS-STOP as the samples before loading. Likewise, any empty wells should be loaded with this solution. This ensures uniform salt concentration across the gel and produces straighter resolution of protein down the length of the gel. This is not just of aesthetic value but becomes important during quantitation procedures. Overlay the top of the slab sandwich with running buffer before attaching the plates to the electrophoresis apparatus.

6. Fill the upper and lower reservoirs of the electrophoresis apparatus as recommended by the manufacturer. Electrophorese samples at a constant voltage (50–80 V) through the stacking gel and subsequently at voltages >50 V, depending on the running time intended. If apparatus has a cooling facility, voltages up to 250 V can be used safely. Otherwise, proteins are equally well resolved by slower electrophoresis (a 7.5% 16-cm gel requires ~800–900 Vh).

7. For protein phosphorylation work, using relatively large amounts of radioactivity, safety is an essential consideration. For this reason, the gel should not be run until the bromophenol blue front is off the bottom edge of the gel. [^{32}P]ATP that has not been incorporated into proteins runs approx 2 cm ahead of the bromophenol blue front and, therefore, would contaminate the large amount of running buffer in the lower reservoir. It is preferable to stop the electrophoresis when the bromophenol blue front is 2.5 cm from the bottom. The gel segment below this front will contain immobilized [^{32}P]ATP, which can be safely disposed of by cutting and discarding this section of gel.

8. Fix and stain gels with Coomassie blue solution (0.1% [w/v] brilliant blue R250, 10% [v/v] acetic acid, 50% [v/v] methanol) for 30 min with shaking. Destain the gel by replacing the Coomassie solution with destaining solution (30% [v/v] methanol, 10% [v/v] acetic acid) and shaking (2–3 h) until the required intensity of protein stain is obtained.

9. Place the gels onto a double fold of moist Whatman 3MM paper, cover with transparent film, dry with heat, and vacuum (a 7.5% gel takes 1–2 h at 70°C).

10. Mount the dried gels on card and tape onto autoradiography film. Pinholes can be made through the film and card to allow accurate alignment of gel and autoradiogram after exposure. Exposure times depend on the efficacy of prelabeling and the amount of protein loaded. With 100 μg protein and reasonable labeling efficacy, a 1–2 h exposure without an intensifying screen is often enough.

3.4.2. Autoradiographic Quantitation of ^{32}P-Labeled Protein

Quantitation of phosphorylation of individual protein bands can be carried out by a number of means. The most direct, but perhaps most tedious method is to cut out the protein bands of interest from the gel. To expedite this, align the autoradiogram and gel using the localizing holes made through the two before exposure. Identify and mark the phosphoprotein bands of interest on the autoradiogram with a pin, and finally cut around the corresponding marks on the gel to excise the gel pieces. Quantitate ^{32}P-radioactivity in the gel pieces by liquid scintillation spectrometry or, if the activity is high enough, by analyzing the dry bands for Cerenkov radiation. If the equipment is available, quantitation is greatly facilitated by densitometric scanning of the autoradiogram, although the linearity of response using this method is dependent on the characteristics of the autoradiographic film. Even more flexible, and with a broader range of linearity of response, are currently available phosphorimager storage devices for direct quantitation of radioactivity on the gel (e.g., Molecular Dynamics, Sunnyvale, CA).

3.4.3. Confirming Protein Phosphorylation

It is important to ensure that the phosphorylation changes observed after given treatments of synaptosomes represent posttranslational modification of proteins as opposed to lipids or nucleic acids. Additionally, it is essential to demonstrate that phosphorylation represents a phosphomonoester bond and not a acyl-phosphate associated with an ATPase or phosphohistidine. This can be carried out by a series of tests on control and stimulated synaptosomal samples that have been stopped with volumes of 100% (w/v) ice-cold trichloracetic acid (TCA) to achieve 5% final TCA. The TCA suspension is divided into aliquots containing approx 300 μg synaptosomal protein and centrifuged in a microfuge for 1 min.

1. A control pellet is washed in 0.5 mL of $0.5M$ NaCl, reprecipitated by addition of 5% (w/v) final TCA, and washed three times with ice-cold acetone before solubilization in 5% (w/v) SDS for resolution on SDS-PAGE. This control is compared with TCA-precipitated samples treated as follows.

2. To ensure that the radioactivity incorporated is not in phospholipid, another pellet is treated the same as the control, but is additionally extracted three times with 1 mL of 2:1 mix of chloroform:methanol before the addition of 10% (w/v) SDS. Labeling in this sample should display no difference compared with the control if the signal is from a nonlipid component.

3. That the ^{32}P-label is not associated with the phosphodiester bond in nucleic acids is confirmed by treating another TCA precipitate the same as the control, and additionally incubating the final 5% (w/v) SDS extract with 10 μg DNase I, DNase II, and RNase A for 10 min at 37°C. The ^{32}P-labeling should be insensitive to this treatment if it is not associated with nucleic acids.

4. Conversely, if the incorporation of ^{32}P-label is sensitive to incubation of the 5% (w/v) SDS extract with 10 μg pronase (obliteration of labeling of large mol-wt bands on SDS-PAGE), phosphorylation of proteins is confirmed.

5. Phosphorylation of proteins can occur as ^{32}P incorporated via phosphomonoester bonds with amino acids with hydroxyl side chains (serine, threonine, or tyrosine) or via acyl-phosphate bonds with amino acids with carboxylic side chains. If the former is the case, radio-labeling should be insensitive to treatment with $0.5M$ NaOH at 4°C or 100% (w/v) TCA at 100°C, but sensitive to $0.5M$ NaOH at 100°C. These agents (0.2 mL) are added to TCA-precipitated pellet instead of $0.5M$ NaCl (*see* step 1) and incubated for 30 min before the addition of 1 mL 100% (w/v) TCA, followed by centrifugation, acetone extraction, and solubilization in 5% (w/v) SDS.

Another consideration in prelabeling studies is delineation of changes in actual phosphate levels on a given protein (indicated by increased ^{32}P-labeling) after a particular stimulus, as opposed to an increase of turnover of existing cold-phosphorylated residues with ^{32}P-label. Reversibility of ^{32}P-labeling of a protein is good evidence that an observed change is owing to an increase of phosphorylation rather than turnover, since a decrease in ^{32}P-labeling would otherwise only occur if the specific activity of [^{32}P]ATP decreased. Likewise, protein dephosphorylation is not likely the result of a turnover effect in the presence of a constant specific activity of synaptosomal [^{32}P]ATP. Another indicator that the incorporation of ^{32}P is not owing to turnover is similar relative increases in phosphorylation of a particular protein after stimulation at different levels of prelabeling.

3.4.4. Synaptosomal Phosphoproteins

Figure 2 shows a typical phosphorylation pattern and time-course following depolarization of Percoll-purified synaptosomes with 30 mM KCl, 1 mM 4AP, and 1 μM ionomycin. A number of proteins clearly increase in phosphorylation, whereas others, with high basal phosphorylation levels, are dephosphorylated. The identity of many of these proteins has been established. Known synaptosomal substrates are reviewed in Table 1 in terms of the kinase or phosphatase systems involved in altering their phosphorylation state.

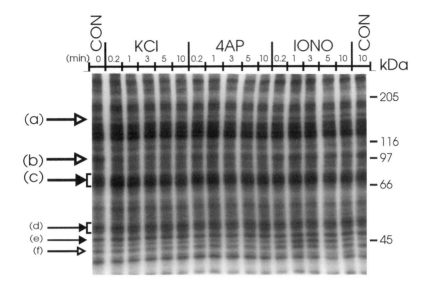

Fig. 2. Autoradiogram of synaptosomal phosphoproteins separated on SDS-PAGE: time-courses of stimulation by KCl (30 mM), 4-aminopyridine (4AP; 1 mM), and ionomycin (IONO; 1 μM). Some phosphoproteins exhibit increases in phosphorylation (solid arrows), whereas others are dephosphorylated (open arrows). Phosphoproteins unequivocally identified by phosphopeptide mapping and other analytical procedures: **(A)** Synaptojanin (P145), **(B)** Dynamin (P96), **(C)** Synapsins Ia/Ib/IIa and MARCKS (P85–75), **(D)** Synapsin IIb, CaMKII β subunit, and Tubulin (P55), **(E)** B50 (GAP-43), and **(F)** Pyruvate dehydrogenase.

Well-characterized phosphoproteins relevant to the following discussion are highlighted in Fig. 2 (solid arrows—phosphorylation, open arrows—dephosphorylation).

3.4.5. Resolution of Synaptosomal Phosphoproteins by Two-Dimensional Electrophoresis

Figure 2 demonstrates that the profile of synaptosomal phosphoproteins resolved by SDS-PAGE is complex. It is conceivable, and indeed likely, that some proteins are comigrating in the one-dimensional gel electrophoresis system used in Fig. 2, which would make quantitation ambiguous. Once a phosphoprotein of interest has been identified by one-dimensional SDS-PAGE, the "purity" of a particular band

Table 1
Synaptosomal Phosphoproteins

Mol wt range reported	Identity (other names)	Ca/Calmodulin-dependent kinases and calcineurin	PKA	PKC	Phosphatase 1 & 2A other modulators
254	Spectrin (Fodrin)	(1)[a]			
~150	?				
139, 140, 145	Synaptojanin (4) (P145)	⇓ (3)[b,g], (5)[b], (6)[b,c], (7)[b,c], (8)[c,e], (9)[c,d,e], (10)[c,d,f]	(8)[h]		(8)[i], (9)[j]
90, 93, 96	Dynamin (Dephosphin, P96)	⇓ (11)[b], (3)[a,b,g], (5)[b], (6)[b,c], (7)[b,c], (8)[c,e], (9)[c,d,e], (12)[c,d], (10)[c,d,f]	(8)[h]	(8)[k]	(8)[i], (9)[j]
83, 87 (9)[j]	MARCKS (87 kDa)	(5)[b], (6)[b,c], (7)[b,c], (13)[a,b,c], (9)[c,d,e], (12)[c,d]	(13)[k,l,m]	(14)[k], (15)[n], (12)[n], (16)[n],	
80, 81, 86, 80, 77, 75	Synapsin Ia/b (Protein I)	(11)[b], (2)[a,b,p], (17)[a,b], (3)[a,b], (18)[a], (5)[b], (14)[a], (6)[b,c], (7)[b,c], (1)[a], (13)[a,b,c], (8)[c,e], (9)[c,d,e], (12)[c,d], (10)[c,d,f]	(17)[h,l], (14)[l], (13)[h,l,p], (19)[a],	(15)[n], (12)[n], (16)[n]	(8)[i], (9)[j]
72	Synapsin IIa (Protein IIIa)	(7)[b,c], (13)[a,b,c]	(14)[l], (20)[k], (13)[h,l,m]		
65	Parafuscin ?	(3)[a,b], (7)[b,c], (13)[a,b,c]	(14)[l], (13)[h,l,p], (21)[l]		
65	?	⇓ (22)[c]			
60, 62, 63, 65, 60	Ca/CAM kinase II, DPH-L, tubulin, Regulatory subunit of PKA (RII)	(2)[a,b,p], (3)[a,b], (23)[a], (18)[a], (6)[b,c], (7)[b,c], (1)[a], (13)[a,b,c], (13)[a,b,c]			
57	Synapsin IIb (Protein IIIb), Tubulin, P57	(2)[a,b,p], (6)[b,c], (13)[a,b,c], (8)[c,e]	(14)[l], (20)[k], (13)[h,l,m]		(8)[i]

54, 55	Regulatory subunit of PKA (R-II)	(3)[a,b], (23)[a,q], (13)[a,b,c]		(13)[h,l], (21)[k,l]
48, 51, 54	Ca/CAM kinase II, DPH-M, tubulin	(2)[a,b,p], (3)[a,b], (23)[a], (18)[a], (6)[b,c], (7)[b,c], (1)[p], (13)[a,b,c],		
45	B50 (GAP-43,FI)	(6)[b,c], (7)[b,c], (1)[p], (13)[a,b,c], 9[c,d,e]	(14)[k]	(13)[h,l]	(9)[j]
43	Actin	(13)[a,b,c]		
41, 42	Pyruvate dehydrogenase	⇓ (3)[a,b], (6)[b,c], (7)[b,c]		(21)[k,l]
35, 36	Succinyl Co-A synthase ?	(6)[b,c,f]		(21)[k,l]

(1) Gower et al. (1986); (2) DeLorenzo et al. (1979); (3) Sieghart et al. (1980); (4) McPherson et al. (1996); (5) Wu et al. (1982); (6) Robinson and Dunkley (1983); (7) Robinson and Dunkley (1985); (8) Robinson and Dunkley (1987); (9) Sihra et al. (1992); (10) Sihra et al. (1993); (11) Krueger et al. (1977); (12) Coffey et al. (1993); (13) Dunkley et al. (1986); (14) Walaas et al. (1983); (15) Wang et al. (1988); (16) Coffey et al. (1994); (17) Huttner and Greengard (1979); (18) Burke and DeLorenzo (1981); (19) Sihra et al. (1989); (20) Browning et al. (1987); (21) Rauch and Roskoski, Jr. (1984); (22) Gomez-Puertas et al. (1991); (23) O'Callaghan et al. (1980).

⇓ Protein dephosphorylation. [a]In vitro Ca-dependent phosphorylation. [b]Synaptosomal Ca-influx stimulated by VER-mediated depolarization. [c]Synaptosomal Ca-influx stimulated by KCl-mediated depolarization. [d]Synaptosomal Ca-influx stimulated by 4AP-mediated depolarization. [e]Synaptosomal Ca-influx with Ca-ionophore (A23187 or ionomycin). [f]Comparison of Ca/KCl and Ba/KCl. [g]No in vitro dephosphorylation. [h]cAMP (4–10 mM). [i]Guanidine. [j]Okadaic acid (0.1–1 µM). [k]In vitro phosphorylation conditions with lysate or purified kinase. [l]cAMP in vitro (2–10 µM). [m]In vitro phosphorylation > Intact synaptosome phosphorylation. [n]PDBU (0.1–1 mM). [o]Forskolin (50 µM). [p]Intact synaptosome phosphorylation > In vitro phosphorylation. [q]Acid labile phosphorylation.

Abbreviations: PKA, cAMP-dependent protein kinase; PKC, protein kinase C; VER, veratridine, KCl, high external [K+] (15–60 mM); 4AP, 4-aminopyridine (0.5–3 mM).

should be ascertained by rerunning samples with alternative acrylamide percentages or with gradients of acrylamide concentration in the resolving gels to obtain greater resolution (Robinson and Dunkley, 1983). Alternatively, several two-dimensional gel systems can be used to resolve synaptosomal phosphoproteins (Table 2). These involve the separation of the original sample in the first dimension using tube gels under conditions (outlined in Table 2) in which resolution is based on the native charge and isoelectric point (pI) of the proteins. This is followed by second dimension separation of the extracted tube gel in the perpendicular direction using slab SDS-PAGE as described in Section 3.4.1. No one system is ideal, and the choice of protocol depends on the intrinsic properties of the phosphoproteins of interest, most notably their isoelectric points. Isoelectric focusing (IEF) is optimal for neutral and acidic proteins and represents a good starting point for analysis beyond SDS-PAGE (O'Farrell, 1975). However, many basic proteins are not resolved well by this system (e.g., the synapsins, Fig. 2, Table 1), and some basic proteins may precipitate at the origin. An alternative that overcomes some of the problems of IEF is nonequilibrium pH gel electrophoresis (NEPHGE) (O'Farrell et al., 1977). The latter method has been used to resolve neutral and basic synaptosomal proteins to greater effect than IEF (Dunkley et al., 1986b). Several other modifications of these systems have been described, but the general principles are similar. Another two-dimensional system that has been of utility in resolving synaptosomal proteins utilizes a "mixed detergent" system that separates on the basis of a combination of principles involved in IEF, NEPHGE, and SDS-PAGE (Imada and Sueoka, 1980).

The fundamental methods for two-dimensional separation are similar for all the aforementioned separation protocols, differing mainly in the resolving gel composition of the first-dimensional electrophoresis in tube gels. Resolving gel compositions and other buffers are summarized in Table 2. After the first dimension, the tube gels are extracted with water pressure and equilibrated with SDS-STOP (Section 3.4.1.) before being placed horizontally on SDS-PAGE slab gels (no sample wells). The tube gels are sealed onto the slab gels using 1% (w/v) agarose melted in SDS-STOP before electrophoresis and analysis, as described in Section 3.4.1.

3.4.6. Phosphoprotein "Fingerprinting" by Limited Proteolysis

Pharmacological manipulation of synaptosomes and analysis by one- and two-dimensional electrophoresis provide a basic knowledge of the kinase and phosphatase systems that target a particular phosphoprotein (Table 1). In many cases, there are multiple phosphorylation sites on a protein, and these may be phosphorylated or dephosphorylated by multiple kinases or phosphatases. In order to obtain information on the specificity of individual phosphorylation sites, these have to be physically resolved. This is possible by using proteases to effect "limited proteolysis" under controlled conditions, in order to generate phosphopeptide maps, which effectively "fingerprint" the protein. Two such systems that have been used in analyzing synaptosomal phosphoproteins are described.

3.4.6.1. PROTEIN FINGERPRINTING: LIMITED PROTEOLYSIS WITH *STAPHYLOCOCCUS AUREUS* V8 PROTEASE (SAP MAP) (CLEVELAND ET AL., 1977)

Reagents

SAP buffer	Stock	100 mL
125 mM Tris, pH 6.8	1M	40 mL
0.1% SDS	10%	1 mL
20% glycerol		20 mL
H_2O		39 mL
0.0025% bromophenol blue		2.5 mg

S. AUREUS V8 PROTEASE. Make 0.5–1 mg/mL with SAP buffer containing 0.5 mg/mL pyronin Y and only 10% (w/v) glycerol.

GEL. 15% Discontinuous SDS-PAGE. Pour a 15% resolving gel leaving 4.5 cm from the bottom of what would be the sample well (i.e., 6–6.5 cm from the top of the plate with comb/well depths of 2 cm). Pour a 3.5% stacking gel and use a wide 11-lane sample well comb.

Table 2
Two-Dimensional Electrophoresis—Tube Gel Compositions and Electrophoresis Conditions

Source	IEF, O'Farrell, 1975	NEPHGE, O'Farrell et al., 1977	Mixed detergent, Imada and Sueoka, 1980
Resolving gel	9M urea 4% Acrylamide/bisacrylamide (28.38/1.62%) 2% Nonidet P-40 H$_2$O 2% Ampholines[a] pH 3.0–10.0[c] 1 μL/mL APS 0.7 μL/mL TEMED	9M urea 4% Acrylamide (28.38/1.02%) 2% Nonidet P-40 H$_2$O 2% Ampholines[c] 2 μL/mL APS 1.4 μL/mL TEMED	9M urea 375 mM Tris 2% Acrylamide/bisacrylamide (30/3%) 0.1% SDS 0.3% Triton CF-10 1.5 μL/mL TEMED 1 μL/mL APS Stacking gel: 125 mM Tris, 1.6% Acrylamide/bisacrylamide (30/3%), 0.1% SDS, 0.3% Triton CF-10, 0.2 μL/mL TEMED, 4 μL/mL APS
Tubes	130 × 2.5 mm (11.5 cm gel)	130 × 2.5 mm (12 cm gel)	160 × 2 mm (11 cm resolve/3 cm stack)
Gel overlay	8M urea	H$_2$O	125 mM Tris, pH 6.8 0.1% SDS 0.3% Triton CF-10
Sample buffer	9.5M urea 2% NP40 2% Ampholines[d] 5% β-mercaptoethanol	9.5M urea 2% NP40 2% Ampholines[a] 5% β-mercaptoethanol	125 mM Tris, pH 6.8 3% SDS 10% Glycerol 2 mM phenylmethylsulfonyl fluoride 2 mM EDTA

Sample overlay	10 µL 9M urea 1% Ampholines[b]	20 µL 8M urea 1% Ampholines[b] (5% NP40 for overlaying SDS samples)	1 mM N-ethylmaleimide 1 mM Iodoacetamide 5% β-Mercaptoethanol 0.005% Bromophenol blue Cathode buffer
Cathode solution	20 mM NaOH (degassed)—upper	20 mM NaOH (degassed)—lower	50 mM Glycine-NaOH, pH 10.5—upper 0.1% SDS 0.3% Triton CF-10
Anode solution	10 mM H_3PO_4—lower	10 mM H_3PO_4—upper	375 mM Tris, pH 9.0—lower 0.1% SDS 0.3% Triton CF-10
Electrophoresis	Prerun: 200 V for 15 min; 300 V for 30 min; 400 V for 30 min 400 V for 12 h; 800 V for 1 h 10,000 > optimal Vh > 5000	Prerun: None 400 V for 4–5 h; optimal 1600–2000 Vh For very basic proteins: 1000 Vh (pH 3.5–10.0) or 1200 Vh (pH 7.0–10.0)	Prerun: None 75–100 V for 2 h until stacked, and then 500 V for 1 h
Optimal pI separation	Neutral and acidic	Neutral and basic	Acidic, neutral, and basic
Synaptosomal examples	Dunkley et al., 1986; Robinson, 1991	Dunkley et al., 1986	Wang et al., 1988

[a] 1.6%, pH 5.0–7.0; 0.4%, pH 3.0–10.0.
[b] 0.8%, pH 5.0–7.0; 0.2%, pH 3.0–10.0.
[c] pH 7.0–10.0 for basic proteins or pH 3.5–10.0 for broad range or pH 8.0–10.0 for overlap with IEF.
[d] Anderson and Anderson, 1977: 2% SDS, 5% β-mercaptoethanol, 10% glycerol.

PRESTAINED MOL-WT MARKERS. The SAP gel is not stained/destained; hence, in order to determine the size of the phosphopeptides, the standards have to be ready-stained with Coomassie blue or other dye for visualization. The quality of prestained standards tends to be variable; thus, prior appraisal is recommended. Standard mixes containing a number of proteins <50 kDa are appropriate and should be solubilized in SDS-STOP.

PROCEDURE

1. Run the initial SDS-PAGE using a 20-lane sample well comb if possible. After electrophoresis, treat the gels according to one the following methods to identify and localize phosphoproteins for mapping:
   a. Seal the gel in polyethylene, and image the phosphoproteins in the "wet" gel using a phosphorimager.
   b. Fix the gels in 30% (v/v) methanol/10% (v/v) acetic acid for 3 min, and then wash in H_2O for 30 min before drying for autoradiography.
   c. Stain and destain gels as usual for SDS-PAGE, and dry for autoradiography.
2. Align the autoradiogram or phosphorimage with the gel, mark the gel bands of interest with a pin and excise the gel pieces.
3. If the primary gel is treated as in: (a) proceed to step 6; (b) proceed to step 5; or (c) proceed to step 4.
4. Reswell the gel pieces by shaking in 1 mL SAP buffer for 15 min. This step also neutralizes residual acetic acid in gel pieces. If the bromophenol blue decolorizes to indicate persistent acid, add small volumes of $1M$ Tris-base and continue shaking until neutrality is indicated by the persistence of the bromophenol blue color.
5. Replace the SAP buffer with 1 mL SDS-STOP and shake for 15 min before boiling for 3 min.
6. Fill the sample wells with SAP buffer. Remove any filter paper backing from the reswollen gel pieces and insert them into the well, ensuring that there are no bubbles trapped. Leave the end wells free for mol-wt standards.

Equalize the volume in each sample well with SAP buffer, leaving a sufficient amount to cover the gel piece. Fill the remainder of each well by overlaying with SDS-PAGE running buffer.

7. Mount the gel onto electrophoresis apparatus and fill upper and lower reservoirs with running buffer. Overlay each gel piece with 20 µL 0.5 mg/mL SAP using a Hamilton-type syringe. An even pink layer of pyronin Y should form over each gel piece.

8. Underlay the running buffer in the end lanes with aliquots of prestained low-mol-wt (<50 kDa) standards in 1X SDS-STOP.

9. Electrophorese at 40 V* until the samples have entered the stacking gel and then increase voltage to 60 V* until samples enter the resolving gel. Subsequently, electrophoresis can be carried out slowly at 60 V or rapidly at 250 V with cooling.

10. At the end of electrophoresis, dry the gel directly, with no further treatment, using low heat (~50°C) and extended time (~2–3 h) to prevent cracking.

Figure 3A shows a typical SAP MAP of a group of phosphoproteins migrating as a 85–75 kDa band on one-dimensional SDS-PAGE. This group includes the synapsins (Ia, Ib, and IIa) as well as **Myritoylated-Alanine-Rich-C-Kinase-Substrate** (MARCKS; 87 kDa) (*see* Table 1). The kinase specificity of the phosphopeptides derived from the synapsins has been well characterized (Huttner and Greengard, 1979; Dunkley et al., 1986), which is valuable for assessing the intrasynaptosomal activity of CaMKII (35-kDa peptide substrate from synapsins Ia and IIa), CaMKI (10-kDa peptide substrate from synapsin Ia, Ib), and PKA (10-kDa peptide substrate from

*Proteolysis occurs while the SAP is in the sample well and the stacking gel. It is therefore critical to determine the amount of time the sample remains in the stacking gel by strictly controlling the length of the stacking gel and the running voltages used during stacking. For reproducible limited proteolysis, it is also important to control the ambient temperature during electrophoresis through the stacking gel.

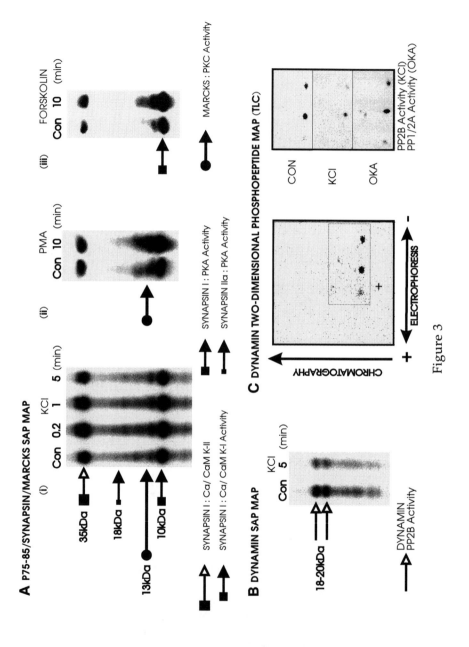

Figure 3

100

synapsin Ia, Ib). Similarly, MARCKS, which is a specific substrate for PKC, is proteolyzed to produce a 13-kDa peptide that reports intrasynaptosomal PKC activity (Wu et al., 1982; Wang et al., 1988). Proteolysis of dynamin (P96, dephosphin, Fig. 2 and Table 1), results in two phosphopeptides, 21 and 19 kDa, (Fig. 3B) (Robinson, 1991b,c) which reflect intrasynaptosomal protein phosphotase 2B (calcineurin) activity (Fig. 3B), as well as the activity of a phorbol-ester-insensitive PKC (Robinson, 1991a,c).

The SAP MAP technique is generally applied to individual bands or spots cut from one- (Huttner and Greengard, 1979) or two-dimensional (Dunkley et al., 1986b; Wang et al.,

---

Fig. 3. (*opposite page*) Phosphopeptide mapping of synaptosomal phosphoproteins: use to monitor specific synaptosomal kinase and phosphatase activities. (**A**) SAP phosphopeptide maps of 85–75 kDa region on SDS-PAGE (Fig. 1) constituting synapsins Ia, Ib, IIa, and MARCKS, obtained after stimulation of synaptosomes with (i) 30 mM KCl, (ii) 100 nM phorbol 12-myristate, 13-acetate (PMA), or (iii) 50 μM forskolin for the time indicated. Con: unstimulated control. The 35-kDa synapsin Ia and Ib phosphopeptide (j and open arrow) reports CaMKII activity and is seen to be transiently phosphorylated (i). The 10-kDa phosphopeptide synapsin Ia and Ib (j) reports CaMKI activity (i) and PKA activity (iii). The 18-kDa peptide (j) phosphopeptide is derived from synapsin IIa and is a good substrate for PKA in vitro (Walaas et al., 1983). The 13-kDa phosphopeptide (d) ([i] and [ii]) is derived from MARCKS and is a specific substrate for PKC, as indicated by PMA stimulation of its phosphorylation (ii). (**B**) SAP phosphopeptide map of 96-kDa band on SDS-PAGE (Fig. 1) constituting dynamin obtained in control conditions (Con) and after depolarization with KCl (30 mM) for 5 min. Two phosphopeptides, 18 and 20 kDa, are dephosphorylated and report protein phosphatase 2B (PP2B, calcineurin) activity. (**C**) Two-dimensional phosphopeptide map/ TLC of a trypsin/thermolysin digest of dynamin (left panel). The dotted-line box delineates areas represented in the right panel showing maps in control conditions (CON) and after treatment of synaptosomes with 30 mM KCl or 1 μM okadaic acid (OKA). Depolarization with KCl causes dephosphorylation of three phosphopeptides and reports protein phosphatase 2B (PP2B, calcineurin) activity. Treatment with OKA increases the phosphorylation of dynamin phosphopeptides above basal conditions (CON) and reports constitutive protein phosphatase 2A (PPA) and/ or protein phosphatase 1 (PP1) activity in synaptosomes.

1988) gels to fingerprint individual proteins. More general screening of phosphorylation changes can be conducted by cutting complete lanes of phosphoproteins resolved by SDS-PAGE and proteolyzing/electrophoresing them in a perpendicular direction to generate an expansive set of phosphopeptides from synaptosomal phosphoproteins (Dunkley et al., 1986b).

3.4.6.2. PROTEIN FINGERPRINTING: TWO-DIMENSIONAL PHOSPHOPEPTIDE PEPTIDE MAPPING BY THIN-LAYER CHROMATOGRAPHY (TLC)

An alternative method of analysis for characterizing phosphoproteins is two-dimensional TLC of trypsin/thermolysin digests of phosphoproteins extracted from gel pieces obtained after SDS-PAGE or two-dimensional electrophoresis. This arguably provides a more definitive phosphopeptide pattern from a phosphoprotein than is possible by one-dimensional SAP phosphopeptide mapping.

REAGENTS

1. 50% (v/v) Methanol/10% (v/v) acetic acid;
2. 50% (v/v) Methanol/50% water;
3. Trypsin (TPCK-treated);
4. Thermolysin;
5. 50 mM NH_4HCO_3, pH 8.0/1 mM dithiothreitol (DTT);
6. Phenol red solution;
7. First-dimension sample and running buffer: 10% (v/v) acetic acid/1% (v/v) pyridine, pH 3.5;
8. Second-dimension running buffer: 37.5% butanol/7.5% acetic acid/25% pyridine/30% (v/v) water;
9. TLC sheets: 20 × 20 cm plastic-backed cellulose (Eastman-Kodak [Rochester, NY] #136 6061).

EQUIPMENT

1. Horizontal electrophoresis tank;
2. Vertical TLC tank;
3. Lyophilizer or "Speed-Vac" centrifuge.

PROCEDURE

1. Excise the band/spot of interest from a well-resolved one- or two-dimensional gel separation and determine the radioactivity by Cerenkov counting.
2. Swell the gel piece first in water and then by shaking in 50% (v/v) methanol/10% (v/v) acetic acid for 6–10 h with frequent solvent changes.
3. Wash the reswollen gel piece with 50% (v/v) methanol/50% water for 2 h, changing the solvent every hour to ensure that all the acid is removed.
4. Clean the gel piece free of any paper debris and lyophilize or dry in a Speed-Vac (Savant).
5. Begin proteolysis by resuspending the lyophilized material in a 50 mM NH_4HCO_3, pH 8.0/1 mM DTT solution containing 50 µg/mL trypsin and 50 µg/mL thermolysin (1 mL final volume), and incubate at 37°C for 24 h.
6. Centrifuge the solution at 13,000g to remove particulate matter and transfer the supernatant into another tube. Add a further 0.5 mL 50 mM NH_4HCO_3, pH 8.0/1 mM DTT to the gel piece, and incubate at 37°C for 2 h. Centrifuge as before, and combine the supernatants. Lyophilize or Speed-Vac dry the combined supernatants.
7. Determine the total extracted phosphopeptide ^{32}P-activity by Cerenkov counting.
8. Resuspend the dried phosphopeptides in 10–20 µL of 10% (v/v) acetic acid/1% (v/v) pyridine (pH 3.5) buffer to about 100 cpm/µL and vortex thoroughly.
9. Apply the resolubilized phosphopeptide in small volumes (0.5 µL) to a 20 × 20 cm plastic-backed cellulose TLC plate (Eastman-Kodak), 3 cm up from the middle of one edge. Dry the spot in between each application. Application of 500–1000 cpm is usually ideal. Spot 0.5 µL of phenol red solution on top of the sample as a migration marker.
10. Wet the plate with 10% (v/v) acetic acid/1% (v/v) pyridine, pH 3.5 buffer, avoiding the sample spot. This can be expedited by blotting the buffer onto the TLC plate with a piece of Whatman 3MM filter paper that has a hole cut at the position of the sample.

11. Blot the TLC sheet almost dry, place in a horizontal electrophoresis tank containing 10% (v/v) acetic acid/1% (v/v) pyridine pH 3.5 buffer, and electrophorese at 400 V for 1.5 h until the phenol red marker migrates 5 cm toward the cathode.
12. Remove plates and dry them in a fume hood before conducting ascending TLC in the perpendicular direction (origin down), using a mobile phase composed of butanol:acetic-acid:water:pyridine in the proportions 15:3:12:10. Chromatograph to within 2 cm of the top of the plate (4 h).
13. Dry the plate and visualize the ^{32}P peptides using a phosphorimager or by autoradiography. Quantitate by densitometry, or alternatively, the area corresponding to the ^{32}P activity can be scratched off using a sharp blade and the radioactivity therein quantitated by liquid scintillation spectrometry.

Figure 3C shows typical phosphopeptide patterns obtained from digests of dynamin (Table 1), which has been obtained from one-dimensional SDS-PAGE gel bands after electrophoresis of phosphoproteins from control, KCl, and okadaic acid (OKA) treated synaptosomes (Fig. 3C). The KCl-evoked dephosphorylation of three phosphopeptides reflects voltage-dependent Ca^{2+}-entry and intrasynaptosomal calcineurin activation. Increases in phosphorylation of these phosphopeptides and the appearance of other phosphopeptides with OKA are indicative of constitutive protein phosphatase 1 and 2A activity in synaptosomes.

3.4.7. Stoichiometry of Protein Phosphorylation
3.4.7.1. BACKGROUND

The analytical techniques considered thus far provide information regarding whether a specific synaptosomal protein is phosphorylated or dephosphorylated after a given stimulation paradigm. To provide evidence that this change in phosphorylation state is of physiological consequence, it is important to show that the stoichiometry of phosphorylation (mol phosphate/mol protein) is significantly altered.

This requires the accurate determination of the total phosphate incorporated into the protein and the molar concentration of protein. These studies are only realistically feasible when an antiserum has been raised against the protein of interest. For antigen, it is therefore necessary to obtain either pure protein or primary sequence information on the basis of which synthetic peptides can be produced. Preparative separation of the protein by gel electrophoresis is one route to obtaining pure protein for use as immunogen. Alternatively, it is possible to obtain primary sequence information from phosphopeptides, produced in tryptic digests of excised gel pieces and further purified using high-performance liquid chromatography (HPLC).

The ^{32}P-labeling of the protein reflects the specific activity of the synaptosomal ATP pool. In determining the mol phosphate content of the protein, incorporation of ^{32}P-label into the phosphoprotein has to be determined in parallel with measurement of the specific ^{32}P activity of synaptosomal ATP following prelabeling.

3.4.7.2. DETERMINATION OF PHOSPHATE INCORPORATION—
SPECIFIC QUANTITATION OF PHOSPHORYLATION
BY IMMUNOPRECIPITATION

The extent of phosphorylation of a protein in a sample of ^{32}P$_i$ prelabeled synaptosomes is best determined quantitatively by immunoprecipitation. In phosphorylation studies, it is essential that protein phosphatase and kinase activities be limited after the samples are stopped. There are numerous immunoprecipitation protocols, some designed to obtain coprecipitation of associated proteins and others for the sole purpose of specifically isolating the protein of interest, with little or no contamination. The latter is appropriate for the present purpose and involves initial disruption of protein/protein interactions using SDS solubilization. Subsequently, since antibodies will not bind antigen in the presence of SDS, the detergent has to be removed. This is expedited by comicellation of SDS with an excess of the nonionic detergent Nonidet P-40 (NP40) (*see also* Chapter 6).

REAGENTS

1. Immunoprecipitation stop (IP-STOP): Make up a 2X stock containing 100 mM Tris, pH 7.4, 10 mM EDTA, 10 mM EGTA, 100 mM NaF, 20 mM $Na_4P_2O_7$, 2 mM Na_3VO_4, and 2% (w/v) SDS.
2. NP40/inhibitor solution: Make up a 2X stock containing 300 mM NaCl, 30 mM HEPES, 2 mM EDTA, 2 mM EGTA, 100 mM NaF, 20 mM $Na_4P_2O_7$, 2 mM Na_3VO_4, 20 μg/mL leupeptin, 20 μg/mL antipain, 4 μg/mL, pepstain, 4 μg/mL chymostatin, 50 KIU trasylol (aprotinin), and 8% (v/v) NP40.
3. *S. aureus* cells (SAC): Precipitation of protein/antibody complexes can be effected with SAC (Pansorbin; Calbiochem, La Jolla, CA), which express surface protein-A or using protein A-conjugated Sepharose CL-4B (Pharmacia). The latter has greater immunoglobulin capacity, but SAC is inexpensive and easier to work with. Thus, apart from immunoprecipitation of very low-abundance phosphoproteins, SAC is appropriate for most purposes provided it is properly prepared.

   Preparation of SAC: Centrifuge the 10% suspension obtained from the manufacturer at 3000g for 5 min and resuspend as a 1% slurry in a buffer containing 140 mM NaCl, 3 mM KCl, 80 mM Na_2HPO_4, 15 mM NaH_2PO_4, 3% (w/v) SDS, and 10% (v/v) β-mercaptoethanol and boil for 20 min. Cool and pellet the SAC by centrifugation at 10,000g for 15 min. Wash the SAC twice by resuspension (to 1% slurry) and recentrifugation in the aforementioned buffer, but with the β-mercaptoethanol omitted. Finally, wash the SAC three times in a standard SAC wash buffer (*see below*). Resuspend the final pellet as a 10% slurry in SAC wash buffer (add 10 mM NaN_3 for storage at 4°C).
4. SAC wash buffer: 150 mM $NaCl_2$, 15 mM HEPES, pH 7.4, 1 mM EDTA, and 1% (v/v) NP4O.

PROCEDURE

1. Synaptosomal incubations (100 μg protein) should be stopped in buffers containing a final concentration of 1% (w/v) SDS. SDS-STOP modified by omitting β-mercapto-

ethanol (disrupts antibody) and adjusting SDS to achieve 1% final may be utilized if samples can be boiled immediately (5 min). Alternatively, to ensure inhibition of phosphorylation changes postlysis, kinase, and phosphatase inhibitors are included in the IP-STOP. Add an equal volume of a 2X concentrated stock of IP-STOP to terminate reactions. Boil the samples to completely solubilize proteins.

2. To sequester the SDS, add an equal volume of 2X NP40/ inhibitor solution to stopped samples.
3. Add 50 μL of prepared SAC to the NP40-treated samples in microfuge tubes. Vortex and centrifuge the suspension for 1 min at 10,000g. Aspirate the supernatant into a fresh tube and discard the pellet. This step "preclears" the sample of proteins that bind nonspecifically to SAC in the absence of antibody.
4. Add antibody to the sample and incubate for 1–2 h at 4°C with shaking. Antibody should be added in excess; for a protein of approx 66 kDa and 1% abundance, 10 μg of affinity-purified IgG added to 100 μg total synaptosomal protein is about 10-fold excess with respect to binding capacity.
5. Add 50 μL SAC and incubate for 1–2 h at 4°C with shaking.
6. Centrifuge the suspension at 10,000g in a microcentrifuge for 1 min. Aspirate and discard the supernatant. Resuspend the pellet in 500 μL SAC-wash buffer with vortexing. Recentrifuge and repeat the wash twice.
7. Resuspend the final washed pellet in 100 μL SDS-STOP and boil samples for 5 min. Centrifuge at 10,000g, and use the supernatants for SDS-PAGE as described in Sections 3.4.1. and 3.4.2.

3.4.7.3. DETERMINATION OF PHOSPHATE INCORPORATION—
SPECIFIC ACTIVITY OF SYNAPTOSOMAL [^{32}P]ATP

1. Take small aliquots of ^{32}P-labeled synaptosomes from each sample before immunoprecipitation and add 100 μL 0.4M perchloric acid (HClO$_4$) to deproteinate the sample. Centrifuge at 10,000g for 10 min at 4°C and aspirate the

supernatant into another tube. Neutralize the supernatant with 10M KOH and recentrifuge at 10,000g for 10
min at 4°C to sediment the $KClO_4$. Aspirate the supernatant for HPLC analysis.

2. Determine the specific activity of the [γ-^{32}P]ATP in the
perchloric acid-extracted supernatant by separating the
nucleotides on reverse-phase HPLC using a C18 column
and absorbance detector (254 nm). Inject 10 μL of the
extracted sample and elute isocratically for 10 min at 1 mL/
min with mobile phase buffer A (100 mM K$^+$-phosphate,
pH 6.0), followed by 10 min at 1 mL/min with buffer B
(75 mM K$^+$-phosphate, pH 6.0/25% [v/v] methanol).
3. Collect [γ-^{32}P]ATP in the ATP peak eluted and determine
the ^{32}P activity therein by liquid scintillation spectrometry.
4. Quantitate the amount of ATP in the eluted peak by comparison of peak areas with pure ATP external standards run
under identical chromatographic conditions as the sample.

3.4.7.4. QUANTITATION OF PHOSPHOPROTEINS BY IMMUNOBLOTTING

The mole amount of the protein of interest in the synaptosomal samples is determined in incubations of unlabeled
synaptosomes carried out in parallel with incubations of ^{32}P-
labeled terminals. Immunoblotting (Western blotting) after
electrotransfer of proteins on one-dimensional gels onto nitrocellulose membranes is perhaps the most convenient method
(Towbin et al., 1979; Lin and Kasamatsu, 1983). However,
the efficiency of electrotransfer and nitrocellulose binding
varies from protein to protein, so it is essential to ensure that
there is a quantitative transfer of the protein of interest. If
highly specific antibody is available, a dot-immunobinding
assay of synaptosomal samples spotted directly onto nitrocellulose is more convenient and perhaps more quantitative
(Jahn et al., 1984; Chapter 5). Pure protein standards should
be run in parallel with the samples. Electrotransferred or dot-
blotted proteins are incubated with antibody and washed in
accordance with the titer and characteristics of the antibody.
^{125}I-Protein A is used as a reporter for quantitation by autoradiography (Sihra et al., 1989).

3.4.8. Translocation of Synaptosomal Phosphoproteins

The change in the phosphorylation state of a protein can alter any number of properties of the protein to lead to functional effects. One property that has provided useful information has been the phosphorylation-dependent physicochemical change that occurs to alter the avidity of an extrinsic protein for membranes (Schiebler et al., 1986; Sihra et al., 1989). Thus, translocation of synapsin I from a particulate (membrane) to a cytosolic compartment provided a demonstration of the phosphorylation-dependent change in the affinity of the protein for synaptic vesicles as a direct consequence of synaptosomal stimulation (Sihra et al., 1989). Using the appropriate equipment, translocation of other such presynaptic proteins can be determined and the stoichiometry of phosphorylation of the translocating protein evaluated.

3.4.8.1. REAGENTS

1. Hypotonic lysis buffer: 20 mM HEPES, pH 7.4, 10 mM $Na_4P_2O_7$, 1 mM EDTA, 1 mM EGTA, 10 µg/mL leupeptin, 10 µg/mL antipain, 1 µg/mL chymostatin, 1 µg/mL pepstatin, and 0.1% (w/v) aprotinin.
2. Saponin buffer: 0.3% (w/v) saponin in HBM containing 10 mM $Na_4P_2O_7$, 3 mM EDTA, 3 mM EGTA, 10 µg/mL leupeptin, 10 µg/mL antipain, 1 µg/mL chymostatin, 1 µg/mL pepstatin, and 0.1% (w/v) aprotinin.

3.4.8.2. EQUIPMENT

Bench-top ultracentrifuge with rotor capable of taking small 1–3-mL tubes (Beckman TL-100 ultracentrifuge with a TLA 100.2 fixed-angle rotor).

3.4.8.3. PROCEDURE

1. Incubate and stimulate/depolarize mock-prelabeled or ^{32}P-labeled synaptosomes as described in Section 2.2. Terminate the incubation by disruption of synaptosomes by either hypotonic lysis or permeabilization with saponin as follows:
   a. Effect hypotonic lysis by adding 10 vol of ice-cold hypotonic lysis buffer.

b. In situations where there is reason to suspect that a change in osmolality and/or dilution owing to the addition of hypotonic lysis buffer could be responsible for the effective translocation of a protein, rather than the physicochemical changes induced by an alteration of phosphorylation state, an alternative method of synaptosomal disruption is necessary. Using a detergent, such as saponin, disruption or lysis of synaptosomes can be obtained without significant dilution. Disrupt synaptosomes after appropriate incubation by the addition of a 1/3 vol of saponin buffer to produce a final concentration of 50 μg saponin/mg synaptosomal protein and place the samples at 4°C. After 10 min in the presence of saponin, add cholesterol (80 μg/mL, final concentration) to quench the effect of saponin.

The exact concentration of saponin required for a given protein concentration of synaptosomes and the lysis time should be determined by prior titration of synaptosomal resuspensions with increasing concentrations of saponin. The optimal concentration is that at which only plasma membranes are disrupted. This can be ensured by determining the solubilization (appearance in the cytosol) of a known intrinsic synaptic vesicle protein, such as synaptophysin (Jahn et al., 1985).

2. Following hypotonic lysis or saponin disruption, fractionate the samples by centrifugation at 200,000g for 20 min (Beckman TL-100 ultracentrifuge with a TLA 100.2 fixed-angle rotor). Centrifuge tubes should be pretreated with BSA to minimize binding of synaptosomal proteins to tube walls. Aspirate the supernatants and freeze immediately in liquid nitrogen. When hypotonic lysis has been used, lyophilize the frozen supernatants. Store supernatant and pellet samples at −70°C until resolubilization.

3. Resolubilize lyophilized supernatants from hypotonic lysis with 0.1% Triton X-100 (v/v in H_2O). Pellets should be resuspended in 0.1% Triton X-100 (v/v in incubation buffer), so that the salt concentration in the supernatant

and pellet resuspensions is identical. For saponin disrupted synaptosomes, supernatants should be used without further treatment and pellets should be resolubilized in buffer containing the same concentration of saponin and cholesterol as that used for disruption.

4. Materials

Leupeptin, antipain, chymostatin, and pepstatin are obtained from Chemicon (Los Angeles, CA). *S. aureus* V8 protease and other proteases are obtained from Miles Laboratories (West Haven, CT). ^{32}P and ^{125}I-protein A are obtained from Du Pont-New England Nuclear (Boston, MA). Sources of other reagent materials and equipment are indicated in the text; otherwise, chemicals of analytical grade are obtained from standard commercial sources.

5. Final Remarks

The intention in this chapter has been to provide a working knowledge of current techniques employed in studying protein phosphorylation and dephosphorylation in synaptosomes. The information, although not exhaustive, should provide enough detail to enable investigators to modify or seek alternatives that are more suited to their specific requirements. On the basis of initial characterization of presynaptic phosphoproteins using these approaches, it becomes feasible to develop more sophisticated means, such as, phosphorylation-site-specific antibodies (Chapter 5), for assaying the phosphorylation/dephosphorylation of a specific phosphoprotein of interest.

Acknowledgments

I would like to thank Jasmina Jovanovic and Frederico Trindade-Ramage for helpful discussion. Thanks are owed to Tristan Piper and Michael Perkinton for their comments and input during the preparation of the manuscript. The author is supported by a University Award from the Wellcome Trust.

References

Abdel-Latif, A. A. (1966) A simple method for isolation of nerve-ending particles from rat brain. *Biochim. Biophys. Acta* **121**, 403–406.

Anderson, L. and Anderson, N. G. (1977) High resolution two-dimensional electrophoresis of human plasma proteins. *Proc. Natl. Acad. Sci. USA* **74**, 5421–5425.

Autilio, L. A., Appel, S. H., Pettis, P., and Gambetti, P. L. (1968) Biochemical studies of synapses in vitro. I. Protein synthesis. *Biochemistry* **7**, 2615–2622.

Babitch, J. A., Breithaupt, T. B., Chiu, T. C., Garadi, R., and Helseth, D. L. (1976) Preparation of chick brain synaptosomes and synaptosomal membranes. *Biochim. Biophys. Acta* **433**, 75–89.

Booth, R. F. and Clark, J. B. (1978) A rapid method for the preparation of relatively pure metabolically competent synaptosomes from rat brain. *Biochem. J.* **176**, 365–370.

Browning, M. D., Huang, C., and Greengard, P. (1987) Similarities between protein IIIa and protein IIIb, two prominent synaptic vesicle-associated phosphoproteins. *J. Neurosci.* **7**, 847–853.

Burke, B., E. and DeLorenzo, R. J. (1981) Ca2+- and calmodulin-stimulated endogenous phosphorylation of neurotubulin. *Proc. Natl. Acad. Sci. USA* **78**, 991–995.

Chao, S. H., Suzuki, Y., Zysk, J. R., and Cheung, W. Y. (1984) Activation of calmodulin by various metal cations as a function of ionic radius. *Mol. Pharmacol.* **26**, 75–82.

Chin, G. J., Shapiro, E., Vogel, S. S., and Schwartz, J. H. (1989) Aplysia synaptosomes. I. Preparation and biochemical and morphological characterization of subcellular membrane fractions. *J. Neurosci.* **9**, 38–48.

Cleveland, D. W., Fischer, S. G., Kirschner, M. W., and Laemmli, U. K. (1977) Peptide mapping by limited proteolysis in sodium dodecyl sulfate and analysis by gel electrophoresis. *J. Biol. Chem.* **252**, 1102–1106.

Coffey, E. T., Herrero, I., Sihra, T. S., Sanchez-Prieto, J., and Nicholls, D. G. (1994) Glutamate exocytosis and MARCKS phosphorylation are enhanced by a metabotropic glutamate receptor coupled to a protein kinase C synergistically activated by diacylglycerol and arachidonic acid (published erratum appears in *J. Neurochem.* 1995 **64(1)**, 471). *J. Neurochem.* **63**, 1303–1310.

Coffey, E. T., Sihra, T. S., and Nicholls, D. G. (1993) Protein kinase C and the regulation of glutamate exocytosis from cerebrocortical synaptosomes. *J. Biol. Chem.* **268**, 21,060–21,065.

De Robertis, E., De Iraldi, A. P., Rodriguez, G., and Gomez, J. (1961) On the isolation of nerve endings and synaptic vesicles. *J. Biophys. Biochem. Cytol.* **9**, 229–235.

Dekker, L. V., de Graan, P. N., De Wit, M., Hens, J. J., and Gispen, W. H. (1990) Depolarization-induced phosphorylation of the protein kinase C substrate B-50 (GAP-43) in rat cortical synaptosomes. *J. Neurochem.* **54**, 1645–1652.

DeLorenzo, R. J. (1982) Calmodulin in neurotransmitter release and synaptic function. *Fed. Proc.* **41**, 2265–2272.

DeLorenzo, R. J., Freedman, S. D., Yohe, W. B., and Maurer, S. C. (1979) Stimulation of Ca2+-dependent neurotransmitter release and presynaptic nerve terminal protein phosphorylation by calmodulin and a calmodulin-like protein isolated from synaptic vesicles. *Proc. Natl. Acad. Sci. USA* **76**, 1838–1842.

Dowdall, M. J. and Whittaker, V. P. (1973) Comparative studies in synaptosome formation: the preparation of synaptosomes from the head ganglion of the squid, *Loligo pealii. J. Neurochem.* **20**, 921–935.

Dunkley, P. R. and Robinson, P. J. (1986) Depolarization-dependent protein phosphorylation in synaptosomes: mechanisms and significance. *Prog. Brain Res.* **69**, 273–293.

Dunkley, P. R., Baker, C. M., and Robinson, P. J. (1986b) Depolarization-dependent protein phosphorylation in rat cortical synaptosomes: characterization of active protein kinases by phosphopeptide analysis of substrates. *J. Neurochem.* **46**, 1692–1703.

Dunkley, P. R., Heath, J. W., Harrison, S. M., Jarvie, P. E., Glenfield, P. J., and Rostas, J. A. (1988) A rapid Percoll gradient procedure for isolation of synaptosomes directly from an S1 fraction: homogeneity and morphology of subcellular fractions. *Brain Res.* **441**, 59–71.

Dunkley, P. R., Jarvie, P. E., Heath, J. W., Kidd, G. J., and Rostas, J. A. (1986a) A rapid method for isolation of synaptosomes on Percoll gradients. *Brain Res.* **372**, 115–129.

Enriquez, J. A., Sanchez-Prieto, J., Muino Blanco, M. T., Hernandez-Yago, J., and Lopez-Perez, M. J. (1990) Rat brain synaptosomes prepared by phase partition. *J. Neurochem.* **55**, 1841–1849.

Forn, J. and Greengard, P. (1978) Depolarizing agents and cyclic nucleotides regulate the phosphorylation of specific neuronal proteins in rat cerebral cortex slices. *Proc. Natl. Acad. Sci. USA* **75**, 5195–5199.

Gomez-Puertas, P., Martinez-Serrano, A., Blanco, P., Satrustegui, J., and Bogonez, E. (1991) Conditions restricting depolarization-dependent calcium influx in synaptosomes reveal a graded response of P96 dephosphorylation and a transient dephosphorylation of P65. *J. Neurochem.* **56**, 2039–2047.

Gower, H., Rodnight, R., and Brammer, M. J. (1986) Ca2+ sensitivity of Ca2+-dependent protein kinase activities toward intrinsic proteins in synaptosomal membrane fragments from rat cerebral tissue. *J. Neurochem.* **46**, 440–447.

Gray, E. G. and Whittaker, V. P. (1962) The isolation of nerve endings from brain: an electron-microscopic study of cell fragments derived by homogenization and centrifugation. *J. Anat.* **96**, 79–88.

Greengard, P. (1987) Neuronal phosphoproteins. Mediators of signal transduction (review). *Mol. Neurobiol.* **1**, 81–119.

Harrison, S. M., Jarvie, P. E., and Dunkley, P. R. (1988) A rapid Percoll gradient procedure for isolation of synaptosomes directly from an S1 fraction: viability of subcellular fractions. *Brain Res.* **441**, 72–80.

Huttner, W. B. and Greengard, P. (1979) Multiple phosphorylation sites in protein I and their differential regulation by cyclic AMP and calcium. *Proc. Natl. Acad. Sci. USA* **76**, 5402–5406.

Huttner, W. B., DeGennaro, L. J., and Greengard, P. (1981) Differential phosphorylation of multiple sites in purified protein I by cyclic AMP-dependent and calcium-dependent protein kinases. *J. Biol. Chem.* **256**, 1482–1488.

Imada, M. and Sueoka, N. (1980) A two-dimensional polyacrylamide gel electrophoresis system for the analysis of mammalian cell surface proteins. *Biochim. Biophys. Acta* **625**, 179–192.

Israel, M., Lazereg, S., Lesbats, B., Manaranche, R., and Morel, N. (1985) Large-scale purification of Torpedo electric organ synaptosomes. *J. Neurochem.* **44**, 1107–1110.

Jahn, R., Schiebler, W., and Greengard, P. (1984) A quantitative dot-immunobinding assay for proteins using nitrocellulose membrane filters. *Proc. Natl. Acad. Sci. USA* **81**, 1684–1687.

Jahn, R., Schiebler, W., Ouimet, C., and Greengard, P. (1985) A 38,000-dalton membrane protein (p38) present in synaptic vesicles. *Proc. Natl. Acad. Sci. USA* **82**, 4137–4141.

Jessell, T. M. and Kandel, E. R. (1993) Synaptic transmission: a bidirectional and self-modifiable form of cell-cell communication (review). *Cell* **72(Suppl.)**, 1–30.

Kauppinen, R. A. and Nicholls, D. G. (1986) Synaptosomal bioenergetics. The role of glycolysis, pyruvate oxidation and responses to hypoglycaemia. *Eur. J. Biochem.* **158**, 159–165.

Kauppinen, R. A., Sihra, T. S., and Nicholls, D. G. (1986) Divalent cation modulation of the ionic permeability of the synaptosomal plasma membrane. *Biochim. Biophys. Acta* **860**, 178–184.

Kelly, R. B. (1993) Storage and release of neurotransmitters (review). *Cell* **72(Suppl.)**, 43–53.

Kennedy, M. B. and Greengard, P. (1981) Two calcium/calmodulin-dependent protein kinases, which are highly concentrated in brain, phosphorylate protein I at distinct sites. *Proc. Natl. Acad. Sci. USA* **78**, 1293–1297.

Kennedy, M. B., McGuinness, T., and Greengard, P. (1983) A calcium/calmodulin-dependent protein kinase from mammalian brain that phosphorylates synapsin I: partial purification and characterization. *J. Neurosci.* **3**, 818–831.

Krueger, B. K., Forn, J., and Greengard, P. (1977) Depolarization-induced phosphorylation of specific proteins, mediated by calcium ion influx, in rat brain synaptosomes. *J. Biol. Chem.* **252**, 2764–2773.

Kyriazi, H. T. and Basford, R. E. (1986) Intractable unphysiologically low adenylate energy charge values in synaptosome fractions: an explanatory hypothesis based on the fraction's heterogeneity. *J. Neurochem.* **47,** 512–528.

Laemmli, U. K. (1970) Cleavage of structural proteins during the assembly of the head of bacteriophage T4. *Nature* **227,** 680–685.

Lin, W. and Kasamatsu, H. (1983) On the electrotransfer of polypeptides from gels to nitrocellulose membranes. *Anal. Biochem.* **128,** 302–311.

Liu, J., Farmer, J. D., Jr., Lane, W. S., Friedman, J., Weissman, I., and Schreiber, S. L. (1991) Calcineurin is a common target of cyclophilin-cyclosporin A and FKBP-FK506 complexes. *Cell* **66,** 807–815.

McMahon, H. T. and Nicholls, D. G. (1991) The bioenergetics of neurotransmitter release (review). *Biochim. Biophys. Acta* **1059,** 243–264.

McPherson, P. S., Garcia, E. P., Slepnev, V. I., David, C., Zhang, X., Grabs, D., Sossin, W. S., Bauerfeind, R., Nemoto, Y., and DeCamilli, P. (1996) A presynaptic inositol-5-phosphatase. *Nature* **379,** 353–357.

Michaelson, D. M. and Avissar, S. (1979) Ca^{2+}-dependent protein phosphorylation of purely cholinergic Torpedo synaptosomes. *J. Biol. Chem.* **254,** 12,542–12,546.

Musgrave, M. A., Madigan, M. A., Bennett, B. M., and Goh, J. W. (1994) Stimulation of postsynaptic and inhibition of presynaptic adenylyl cyclase activity by metabotropic glutamate receptor activation. *J. Neurochem.* **62,** 2316–2324.

Nagy, A. and Delgado-Escueta, A. V. (1984) Rapid preparation of synaptosomes from mammalian brain using nontoxic iso-osmotic gradient material (Percoll). *J. Neurochem.* **43,** 1114–1123.

Nicholls, D. G. (1978) Calcium transport and proton electrochemical potential gradient in mitochondria from guinea-pig cerebral cortex and rat heart. *Biochem. J.* **170,** 511–522.

Nicholls, D. G., Sihra, T. S., and Sanchez-Prieto, J. (1987) The role of the plasma membrane and intracellular organelles in synaptosomal calcium regulation (review). *Soc. Gen. Physiol. Ser.* **42,** 31–43.

Nichols, R. A., Suplick, G. R., and Brown, J. M. (1994) Calcineurin-mediated protein dephosphorylation in brain nerve terminals regulates the release of glutamate. *J. Biol. Chem.* **269,** 23,817–23,823.

Nordmann, J. J., Desmazes, J. P., and Georgescauld, D. (1982) The relationship between the membrane potential of neurosecretory nerve endings, as measured by a voltage-sensitive dye, and the release of neurohypophysial hormones. *Neuroscience* **7,** 731–737.

O'Callaghan, J. P., Dunn, L. A., and Lovenberg, W. (1980) Calcium-regulated phosphorylation in synaptosomal cytosol: dependence on calmodulin. *Proc. Natl. Acad. Sci. USA* **77,** 5812–5816.

O'Farrell, P. H. (1975) High resolution two-dimensional electrophoresis of proteins. *J. Biol. Chem.* **250,** 4007–4021.

O'Farrell, P. Z., Goodman, H. M., and O'Farrell, P. H. (1977) High resolution two-dimensional electrophoresis of basic as well as acidic proteins. *Cell* **12,** 1133–1141.

Pant, H. C., Pollard, H. B., Pappas, G. D., and Gainer, H. (1979) Phosphorylation of specific, distinct proteins in synaptosomes and axons from squid nervous system. *Proc. Natl. Acad. Sci. USA* **76,** 6071–6075.

Rapier, C., Lunt, G. G., and Wonnacott, S. (1990) Nicotinic modulation of [³H]dopamine release from striatal synaptosomes: pharmacological characterization. *J. Neurochem.* **54,** 937–945.

Rauch, N. and Roskoski, R., Jr. (1984) Characterization of cyclic AMP-dependent phosphorylation of neuronal membrane proteins. *J. Neurochem.* **43,** 755–762.

Robinson, P. J. (1991a) The role of protein kinase C and its neuronal substrates dephosphin, B-50, and MARCKS in neurotransmitter release (review). *Mol. Neurobiol.* **5,** 87–130.

Robinson, P. J. (1991b) Dephosphin, a 96,000 Da substrate of protein kinase C in synaptosomal cytosol, is phosphorylated in intact synaptosomes. *FEBS Lett.* **282,** 388–392.

Robinson, P. J. (1991c) Characterization of dephosphin: 96,000 and 93,000 dalton substrates of protein kinase C in rat and human brain synaptosomal cytosol and synaptic vesicles. *Neurosci. Res. Commun.* **9,** 167–176.

Robinson, P. J. (1992a) Potencies of protein kinase C inhibitors are dependent on the activators used to stimulate the enzyme. *Biochem. Pharmacol.* **44,** 1325–1334.

Robinson, P. J. (1992b) Differential stimulation of protein kinase C activity by phorbol ester or calcium/phosphatidylserine in vitro and in intact synaptosomes. *J. Biol. Chem.* **267,** 21,637–21,644.

Robinson, P. J. and Dunkley, P. R. (1983) Depolarization-dependent protein phosphorylation in rat cortical synaptosomes: factors determining the magnitude of the response. *J. Neurochem.* **41,** 909–918.

Robinson, P. J. and Dunkley, P. R. (1985) Depolarization-dependent protein phosphorylation and dephosphorylation in rat cortical synaptosomes is modulated by calcium. *J. Neurochem.* **44,** 338–348.

Robinson, P. J. and Dunkley, P. R. (1987) Altered protein phosphorylation in intact rat cortical synaptosomes after in vivo administration of fluphenazine. *Biochem. Pharmacol.* **36,** 2203–2208.

Robinson, P. J., Jarvie, P. E., and Dunkley, P. R. (1984) Depolarization-dependent protein phosphorylation in rat cortical synaptosomes is inhibited by fluphenazine at a step after calcium entry. *J. Neurochem.* **43,** 659–667.

Salamin, A., Deshusses, J., and Straub, R. W. (1981) Phosphate ion transport in rabbit brain synaptosomes. *J. Neurochem.* **37,** 1419–1424.

Santiapillai, N. F., Gray, S. R., Phillips, R. E., and Richardson, P. J. (1989) Isolation of nerve terminals from crustacean muscle. *J. Neurochem.* **53,** 1527–1535.

Schiebler, W., Jahn, R., Doucet, J. P., Rothlein, J., and Greengard, P. (1986) Characterization of synapsin I binding to small synaptic vesicles (published erratum appears in *J. Biol. Chem.* 1986; **261(26)**, 12,428). *J. Biol. Chem.* **261,** 8383–8390.

Sieghart, W., Forn, J., and Greengard, P. (1979) Ca2+ and cyclic AMP regulate phosphorylation of same two membrane-associated proteins specific to nerve tissue. *Proc. Natl. Acad. Sci. USA* **76,** 2475–2479.

Sieghart, W., Schulman, H., and Greengard, P. (1980) Neuronal localization of Ca^{2+}-dependent protein phosphorylation in brain. *J. Neurochem.* **34,** 548–553.

Sihra, T. S. and Nichols, R. A. (1993) Mechanisms in the regulation of neurotransmitter release from brain nerve terminals: current hypotheses (review). *Neurochem. Res.* **18,** 47–58.

Sihra, T. S. and Pearson, H. (1995) Ca/Calmodulin-dependent kinase II inhibitor KN-62 attenuates glutamate release by inhibiting voltage-dependent Ca-channels. *Neuropharmacology* **34,** 731–741.

Sihra, T. S., Bogonez, E., and Nicholls, D. G. (1992) Localized Ca2+ entry preferentially effects protein dephosphorylation, phosphorylation, and glutamate release. *J. Biol. Chem.* **267,** 1983–1989.

Sihra, T. S., Piomelli, D., and Nichols, R. A. (1993) Barium evokes glutamate release from rat brain synaptosomes by membrane depolarization: involvement of K^+, Na^+, and Ca^{2+} channels. *J. Neurochem.* **61,** 1220–1230.

Sihra, T. S., Scott, I. G., and Nicholls, D. G. (1984) Ionophore A23187, verapamil, protonophores, and veratridine influence the release of gamma-aminobutyric acid from synaptosomes by modulation of the plasma membrane potential rather than the cytosolic calcium. *J. Neurochem.* **43,** 1624–1630.

Sihra, T. S., Wang, J. K., Gorelick, F. S., and Greengard, P. (1989) Translocation of synapsin I in response to depolarization of isolated nerve terminals. *Proc. Natl. Acad. Sci. USA* **86,** 8108–8112.

Sim, A. T., Dunkley, P. R., Jarvie, P. E., and Rostas, J. A. (1991) Modulation of synaptosomal protein phosphorylation/dephosphorylation by calcium is antagonized by inhibition of protein phosphatases with okadaic acid. *Neurosci. Lett.* **126,** 203–206.

Snelling, R. and Nicholls, D. (1984) The calmodulin antagonists, trifluoperazine and R24571, depolarize the mitochondria within guinea pig cerebral cortical synaptosomes. *J. Neurochem.* **42,** 1552–1557.

Tamir, H., Rapport, M. M., Roizin, L., Huang, Y. L., and Liu, J. C. (1974) Preparation of synaptosomes and vesicles with sodium diatrizoate. *J. Neurochem.* **23,** 943–949.

Terrian, D. M., Johnston, D., Claiborne, B. J., Ansah-Yiadom, R., Strittmatter, W. J., and Rea, M. A. (1988) Glutamate and dynorphin release from a subcellular fraction enriched in hippocampal mossy fiber synaptosomes. *Brain Res. Bull.* **21,** 343–351.

Thorne, B., Wonnacott, S., and Dunkley, P. R. (1991) Isolation of hippocampal synaptosomes on Percoll gradients: cholinergic markers and
ligand binding sites. *J. Neurochem.* **56,** 479–484.
Tibbs, G. R., Barrie, A. P., Van Mieghem, F. J., McMahon, H. T., and
Nicholls, D. G. (1989) Repetitive action potentials in isolated nerve
terminals in the presence of 4-aminopyridine: effects on cytosolic
free Ca^{2+} and glutamate release. *J. Neurochem.* **53,** 1693–1699.
Tokumitsu, H., Chijiwa, T., Hagiwara, M., Mizutani, A., Terasawa, M.,
and Hidaka, H. (1990) KN-62, 1-[N,O-bis(5-isoquinolinesulfonyl)-
N-methyl-L-tyrosyl]-4-phenylpiperazine, a specific inhibitor of
Ca^{2+}/calmodulin-dependent protein kinase II. *J. Biol. Chem.* **265,**
4315–4320.
Towbin, H., Staehelin, T., and Gordon, J. (1979) Electrophoretic transfer
of proteins from polyacrylamide gels to nitrocellulose sheets: procedure and some applications. *Proc. Natl. Acad. Sci. USA* **76,** 4350–4354.
Turner, T. J., Adams, M. E., and Dunlap, K. (1992) Calcium channels
coupled to glutamate release identified by omega-Aga-IVA. *Science* **258,** 310–313.
Verhage, M., McMahon, H. T., Ghijsen, W. E., Boomsma, F., Scholten,
G., Wiegant, V. M., and Nicholls, D. G. (1991) Differential release
of amino acids, neuropeptides, and catecholamines from isolated
nerve terminals. *Neuron* **6,** 517–524.
Verity, M. A. (1972) Cation modulation of synaptosomal respiration. *J.
Neurochem.* **19,** 1305–1317.
Walaas, S. I. and Greengard, P. (1991) Protein phosphorylation and neuronal function (review). *Pharmacol. Rev.* **43,** 299–349.
Walaas, S. I., Nairn, A. C., and Greengard, P. (1983) Regional distribu-
tion of calcium- and cyclic adenosine 3':5'-monophosphate-regu-
lated protein phosphorylation systems in mammalian brain. I.
Particulate systems. *J. Neurosci.* **3,** 291–301.
Wang, J. K., Walaas, S. I., and Greengard, P. (1988) Protein phosphorylation in nerve terminals: comparison of calcium/calmodulin-
dependent and calcium/diacylglycerol-dependent systems. *J.
Neurosci.* **8,** 281–288.
Wang, J. K., Walaas, S. I., Sihra, T. S., Aderem, A., and Greengard, P.
(1989) Phosphorylation and associated translocation of the 87-kDa
protein, a major protein kinase C substrate, in isolated nerve terminals. *Proc. Natl. Acad. Sci. USA* **86,** 2253–2256.
Whittaker, V. P. (1968) The morphology of fractions of rat forebrain synaptosomes separated on continuous sucrose density gradients.
Biochem. J. **106,** 412–417.
Whittaker, V. P. (1969) The synaptosome, in *Handbook of Neurochemistry,*
vol. 2 (Lajtha, A., ed.), Plenum, New York, pp. 327–364.
Whittaker, V. P. (1988) Synaptosome preparations (letter). *J. Neurochem.*
50, 324,325.

Whittaker, V. P. and Greengard, P. (1971) The isolation of synaptosomes from the brain of a teleost fish, *Centriopristes striatus. J. Neurochem.* **18,** 173–176.

Whittaker, V. P., Michaelson, I. A., and Kirkland, R. J. A. (1964) The separation of synaptic vesicles from nerve-ending particles ("synaptosomes"). *Biochem. J.* **90,** 293–305.

Wu, W. C., Walaas, S. I., Nairn, A. C., and Greengard, P. (1982) Calcium/phospholipid regulates phosphorylation of a Mr "87k" substrate protein in brain synaptosomes. *Proc. Natl. Acad. Sci. USA* **79,** 5249–5253.

Protein Kinase
and Phosphatase Inhibitors

Applications in Neuroscience

Hugh C. Hemmings, Jr.

1. Introduction

Extracellular signals, including neurotransmitters, hormones, cytokines, and growth factors, produce many of their physiological effects by regulating the state of phosphorylation of specific proteins in target cells (Walaas and Greengard, 1991). Inhibitors of protein kinases and protein phosphatases are useful tools in evaluating the physiological roles of protein phosphorylation signaling mechanisms, and may ultimately find therapeutic applications. Available inhibitors include natural and synthetic compounds, peptides, and proteins (MacKintosh and MacKintosh, 1994). Natural inhibitors include endogenous protein inhibitors of protein kinases (e.g., the protein kinase inhibitor [PKI] of cyclic AMP [cAMP]-dependent protein kinase [PKA]) and protein phosphatases (e.g., inhibitor-1, inhibitor-2, and DARPP-32), as well as toxins produced by bacteria, fungi, plants, dinoflagellates, and insects. Synthetic inhibitors include peptide analogs derived from intramolecular pseudosubstrate inhibitory domains of protein kinases and phosphatases, derivatives of natural toxins, and analogs and inhibitors of endogenous activators (e.g., cAMP analogs) or substrates (e.g., ATP analogs). This chapter will review the most widely available inhibitors with demonstrated utility, including their basic

From: *Neuromethods, Vol. 30: Regulatory Protein Modification: Techniques and Protocols*
Ed: H. C. Hemmings, Jr. Humana Press Inc.

properties and applications relevant to their use in neuro-science research.

Several important principles are relevant to the use of protein kinase and phosphatase inhibitors:

1. Inhibitor potency can be measured by IC_{50} values (inhibitor concentration yielding 50% inhibition), which are dependent on assay conditions (e.g., enzyme and substrate concentrations), or by K_i values, which are concentration-independent. The inhibitory constant (K_i) is the standard for measuring inhibitor potency, although IC_{50} values are commonly used.

2. The IC_{50} for inhibitors with high affinity binding (low K_i values) depends on the concentration of protein kinase or phosphatase in the assay, i.e., higher inhibitor concentrations are required for enzyme inhibition at higher enzyme concentrations. Therefore, at least stoichiometric amounts of inhibitor are required of even the most potent enzyme inhibitors, such that micromolar concentrations will be required to inhibit an enzyme present at micromolar concentrations in spite of a K_i value in the nanomolar range. Use of inhibitors at such high concentrations may, in some cases, limit their specificity.

3. The mechanism of inhibition may influence the appropriate use of protein kinase and phosphatase inhibitors. For example, the efficacy of protein kinase inhibitors that are competitive antagonists of ATP binding is affected by the ATP concentration in the assay, which is millimolar in intact cells. Noncompetitive inhibitors are not subject to this limitation and are frequently effective at much lower concentrations.

4. Owing to inherent structural similarities among various protein kinases and phosphatases, no inhibitors are absolutely specific and the selectivity (difference in IC_{50} or K_i values) of a number of inhibitors for different enzymes can be very low. In some cases, this limitation can be overcome by comparing two inhibitors with different selectivities for various enzymes or by the use of ago-

nist/antagonist pairs. Evidence for the involvement of a specific protein kinase in a physiological process can be strengthened by observing opposite effects of an activator (e.g., cAMP) and an inhibitor (e.g., [Rp]cAMPS) that act at the same site on the enzyme.

5. Many inhibitors have nonspecific effects unrelated to protein kinase or phosphatase inhibition. This problem can be minimized by employing more than one inhibitor to achieve the same result and by employing structurally related inactive compounds that are available as controls for a number of inhibitors.

6. The ability of inhibitors to enter cells, the technique of application, and metabolism are important factors that affect inhibitor efficacy when used on intact cells.

Application of these general principles will enhance the utility of the protein kinase and phosphatase inhibitors.

2. Protein Kinase Inhibitors

All known members of the protein-serine/threonine kinase and protein-tyrosine kinase families share a related catalytic domain of 250–300 amino acids, which includes the conserved ATP binding region and catalytic residues, and are grouped together in a large protein superfamily (Hanks et al., 1988). The protein-serine/threonine kinases are further subdivided into subfamilies according to their regulators (e.g., cyclic nucleotide-dependent, Ca^{2+}/calmodulin-dependent, or phosphatidylserine-regulated protein kinases); these subfamilies usually exhibit structural similarities within their regulatory domains as well. Because of the conserved protein kinase catalytic domain structure, many protein kinase inhibitors that interact with the catalytic domain (e.g., ATP analogs) lack sufficient specificity for use in physiological experiments. The protein substrate specificity of protein kinases is largely determined by the local phosphorylation site sequence, which has given rise to the concept of a consensus phosphorylation site sequence for each protein kinase. However, not all good substrates conform to these consensus sequences, and there can be considerable overlap in the abilities of different protein kinases to phosphorylate a specific peptide because of many shared specificity determinants (Kemp and Pearson, 1991). Thus, peptide substrate-based inhibitors afford increased, but not absolute, specificity in the design of protein kinase inhibitors compared to ATP analogs. Protein kinase inhibitors that interact with the regulatory domains offer reasonable specificity between protein kinase subfamilies, but limited specificity between members of the same subfamily. Subtle differences in active site behavior between closely related enzymes may soon allow the generation of enzyme-targeted inhibitors that exploit these differences. For example, a peptide-based affinity inactivator that distinguished completely between two closely related protein-serine/threonine kinases has been described recently (Yan et al., 1996). When evaluating protein kinase inhibitors, these limitations should be borne in mind, as well as the fact that most inhibitors have not been tested on a wide range of protein kinases and their isoforms.

2.1. Cyclic Nucleotide-Dependent Protein Kinase Inhibitors

2.1.1. Protein Kinase Inhibitor of PKA

The inhibitor protein of PKA (PKI) is among the most potent and specific of the known protein kinase inhibitors. PKI has been extensively characterized and widely used to study PKA-mediated processes (Walsh and Glass, 1991). A synthetic peptide consisting of residues 5–22 of the 75 amino acid protein (PKI[5–22]amide; TTYADFIASGRTGRRNAI-amide) possesses ~10% of the inhibitory potency of the intact protein (K_i = 1.6 nM) (Glass et al., 1989). The peptide contains a pseudosubstrate domain consisting of RRNAI, but additional structural features increase binding affinity four orders of magnitude over that of any known substrate for PKA. This extreme potency allows considerable specificity for inhibition of PKA by PKI, which only weakly inhibits the closely homologous cGMP-dependent protein kinase (PKG;

K_i = 100 μM). Some caution is required when using shorter peptide derivatives of PKI, since at very high concentrations, they can inhibit PKG. Although PKG possess an analogous pseudosubstrate region in its regulatory domain, the corresponding peptide does not act as a potent inhibitor of the enzyme.

PKI and its peptide derivatives are extremely useful in characterizing protein kinases in tissue extracts in vitro. Inhibition of phosphorylation by PKI is strong evidence for a PKA-mediated reaction. The lack of membrane permeability of PKI peptides has been overcome in a number of whole-cell studies by microinjection of PKI into cells; PKI delivery by liposome fusion and plasmid expression has also been achieved.

2.1.1.1. Use of PKI in Cell Extracts

The high potency and specificity of purified intact PKI and its peptide derivatives make their use in vitro relatively straightforward. Despite this high degree of specificity, indiscriminate use can result in some inhibition of PKG, especially when a high concentration of PKA in the assay necessitates a high concentration of PKI peptide. For example, the concentration of the catalytic subunit of PKA in cells is about 0.3–1 μM (Hoffmann et al., 1977). Full inhibition of PKA requires essentially stoichiometric amounts of PKI peptide (i.e., >1 μM), but a 70-fold excess (70 μM) would inhibit PKG by about 50% (Glass et al., 1986). Thus, an effective concentration of PKI(5–22)amide is usually 10 μM or less, depending on the dilution factor of the tissue extract. A stock solution (1 mM in H_2O) can be prepared and used at a 1:100 dilution to allow unequivocal identification of proteins phosphorylated by PKA. PKI can also be used to minimize background protein kinase activity owing to active free PKA catalytic subunit in assays of other protein kinases (Hemmings and Adamo, 1996).

2.1.1.2. Use of PKI in Intact Cells

The membrane permeability of PKI and its peptide analogs is limited by its high polarity. Despite this, high extracellular concentrations (1 mM) of PKI peptides have been found to enter cells and inhibit PKA (Buchler et al., 1988). Alternatively, cell-permeabilization has been used to study the roles of PKA in insulin secretion from rat islets of Langerhans using PKI(6–22) (Basudev et al., 1995). However, most studies of PKI effects in intact cells have employed microinjection techniques (e.g., Adams and Levitan, 1982; Castellucci et al., 1982). These studies employ injection of an excessive amount of peptide to obtain an all-or-nothing response (Walsh and Glass, 1991); this avoids potential problems because of peptide binding to the glass injection pipet and intracellular peptide destruction. Pipets are filled with a concentrated peptide solution that is then applied to the cell interior by diffusion, pressure injection, or electrophoresis. Intracellular injection can be confirmed by coinjection of a dye or antibody. Intracellular dialysis of neurons subjected to whole-cell electrophysiological recording with 10 μM PKI(5–24) (Surmeier et al., 1995) or with 0.1–1 μg/mL PKI (Chen et al., 1988) resulted in inhibition of PKA-mediated responses. The short half-life of injected PKI or PKI peptides in cells can be partially overcome by the use of a protease-resistant modified form of PKI(6–24) (Fernandez et al., 1991).

Several studies have taken alternative approaches to the introduction of PKI into intact cells. PKI-containing liposomes have been used as an alternative to microinjection in isolated cardiac myocytes (Reisine et al., 1986) and AtT-20 pituitary tumor cells (Reisine et al., 1985), which should be applicable to neuronal cells as well. The molecular cloning of PKI has led to plasmid expression as a means of delivering PKI into cells. A vector containing a synthetic gene for PKI(1–31) has been used for the transient transfection of JEG-3 cells; an inactive mutant form of PKI(1–31) inserted into a similar vector provided a negative control (Grove et al., 1989). A vector containing PKI(1–22) has been used successfully to inhibit PKA-dependent transcription of the enkephalin gene (Huggenvik et al., 1991). A vector containing the full-length PKI protein has been used successfully by transient transfection to study cAMP-dependent regulation of the prolactin gene in GH3 pituitary cells (Day et al., 1989), of tyrosine hydroxylase expression in human neuroblastoma cells (Kim et al., 1994), and of the enkephalin gene in CHO cells (Olsen and Uhler, 1991).

2.1.2. Isoquinolinesulfonamide Inhibitors
(H-8, H-88, H-89)

A number of isoquinolinesulfonamide derivatives have been synthesized and analyzed as protein kinase inhibitors (Hidaka and Kobayashi, 1992). H-8 (N-[2-methylamino ethyl]-5-isoquinolinesulfonamide), an isoquinoline derivative of the calmodulin antagonist W-7 (naphthalenesulfonamide), is a potent and relatively selective inhibitor of PKA and PKG (Table 1). H-88 (N-[2-cinnamylaminoethyl]-5-isoquino-linesulfonamide) also inhibits both PKA and PKG, but H-89 (N-[2-p-bromocinnamylamino ethyl]-5-isoquinolinesulfona-mide), a brominated derivative of H-88, inhibits PKA with a 10-fold lower K_i than PKG. All three inhibitors inhibit PKA competitively with respect to ATP. H-85 (N-[2–formyl-p-chlorocinnamylamino ethyl]-5-isoquinolinesulfonamide) is a useful negative control for H-88 or H-89, since it is a relatively weak inhibitor of PKA and PKG, but shows comparable inhibition of other protein kinases. The principal advantage of these inhibitors is their cell membrane permeability; their selectivity is limited compared to PKI (i.e., 10-fold selectivity of H-89 for PKA vs PKG, but 600-fold for PKA vs other protein kinases). In carefully chosen applications, these inhibitors can be useful in demonstrating a role for PKA and/or PKG. Examples include the use of H-89 (20 μM), but not H-85 (30 μM), to block forskolin-induced, but not nerve growth factor (NGF)-induced, protein phosphorylation and neurite outgrowth in PC12 pheochromocytoma cells (Chijiwa et al., 1990). H-89 had no effects on adenylyl cyclase or cyclic nucleotide phosphodiesterase activities up to 100 μM; however, several isoquinolinesulfonamides (H-7, H-8, and HA-1004) at low micromolar concentrations inhibited μ opioid receptor binding (Aloyo, 1995) or imipramine binding and serotonin uptake in platelets (Helmeste and Tang, 1994).

2.1.2.1. USE OF ISOQUINOLINESULFONAMIDE INHIBITORS

These compounds are prepared as 10-mM stock solutions in dimethylsulfoxide (DMSO) or ethanol. Adequate kinase inhibition is facilitated by pretreatment of intact cells

Table 1
Inhibition Constants of Isoquinolinesulfonamide
and Napthalenesulfonamide Compounds for Various Protein Kinases[a]

Inhibitor	PKA[b]	PKG	CaMKII	MLCK	PKC	CKI	CKII
A-3	4.3	3.8	—	7	47	80	5.1
ML-7	21	—	—	0.3	42	—	—
ML-9	32	—	—	3.8	54	—	—
H-7	3.0	5.8	160	97	6.0	100	780
H-8	1.2	0.5	—	68	15	133	950
H-9	1.9	0.9	60	70	18	110	>300
HA-1004	2.3	1.3	13	150	40	—	—
H-88	0.4	0.8	70	50	80	60	100
H-89	0.05	0.5	30	30	30	40	140
H-85	>100	>100	47	28	>100	50	>100
KN-62	>100	—	0.9	>100	>100	>100	—
KN-04	>100	—	>100	>100	>100	>100	—
KN-93	>30	—	0.4	>30	>30	>30	—
KN-92	>100	—	>100	>100	>100	>100	—
CKI-6	>1000	—	840	—	>1000	50	>300
CKI-7	550	—	195	—	>1000	9.5	90
CKI-8	80	260	—	25	>100	—	—

[a]Values are for K_i (μM). Modified from Hidaka and Kobayashi (1992), with permission.
[b]Abbreviations: PKA, cAMP-dependent protein kinase; PKG, cGMP-dependent protein kinase; CaMKII, Ca^{2+}/calmodulin-dependent protein kinase II; MLCK, myosin light-chain kinase; PKC, protein kinase C; CKI, casein kinase I; CKII, casein kinase II.

for 30 min prior to stimulation to allow adequate diffusion into the cell. The concentration of inhibitor required depends on the concentrations of protein kinase and ATP present in the cell. For external applications, 20 µM H-89 has been shown to be effective in cell culture (Chijiwa et al., 1990), whereas 1–10 µM was effective in superfused whole-cell recordings (Surmeier et al., 1995). The effects of H-89 in whole-cell recordings are reversible with washing, which provides a useful control.

2.1.3. Cyclic Nucleotide Analog Inhibitors

Cyclic nucleotide analogs have been synthesized in an effort to produce phosphodiesterase-resistant activators and inhibitors of PKA and PKG (Botelho et al., 1988). These compounds interact with the cyclic nucleotide binding sites to prevent cAMP- or cGMP-induced activation of PKA and/or PKG. On cAMP binding to the regulatory (R) subunit dimer of PKA, the two bound catalytic (C) subunits dissociate and become active. Since these compounds block the interaction of cAMP with the R-subunit of PKA in contrast to other inhibitors that interact with the C-subunit, they may also affect nonphosphorylation-dependent actions of the R-subunit. These agents are inherently more specific for PKA in their actions compared to inhibitors, such as the isoquinolinesulfonamides, that interact with the structurally similar ATP binding sites. The Rp-diastereomers of cAMP and cGMP phosphorothioate analogs (with an equatorial sulfur substituted for phosphorus) are competitive inhibitors of PKA or PKG activation, whereas the Sp-diastereomers (with an axial sulfur substituted for phosphorus) are activators of PKA and PKG (Butt et al., 1990; Dostmann et al., 1990). The Rp analogs stabilize the holoenzyme form of PKA (R_2C_2) relative to the dissociated active form ($R_2 + 2C$). (Rp)-Adenosine 3':5'-monophosphorothioate [(Rp)-cAMPS] is a competitive inhibitor of type I and type II PKA that is resistant to phosphodiesterase. Use of this compound has been limited by its low membrane permeability and potency ($IC_{50} = 10$ µM). The 8-bromo and 8-chloro derivatives of (Rp)-cAMPS are sig-

Table 2
Inhibition Constants of Cyclic Nucleotide Analogs[a]

Cyclic nucleotide	PKA	PKG
(Rp)-cAMPS	3.7	53
(Rp)-cGMPS	20	20
(Rp)-8–Cl-cGMPS	100	1.5
(Rp)-8–CPT-cGMPS	8.3	0.5
(Rp)-8–Br-PET-cGMPS	11	0.03

[a]Values are for K_i (μM). Data taken from Butt et al. (1990,1994,1995).

nificantly more membrane-permeant than (Rp)-cAMPS, and are relatively selective for inhibition of type I vs type II PKA. The 8-(4-chlorophenylthio) derivative of (Rp)-cAMPS [(Rp)-8-CPT-cAMPS] is extremely lipophilic and membrane permeable, and is site-selective for site B of type II PKA; its PKA vs PKG selectivity is unknown. (Rp)-8-Piperidino-cAMPS [(Rp)-8-PIP-cAMPS] discriminates completely between site A and site B of PKA types I and II (Øgreid et al., 1994). Some of the Rp-diastereomers exhibit partial agonist activity (Butt et al., 1990); however, (Rp)-8-Br-cAMPS and (Rp)-8-Cl-cAMPS are potent inhibitors that exhibit little activation of PKA type I. These two analogs are therefore the most useful for studies in intact cells. The Sp-diastereomers of these compounds are also available as activators.

A major difficulty with these compounds is their poor selectivity between PKA and PKG owing to the structural similarities between cAMP and cGMP and their target enzymes PKA and PKG (Table 2). (Rp)-cAMPS is a relatively selective inhibitor of PKA, but is membrane-impermeant. (Rp)-cGMPS is equipotent as an inhibitor of PKA and PKG; the cGMP derivative (Rp)-8-Cl-cGMPS is a selective inhibitor of PKG compared to (Rp)-cGMPS, but neither one is adequately membrane-permeant. The 8-(4-chlorophenylthio) derivative of (Rp)-cGMPS [(Rp)-8-CPT-cGMPS] is membrane-permeant and somewhat selective for PKG compared to PKA. This compound has been used to selectively inhibit PKG in intact platelets, which contain high levels of both PKA and PKG (Butt et al., 1994). (Rp)-8-Bromo-βphenyl-1,N-etheno-cGMPS [(Rp)-8-Br-PET-cGMPS] is a new cGMP analog with high potency (K_i = 0.03 μM), membrane permeability, and selectivity (PKA K_i = 10 μM) for inhibition of PKG. Although it is not hydrolyzed by cGMP-specific phosphodiesterase, it does inhibit this enzyme, which might act to counteract its antagonism of PKG (Butt et al., 1995). Although there are no data on the effects of these compounds on protein kinases other than PKA or PKG, it is unlikely that they would be inhibitors based on their mechanism of action. Other than (Rp)-8-Br-PET-cGMPS, the Rp cyclic phosphorothioate analogs have been shown to have no significant effects on the activities of three major forms of cyclic nucleotide phosphodiesterase (Erneux et al., 1986).

2.1.3.1. USE OF CYCLIC NUCLEOTIDE ANALOG INHIBITORS

To allow adequate membrane permeation, intact cells should be preincubated with these compounds before protein kinase activation. High extracellular concentrations are often required, even for the more lipophilic derivatives, to ensure adequate intracellular concentrations. Concentrated stock solutions can be made in DMSO, or working solutions up to 25 mM of even the more lipophilic analogs can be made in aqueous buffers. In platelets, a 5-min preincubation with 1 mM (Rp)-8-CPT-cGMPS selectively inhibited PKG over PKA (Butt et al., 1994), whereas 0.1 mM was effective in superfused neurons (Surmeier et al., 1995). In contrast, the relatively hydrophilic derivative (Rp)-cAMPS at 1 μM inhibited VIP-induced smooth muscle relaxation in isolated cells and in muscle strips (Grider, 1993). (Rp)-8-Br-cAMPS and (Rp)-8-Cl-cAMPS inhibited various cAMP-dependent processes in intact 3T3 fibroblasts, IPC leukemia cells, and rat hepatocytes with IC_{50} values from 50–200 μM (Gjertsen et al., 1995). In human platelets, 100 μM (Rp)-8-Br-PET-cGMPS antagonized activation of PKG without affecting activation of PKA; it also inhibited NO-induced vasorelaxation (Erneux et al., 1986).

2.2. Ca²⁺/Calmodulin-Dependent Protein Kinase Inhibitors

The Ca^{2+}/calmodulin-dependent protein kinase (CaMK) family consists of four members with strong sequence homology in their amino-terminal kinase catalytic domain, significant homology in their central regulatory domains (which include the autoinhibitory and calmodulin-binding domains), and variable carboxy-terminal domains (Nairn and Picciotto, 1994). CaMKII, which is highly concentrated in nervous tissue, is the most extensively characterized member of this family (Braun and Schulman, 1995). CaMKII is able to phosphorylate a variety of substrates involved in regulating diverse physiological processes and is therefore termed the multifunctional CaMK. CaMKI and CaMKIV also appear to phosphorylate multiple substrates, although much less is known about these two enzymes. In contrast, CaMKIII, or EF-2 kinase, is a specific protein kinase that regulates protein synthesis by phosphorylating elongation factor-2. Other specific CaMKs include myosin light-chain kinase (MLCK), which regulates the actin-myosin interaction by phosphorylating the myosin light-chain in smooth muscle, and phosphorylase kinase, which regulates glycogen metabolism by phosphorylation of phosphorylase b.

2.2.1. Isoquinolinesulfonamide and Methoxybenzenesulfonamide Inhibitors (KN-62 and KN-93)

CaMKII is an abundant and ubiquitous multifunctional protein kinase involved in the regulation of neurotranmitter synthesis and release (Braun and Schulman, 1995). The isoquinolinesulfonamide derivative KN-62 (1-[N,O-bis 1,5-isoquinolinesulfonyl-N-methyl-L-tyrosyl]-4-phenylpiperazine) is a potent cell-permeable inhibitor of CaMKII (Tokumitsu et al., 1990) that was originally reported to be highly selective for CaMKII compared to the other protein kinases (Table 1). Subsequent studies indicated that KN-62 is a potent inhibitor of CaMKI (K_i = 0.8 μM) in addition to

CaMKII (K_i = 0.9 μM) (Okazaki et al., 1994; CaMKV is equivalent to CaMKI). Okazaki et al. (1994) claim that inhibition of CaMKIV (also called CaMKGR) is negligible (IC_{50} > 10 μM); however, Enslen et al. (1994) have reported similar IC_{50} values (0.5–2 μM) for inhibition of CaMKII and CaMKIV by KN-62. This is consistent with their highly conserved Ca^{2+}/calmodulin-binding domains, with which KN-62 interacts. Inhibition by KN-62 thus provides strong support for the involvement of one or more of the three multifunctional Ca^{2+}/calmodulin-dependent protein kinases (CaMKI, CaMKII, or CaMKIV), of which CaMKII is the most abundant and ubiquitous, but does not specifically implicate CaMKII. Inhibition of CaMKII by KN-62 is competitive with respect to calmodulin and noncompetitive with respect to ATP. The calmodulin-independent autophosphorylated form of CaMKII is not inhibited by KN-62, which disrupts the interaction of calmodulin with CaMKII by binding to its calmodulin binding site and preventing calmodulin-dependent activation. KN-62 has also been shown to block voltage-dependent Ca^{2+} channels (IC_{50} = 0.3 μM) with similar potency to its inhibition of CaMKII (Li et al., 1992). Thus, experiments in intact cells must take into account possible effects of KN-62 on Ca^{2+} influx (Sihra and Pearson, 1995). The methoxybenzenesulfonamide derivative KN-93 (2-[N-2-hydroxyethyl-N-4-methoxybenzenesulfonyl]amino-N-[4-chlorocinnamyl]-N-methylbenzylamine) inhibits CaMKII with comparable potency (IC_{50} = 0.4 μM) by a similar mechanism with the advantage of improved water solubility (Sumi et al., 1991), but may have a nonspecific protonophoric effect (Mamiya et al., 1993). KN-92 (2-[N-4-methoxybenzenesulfonyl]amino-N-[4-chlorocinnamyl]-N-methoxybenzylamine) is an inactive analog of KN-93.

2.2.1.1. Use of KN-62 and KN-93

The physological role of CaMKII, the predominant cellular Ca^{2+}/calmodulin-dependent protein kinase, has been probed using KN-62. KN-62 and its inactive analog KN-04 (N-(1-[N-methyl-p-5-isoquinolinesulfonyl benzyl]-2-[4-

phenylpiperazine]ethyl)-5-isoquinolinesulfonamide; Wenham et al., 1992) are prepared as 10-mM stock solutions in DMSO. KN-62 (0.1–1 μM), but not KN-04 (1 μM), reversibly inhibited the rate of beating of cultured fetal mouse cardiac myocytes (Okazaki et al., 1994); concentrations >10 μM were not completely soluble in buffer. Treatment of PC12 cells with 10 μM KN-62 prevented A-23187-induced autophosphorylation of CaMKII, immediate early gene transcription (Enslen and Soderling, 1994), and KCl-induced activation and phosphorylation of tyrosine hydroxylase (Ishii et al., 1991). Inhibition by KN-62 of GABA release from synaptosomes (Sitges et al., 1995), Ca^{2+}-induced AMPA-type glutamate receptor phosphorylation in cultured hippocampal neurons (Tan et al., 1994), enhanced AMPA receptor function in hippocampal slices (Wyllie and Nicoll, 1994), modulation of neurotransmitter release at the crayfish neuromuscular junction (Noronha and Mercier, 1995), and long-term potentiation in rat hippocampal CA1 cells (Ito et al., 1991) has been used to implicate CaMKII (and, although not usually stated, possibly CaMKI and/or CaMKIV) in these processes. KN-62 (0.5–10 μM) has also been shown to protect fetal cortical neuronal cultures from NMDA- and hypoxia/hypoglycemia-induced cell death (Hajimohammadreza et al., 1995), and to promote cultured cerebellar granule cell survival (Hack et al., 1993). The interpretation of some of these results is obscured by the effects of KN-62 on Ca^{2+} channels. KN-93, which is prepared as a 10-mM stock solution in water, inhibited dopamine synthesis (IC$_{50}$ = 0.37 μM) and tyrosine hydroxylase phosphorylation by CaMKII in PC12 cells (Sumi et al., 1991), and insulin release from pancreatic β cells (Niki et al., 1993). Evidence in support of a specific role of CaMKII in a cellular process can be obtained by parallel experiments comparing the intracellular injection of CaMKII with that of an inhibitor. For example, in *Euhadra* neurons, injection of CaMKII suppressed a quisqualate-induced K$^+$ current, which was enhanced by KN-62 (IC$_{50}$ ~50 μM; Onozuka et al., 1994).

2.2.2. MLCK Inhibitors (ML-7, ML-9, and Wortmannin)

MLCK is a specific Ca^{2+}/calmodulin-dependent protein kinase involved in the regulation of muscle contraction (Nairn and Picciotto, 1994). A derivative of the naphthalenesulfonamide calmodulin antagonist W-7 (ML-9; 1-[5-chloronaphthalene-1-sulfonyl]-1H-hexahydro-1,4-diazepine) was found to inhibit MLCK competitively with respect to ATP, but not with respect to calmodulin or the substrate myosin light-chain (Saitoh et al., 1987). A catalytically active proteolytic fragment of MLCK, which is Ca^{2+}/calmodulin-independent was also inhibited by ML-9. Although ML-9 is not structurally related to ATP, its mechanism of action suggests that it might not be a specific inhibitor. Of the limited number of protein kinases tested, ML-9 exhibited only an eightfold selectivity for inhibition of MLCK over PKA (Table 1). The iodinated analog of ML-9, ML-7 (1-[5-iodonaphthalene-1-sulfonyl]-1H-hexahydro-1,4-diazepine) is more potent and selective for MLCK (70-fold compared to PKA). Wortmannin is a fungal metabolite that irreversibly inhibits MLCK (IC_{50} = 1.9 μM; Nakanishi et al., 1992), but has been shown to be more potent as a phosphatidylinositol 3-kinase inhibitor (IC_{50} = 3 nM; Powis et al., 1994).

2.2.2.1. USE OF MLCK INHIBITORS

MLCK inhibitors have been useful in studying the physiological role of MLCK in smooth muscle and nonmuscle tissues, including neural tissue, where it is present in neurons and glia (Edelman et al., 1992; Akasu et al., 1993). ML-9 and ML-7 are prepared as stock solutions (10–50 mM) in DMSO. ML-9 applied extracellularly (30 min preincubation) blocked myosin light-chain phosphorylation (Ishikawa et al., 1988) and KCl-evoked contraction of vascular smooth muscle (IC_{50} = 18 μM), as well as of skinned smooth muscle fibers (Saitoh et al., 1987). Extracellular application of ML-9 selectively inhibited Ca^{2+}-dependent phosphorylation of myosin light-chain in human platelets (IC_{50} = 12 μM) without affecting PKC-mediated phosphorylation of a 40-kDa protein (Saitoh et al., 1986). ML-7 (3–5 μM) and ML-9 (5–10 μM) selectively inhibited growth cone mobility in goldfish retina explant cultures, in which calphostin C (0.1–1 μM), chelerythrine (up to 50 μM), and KN-62 (up to 50 μM) were without effect (Jian et al., 1994). ML-9 (10–100 μM) and the nonselective inhibitor KT5926 (1–25 μM) (*see* Section 2.3) induced actin breakdown and process growth in cultured astrocytes (Baorto et al., 1992). ML-9 (50 μM) also inhibited insulin release from rat pancreatic islets (Niki et al., 1993) and dopamine release from PC12 cells (Nagatsu et al., 1987).

Wortmannin is prepared as a 10-mM stock solution in DMSO. Effective inhibition requires preincubation with the drug for 30 min (Onozuka et al., 1994; Tokimasa et al., 1995). In combination with a pseudosubstrate peptide inhibitor (Section 2.2.3.), wortmannin has been used to implicate MLCK in ATP-dependent priming of Ca^{2+}-induced exocytosis in permeabilized adrenal chromaffin cells (Kumakura et al., 1994), in acetylcholine release from rat superior cervical ganglion neurons (Mochida et al., 1994), and in reduction of the bullfrog sympathetic neuron M-type K^+ current (Akasu et al., 1993). A role for MLCK in Ca^{2+}-dependent catecholamine release from bovine adrenal chromaffin cells was also suggested by its sensitivity to inhibition (IC_{50} = 1 μM) by wortmannin (Ohara-Imaizumi et al., 1992). However, caution must be used in interpreting uncorroborated results obtained with wortmannin alone given its 600-fold greater potency as a phosphatidylinositol 3-kinase inhibitor.

2.2.3. Pseudosubstrate Peptide Inhibitors of Ca^{2+}/Calmodulin-Dependent Protein Kinases

A number of protein kinases are subject to intrasteric regulation of their catalytic activity by autoinhibitory domains that mimic protein substrate phosphorylation site sequences (Soderling, 1990; Kemp et al., 1991). Synthetic peptides based on pseudosubstrate autoinhibitory domains inhibit their respective kinase activities competitively with respect to protein substrate and/or ATP. These reagents have been useful in studying the physiological roles of protein kinases in cell extracts and in intact cells by microinjection, transfection, or permeabilization. Since many protein kinases share similar phosphorylation site amino acid determinants for substrate recognition, inhibitors based on pseudosubstrate domains exhibit some overlap in their specificities (Table 3; Smith et al., 1990; Hvalby et al., 1994).

Autoinhibitory domain peptides are available for two Ca^{2+}/calmodulin-dependent protein kinases: CaMKII and MLCK. The peptide CaMKII(290–309) was originally identified as an inhibitor of CaMKII; inhibition is competitive with respect to peptide substrate and noncompetitive with respect to ATP (Payne et al., 1988). An amino-terminal extension to give CaMKII(281–309) increased potency 10-fold and changed the mechanism to noncompetitive with respect to peptide substrate and competitive with respect to ATP (Colbran et al., 1989). These peptides also act as calmodulin antagonists, since they bind Ca^{2+}/calmodulin. The critical determinants of the autoinhibitory domain are contained in residues 281–302; elimination of residues carboxy-terminal to residue 302 does not affect kinase inhibition, but disrupts calmodulin binding (Smith et al., 1992). Autophosphorylation of Thr^{286} renders the intact enzyme partially calmodulin-independent; in order to avoid phosphorylation of this residue in peptide analogs, it can be substituted with alanine with no loss in inhibitory potency. [Ala^{286}]CaMKII(281–302) is a potent inhibitor of CaMKII, but is poorly selective, since it also inhibits PKC with a sixfold higher K_i (Table 4). CaMKII(281–309) exhibits slightly more selectivity for a catalytically active proteolytic fragment of CaMKII (Table 3), but this specificity was less apparent with the intact enzyme (Smith et al., 1990). A nonphosphorylatable analog of the CaMKII substrate autocamtide-2, termed autocamtide-2-related inhibitory peptide (AIP), is a potent ($IC_{50} = 40$ nM) and highly specific CaMKII inhibitor (Ishida et al., 1995). Furthermore, AIP did not inhibit CaMKIV, whereas [Ala^{286}]CaMKII(281–302) showed significant inhibition of PKC and CaMKIV at 30 μM, which was required to inhibit CaMKII completely.

Table 3

Amino Acid Sequences of Autoinhibitory Domain Peptides and PKI Peptide, and IC_{50} Values for Protein Kinase Catalytic Domains[a]

Peptides	Sequences	IC_{50} values for protein kinase catalytic domain, μM			
		CaMKII	MLCK	PKC	PKA
CaMKII(281–309)	MHRQETVDCLKKFNARRKLKGAILTTMLA	0.1	60	13	>60
AIP	KKALRQEAVDAL	0.04	—	~20	>20
MLCK(480–501)	AKKLSKDRMKKYMARRKWQKTG	0.9	0.1	0.1	>60
PKC(19–36)	RFARKGALRQKNVHEVKN	4.9	60	0.3	>60
PKI Peptide	IAAGRTGRRQAIHDILVAA	>60	>60	>60	0.2

[a]Multiple basic residues important as substrate specificity determinants are indicated by underlining. Data taken from Smith et al. (1990) and Ishida et al. (1995). AIP, autocamtide-2–related inhibitory peptide.

Table 4

Comparison of Inhibition of PKC and CaMKII
and Induction of LTP by Protein Kinase Inhibitor Peptides[a]

	PKC K_i, μM	CaMKII K_i, μM	LTP IC_{50}, μM
PKC(19–36)	0.28	5.8	≈30
[Ala^{286}]CaMKII			
(281–302)	9.1	1.6	≈1000
Potency ratio	32	0.28	33

[a]Potency ratio indicates potency of PKC(19–36) relative to that of [Ala^{286}]CaMKII(281–302) in inhibiting PKC, CaMKII, or LTP. Data taken from Hvalby et al. (1994).

A number of MLCK autoinhibitory domain peptides are available (Kemp et al., 1991) that inhibit MLCK by acting both as calmodulin antagonists and as substrate antagonists. The peptide MLCK(480–501) is a potent inhibitor of proteolyzed MLCK, CaMKII, and PKC, and is therefore a nonspecific inhibitor (Table 3).

2.2.3.1. USE OF PSEUDOSUBSTRATE PEPTIDE INHIBITORS
OF CA^{2+}/CALMODULIN-DEPENDENT PROTEIN KINASES

Peptide inhibitors can be used to probe the roles of protein kinases in situations in which membrane permeability is not required. Several peptide inhibitors are available commercially; others must be custom-synthesized. Intracellular injection of pseudosubstrate peptide inhibitors has been used to implicate CaMKII (Malenka et al., 1989; Wang and Kelly, 1995), PKC (Wang and Feng, 1992), or both (Malinow et al., 1989) in the induction of long-term potentiation in hippocampal CA1 neurons. However, these studies did not take into account the poor specificity of these inhibitors. A comparison of the relative potencies of [Ala^{286}]CaMKII(281–302) and PKC(19–36), an inhibitor of PKC, in the induction of long-term potentiation (Table 4), indicated that the inhibitory effect of each peptide was attributable to inhibition of PKC (Hvalby et al., 1994).

For intracellular injections, micropipets are filled with peptide solutions of various concentrations, which enter impaled cells by diffusion (Malinow et al., 1989; Hvalby et al., 1994; Wang and Kelly, 1995), pressure injection (McCarron et al., 1992), or intracellular perfusion (Akasu et al., 1993). Most studies required millimolar peptide concentrations in micropipets to achieve effective intracellular concentrations by diffusion (Malinow et al., 1989). The poor selectivity of these inhibitors can make interpretation of results difficult unless careful controls are carried out. This can be achieved by comparing the relative potencies of peptides with different potencies toward specific protein kinases (Hvalby et al., 1994). This approach was used to show that CaMKII(273–302), but not PKC(19–36), inhibited Ca^{2+}-induced enhancement of L-type Ca^{2+} currents in rabbit ventricular myocytes (Anderson et al., 1994), or the effects of muscarinic agonists on hippocampal neuron excitability (Müller et al., 1992). Alternatively, parallel studies employing injection of inactive or constitutively active protein kinase preparations can be used. For example, the nonselective MLCK pseudosubstrate peptide was shown to inhibit contraction when microinjected into smooth muscle cells, whereas injection of a constitutively active proteolytic fragment of MLCK induced Ca^{2+}-independent contraction, which was blocked by the peptide (Itoh et al., 1989). Intracellular application of MLCK(783–804) (30–100 μM) into bullfrog sympathetic neurons reversibly reduced M-type K^+ current, which was enhanced by application of the active fragment of MLCK, whereas application of an inactive MLCK fragment was without effect (Akasu et al., 1993); pretreatment of these cells for 30 min with wortmannin at 0.1–1 μM also inhibited this current (Tokimasa et al., 1995).

Several alternatives to intracellular injection exist for introducing peptide inhibitors into cells. Membrane permeablization of myocytes was used to introduce 10 μM CaMKII(281–309), which inhibited MLCK phosphorylation (Tansey et al., 1994), or 10 μM CaMKII(273–302), which inhibited sarcoplasmic Ca^{2+} transport (Mattiazzi et al., 1994). Other methods of inhibitor introduction include the production of transgenic flies carrying a synthetic minigene encoding CaMKII(273–302) (Griffith et al., 1993), or freeze-thaw intro-

Table 5
Inhibition Constants of Nonselective Inhibitor Compounds[a]

Compound	PKC	PKA	PKG	MLCK	CaMKII
K252a	25	18	20	20	1.8
K252b	20	90	100	150	—
KT5720	>2000	60	>2000	>2000	—
KT5823	4000	>10,000	234	>10,000	—
KT5926	720	1200	160	18	4.4
Staurosporine	0.7	7	8.5	1.3	20

[a]Values are for K_i (nM). Abbreviations as in Table 1. Data taken from Hidaka and Kobayashi (1992) and Kase et al. (1987).

duction of [Ala^{286}]CaMKII(281–302) into rat brain synaptosomes to inhibit glutamate release (Nichols et al., 1990).

2.3. Nonselective Protein Kinase Inhibitors (Staurosporine and Analogs)

Several nonselective protein kinase inhibitors have been identified by screening microbial natural products. Staurosporine is an indolecarbazole natural product isolated from *Streptomyces* that is an extremely potent, but nonselective, protein kinase inhibitor (Table 5). The limited specificity of staurosporine is attributed to its inhibitory interaction with the ATP-binding site of the protein kinase catalytic domain. Its seven-hydroxy derivative, UCN-01, exhibits somewhat more specificity for PKC (Tamaoki, 1991). Staurosporine also inhibits S6 kinase, phosphorylase kinase, and protein-tyrosine kinases, including pp60^{v-src}, insulin receptor kinase, and epidermal growth factor (EGF) receptor kinase.

K-252a is an indolecarbazole alkaloid isolated from *Nocardiopsis* that is a potent inhibitor of PKC, PKA, PKG, CaMKII, and MLCK (Table 5; Kase et al., 1987). Inhibition is competitive with respect to ATP. Recently, K-252a has also been shown to inhibit NGF-induced *trk* proto-oncogene protein-tyrosine kinase activity in PC12 cells (Berg et al., 1992; Nye et al., 1992); effects on calmodulin-dependent enzymes and cyclic nucleotide phosphodiesterase have also been reported. Chemical modification of K-252a has yielded a number of derivatives with slightly different inhibitory specificities and some improved selectivity. For example, 1 μM KT5823 has been used to inhibit the nitric oxide (NO)-mediated component of VIP-induced smooth muscle relaxation selectively (Grider, 1993). K-252b is a nontoxic hydrophilic membrane-impermeant derivative of K-252a that is also an inhibitor of PKC, PKA, and PKG. Although K-252a and K-252b have similar in vitro effects on protein kinase activity, differences in their effects on intact cells have been reported that may suggest ectokinase involvement (Knüsel and Hefti, 1992).

2.3.1. Use of Nonselective Antibiotic Protein Kinase Inhibitors

These compounds are becoming less useful as more selective inhibitors are developed. A number of studies have made use of these compounds to investigate the mechanism of NGF receptor signal transduction in PC12 cells (Knüsel and Hefti, 1992), which appears to involve a direct inhibitory action on trk neurotrophin receptor proteins, and possibly a previously unidentified KT5926-sensitive protein kinase (Teng and Greene, 1994). K-252a and its derivatives staurosporine and UCN-01 are prepared as stock solutions in DMSO and diluted into buffer before use.

2.4. PKC Inhibitors

PKC is an important family of protein-serine/theonine kinases that regulate numerous neuronal functions (Tanaka and Nishizuka, 1994). PKC exists as a family of structurally related isoforms distinguished by their regulatory domains and cofactor dependence (Hug and Sarre, 1993). At least 11 closely related isoforms exist in mammals; they differ in their tissue distributions, regional distributions in brain, subcellular localizations, and sensitivities to various activators, all of which suggests that different isoforms control specific physiological functions. These isoforms have been divided into three groups of genes based on their nucleotide sequences. The conventional PKCs (cPKC) include PKCα, PKCβ1 and β2 (related by alternative splicing of the same primary transcript), and PKCγ, and are activated by Ca^{2+}, diacylglycerol (or phorbol esters), and phosphatidylserine. The new PKCs (nPKC) include PKCδ, PKCε, PKCη, PKCθ, and PKCμ, and are insensitive to Ca^{2+}, but activated by diacylglycerol and phosphatidylserine. The atypical PKCs (aPKC) are poorly characterized, but include PKCζ and PKCλ; these isoforms are insensitive to Ca^{2+} or diacylglycerol. Most PKC inhibitors have not been tested against the individual PKC isoforms, but to a mixture of cPKC isoforms, so the pharmacological differences of the isoforms remain to be defined.

2.4.1. Poorly Selective PKC Inhibitors

A number of PKC inhibitors have been identified, including W-7, phenothiazines, calmidazolium, adriamycin, amiloride, tamoxifen, polymixin B, spermine, 1,12-diaminodecane, dibucaine, palmitoylcarnitine, alkyl lysophospholipids, gangliosides, arachidonic acid, aminoacridines, sangivamycin, CP-46,665-1, H-7, K-252a, staurosporine, and sphingosine. Many of these inhibitors are weak and/or have poor specificity, and have been supplanted by more potent and specific compounds. In spite of their limitations, some of these compounds have been useful as probes to investigate the role of PKC in cellular regulation. H-7, K-252a, staurosporine (Tables 1 and 5), and sphingosine (Hannun et al., 1991) are all potent, but nonselective, membrane-permeant inhibitors of PKC that have been useful in investigating the physiological role of PKC in intact cells. Despite claims to the contrary, the isoquinolinesulfonamide H-7 lacks any useful selectivity between PKC, PKA, and PKG. This limitation can be minimized by comparing the effect of H-7 to that of a second compound, such as HA-1004, which inhibits PKA and PKG effectively, but is weak against PKC (Table 1). Use of these inhibitors requires careful interpretation of the results, since other targets may be affected; inhibition of a cellular response by such nonselective inhibitors can be considered necessary, but insufficient, in demonstrating that a response is mediated by PKC. H-7 supplied by Sigma (St. Louis, MO) has been shown by to be an isomer of H-7 (Iso-H-7) that is significantly less potent as a PKC inhibitor (Quick et al., 1992). The use of isoquinolinesulfonamides, K-252a, and staurosporine is described in Sections 2.1.2. and 2.3.

2.4.1.1. SPHINGOSINE

Sphingosine is a naturally occurring component of sphingolipids that is a potent inhibitor of PKC (Hannun et al., 1991). It acts as a competitive antagonist of diacylglycerol and phorbol ester binding to the regulatory domain of cPKC. Use of sphingosine is limited by its potential cytotoxicity, metabolism to inactive compounds, and effects on other targets. Although sphingosine does not inhibit PKA, it stimulates the EGF receptor tyrosine kinase, inhibits CaMKII, CaMKIII, MLCK, and Na^+,K^+-ATPase activity, and inhibits factor VII binding in the same concentration range required for PKC inhibition. Higher concentrations affect thyrotropin-releasing hormone binding, a CaMK, and pp60$^{v\text{-}src}$. The potencies of sphingosine and certain other PKC inhibitors can be affected by the PKC activator employed (Robinson, 1992).

2.4.1.2. USE OF SPHINGOSINE

Sphingosine is prepared as a stock solution of 100 mM in 95% (v/v) ethanol, which is diluted to 1 mM into an aqueous solution containing 1 mM fatty acid-free bovine serum albumin to yield the working stock solution, and stored at –20°C. The albumin complex helps minimize cytotoxicity and increase solubility. Since sphingosine is protein-bound, highly lipophilic, and partitions into membranes, its effective concentration in intact cell preparations depends on the cell density and the presence of serum proteins. For example, sphingosine inhibits PKC in the 1–20 μM range in washed human platelets, whereas in the presence of serum proteins, 100–200 μM sphingosine is required (Hannun et al., 1991). Useful controls for the effects of sphingosine include the inactive analog N-acetylsphingosine and the ability to override the sphingosine effect with exogenous diacylglycerol.

2.4.2. Calphostin C

Calphostin C (UCN-1028C), a lipophilic perylenequinone antibiotic compound isolated from *Cladosporium*, is a potent and specific inhibitor of PKC (Kobayashi et al., 1989; Tamaoki, 1991). Calphostin C interacts with the regulatory domain of PKC to inhibit phorbol ester and diacylglycerol binding; its activity is not affected by Ca^{2+} or phospholipid. It does not inhibit the spontaneously active proteolytic catalytic fragment of PKC (Tamaoki, 1991). Taken together, these findings provide a rational basis for its selectivity as a PKC inhibitor. The mechanism of calphostin C appears to involve photoactivation to yield singlet oxygen and a short-lived species that reacts with PKC, since its inhibitory activity depends on light activation in the presence of PKC. The IC_{50} for PKC activity in the dark was >1 μM, but in the light was 50 nM (Bruns et al., 1991). IC_{50} values for inhibition of other protein kinases tested were >5 μM (Table 6; Kobayashi et al., 1989). Calphostin C has been shown to be 25 times more potent in inhibiting cytosolic compared to membrane-derived PKC (Budworth and Gescher, 1995). The photoreactivity of calphostin C may lead to irreversible modification of other membrane proteins in addition to PKC (Wang et al., 1993).

2.4.2.1. USE OF CALPHOSTIN C

Calphostin C is prepared as a stock solution in DMSO. It was cytotoxic to cultured cells at submicromolar concentrations, but was able to antagonize the phorbol ester-induced downregulation of EGF binding in A431 cells at fivefold lower concentrations (Bruns et al., 1991). In most whole-cell applications, calphostin C is effective at 1 μM, although lower concentrations minimize nonspecific toxic effects. Continuous culture of hippocampal neurons in the presence of 0.5 μM calphostin C inhibited neurite initiation and axon branching without affecting other aspects of neurite development (Cabell and Audesirk, 1993). Calphostin C has been used to demonstrate a role for PKC in the induction of c-*Fos* expression in cultured striatal neurons by dopamine (Simpson and Morris, 1995), the homologous desensitization

Table 6

IC$_{50}$ Values for Staurosporine-Derived Bisindolylmaleimides, Indolecarbazoles, and Calphostin C

Compound	PKC	PKA	PKG	CaMKII	PK	MLCK	PTK
Staurosporine[a,d]	9	120	18	40	0.5	10	0.4
GF 109203X/Gö 6850[b]	10	2000	—	—	700	—	>65,000
Ro 31-8220[a]	10	1500	—	17,000	—	—	—
Ro 31-8425[c]	7.6	2800	—	19,000	1300	—	—
Gö 6976[d]	20	>100,000	6200	—	—	5800	>10,000
Calphostin C[e]	50	>50,000	>25,000	—	—	>5000	—

Data (in nM) taken from [a]Davis et al. (1989), [b]Toullec et al. (1991), [c]Muid et al. (1991), [d]Gschwendt et al. (1995), and [e]Tamaoki (1991). Abbreviations: PK, phosphorylase kinase; PTK, protein tyrosine kinase; otherwise as in Table 1.

146

of δ opioid receptors expressed in *Xenopus* oocytes (0.1 μ*M* bath application; Ueda et al., 1995), the survival-promoting effect of brain-derived neurotrophic factor in cultured cerebellar granule cells (Zirrgiebel et al., 1995), the induction of long-term potentiation mediated by NMDA receptors in rat hippocampal slices (Lopez-Molina et al., 1993), regulating presynaptic voltage-dependent Ca^{2+} channels in rat hippocampal synaptosomes (Bartschat and Rhodes, 1995), ethanol inhibition of kainate receptors (Dildy-Mayfield and Harris, 1995), modulating the Ca^{2+} sensitivity of NO synthase in rat cerebellar slices (Okada, 1995), glutamate-stimulated protein phosphorylation in cerebellar granule cells (Eboli et al., 1994), mediating NGF-induced neuropeptide Y expression in PC12 cells (Balbi and Allen, 1994), attenuating NMDA-induced increases in intracellular Ca^{2+} in cultured rat cerebellar granule cells (Snell et al., 1994), regulating tyrosine hydroxylase expression in cultured hypothalamic neurons (Kedzierki et al., 1994), the activation of MAP kinase by glutamate receptors in cultured rat hippocampal neurons (Kurino et al., 1995), long-term depression in striatal slices (Calabresi et al., 1994) and cultured Purkinje cells (Linden and Connery, 1991), and preventing glutamate toxicity in cultured cerebellar neurons (Felipo et al., 1993).

2.4.3. Chelerythrine

The benzophenanthridine alkaloid chelerythrine is a potent inhibitor of PKC (IC_{50} = 0.66 μ*M*), with excellent selectivity for PKC among the other protein kinases tested: PKA (IC_{50} = 170 μ*M*), CaMKI (IC_{50} > 100 μ*M*), and a lymphoma protein-tyrosine kinase (IC_{50} = 100 μ*M*) (Herbert et al., 1990). Chelerythrine interacts with the catalytic domain of PKC, which it inhibits competitively with respect to protein substrate (K_i = 0.7 μ*M*) and noncompetitively with respect to ATP (K_i = 0.8 μ*M*). Chelerythrine has not been tested against a wide variety of protein kinases and other nonkinase enzymes, although it has been reported to inhibit alanine aminotransferase (IC_{50} = 4 μ*M*; Walterova et al., 1981).

2.4.3.1. USE OF CHELERYTHRINE

Chelerythrine is soluble in water up to 5 mg/mL, but becomes insoluble in the presence of substantial salt concentrations. It is conveniently prepared as a stock solution of 10 mM in DMSO, and is usually effective in whole-cells at 1 μM. Chelerythrine was cytotoxic to L-1210 tumor cells (IC_{50} = 0.53 μM), as were other PKC inhibitors (Herbert et al., 1990). Chelerythrine (1 μM applied 10 min before treatment) inhibited PKC translocation and cytotoxicity in differentiated PC12 cells subjected to cyanide-induced hypoxia (Pavlakovic et al., 1995). Intrathecal injection of chelerythrine or GF 109203X (*see* Section 2.4.5.1.) reduced nociceptive responses to chemical or thermal stimuli in rats (Yashpal et al., 1995). Chelerythrine has also been used to implicate PKC in regulating neurite growth in cultured rat sympathetic neurons (Campenot et al., 1994).

2.4.4. Pseudosubstrate Peptide PKC Inhibitors

An autoinhibitory pseudosubstrate domain of cPKC was identified from the cPKC amino acid sequence and by the ability of a synthetic peptide based on this sequence to inhibit PKC activity (Kemp et al., 1991). PKC(19–36) binds the protein substrate binding site and inhibits PKC with an IC_{50} value of 0.18–0.3 μM (Tables 3 and 4; Kemp et al., 1991). This peptide does not inhibit nPKC activity (Majumdar et al., 1993). Although these peptides are generally unable to enter cells, the N-myristoylated derivative of PKC(20–28) inhibited phosphorylation of the PKC substrate MARCKS in fibroblasts (Eichholtz et al., 1993). Problems with specificity limit the use of these inhibitors as discussed in Section 2.2.3. An inactive peptide analog of PKC(19–36) ([Glu^{27}]PKC(19–36) is available as a negative control reagent (Malinow et al., 1989).

2.4.4.1. USE OF PSEUDOSUBSTRATE PEPTIDE PKC INHIBITORS

PKC pseudosubstrate peptides have been used in a number of studies to examine the physiological role of PKC in neuronal cells. Their use in the study of long-term potentiation is discussed in Section 2.2.3. Intracellular injection of PKC(19–36) or the shorter analog PKC(19–31) has also been used to demonstrate a role for PKC in mediating substance P-induced inhibition of an inward rectifier K^+ channel in nucleus basalis neurons (Takano et al., 1995), long-term depression of cerebellar Purkinje cells (Crepel et al., 1994), neurosteroid-induced Ca^{2+} channel inhibition in hippocampal CA1 neurons (ffrench-Mullen et al., 1994), modulation of AMPA/kainate receptors in cultured hippocampal neurons (Wang et al., 1994a), enhancement of N- and L-type Ca^{2+} channel currents in frog sympathetic neurons (Yang and Tsien, 1993), and arachidonic acid-induced suppression of Ca^{2+} current in hippocampus (Keyser and Alger, 1990). Pseudosubstrate peptide injection has been used to rule out a role for PKC in mediating potentiation of the response to somatostatin by substance P in chick sympathetic ganglia (Golard et al., 1994), diacylglycerol-induced block of voltage-dependent K^+ channels (Bowlby and Levitan, 1995), phorbol ester-induced enhancement of Ca^{2+} channels in neuroblastoma cells (Reeve et al., 1995), and norepinephrine-induced Ca^{2+} current inhibition in rat sympathetic neurons (Abrahams and Schofield, 1992). In most of these studies, peptide concentrations of 1–20 μM are adequate to inhibit PKC, although higher pipet concentrations are necessary if peptide entry is diffusion-limited. PKC(19–31) and staurosporine have been used to demonstrate that PKC is modulatory, but not obligatory for exocytosis from digitonin-permeabilized bovine adrenal chromaffin cells (Terbush and Holz, 1990). PKC(19–36) has been used in vitro to identify PKC-dependent protein kinase activity in homogenates in the study of long-term potentiation in rat hippocampus (Klann et al., 1991,1992) and of anesthetic activation of PKC in rat brain synaptosomes (Hemmings and Adamo, 1996).

2.4.5. ATP-Competitive Inhibitors of PKC

The isoquinolinesulfonamide H-7 (Table 1) has been used as an inhibitor of PKC, but it is not selective or potent. Selective ATP-competitive PKC inhibitors have been designed based on the extremely potent, but nonselective, protein kinase inhibitor staurosporine (Section 2.3.). Staurosporine is the most potent inhibitor of PKC identified, but is nonspecific in inhibiting a number of other protein kinases (Table 6; Rüegg and Burgess, 1989). Synthesis of staurosporine-related bisindolylmaleimides and indolecarbazoles has provided inhibitors with improved selectivity and similar potency compared to staurosporine. In a novel approach to designing PKC inhibitors, isozyme-specific inhibition of PKCβ2 by in vitro-selected RNA aptamers has been demonstrated (Conrad et al., 1994).

2.4.5.1. BISINDOLYLMALEIMIDES

The bisindolylmaleimides are structural analogs of staurosporine with similar high potency, but improved specificity for inhibition of PKC (Conrad et al., 1994). The most extensively characterized member of this family (GF 109203X or Gö 6850; bisindolylmaleimide I; 2-[1-3-dimethylamino-propyl-1H-indol-3–yl]-3[1H-indol-3–yl]-maleimide) was 70-fold selective for PKC over other protein-serine/threonine kinases tested (Table 6). Similar IC_{50} values were obtained for purified brain PKC and the separated isoforms PKCα, PKCβ1, PKCβ2, and PKCγ (Table 7); the IC_{50} values for three receptor tyrosine kinases were all 65 μM or higher. Like staurosporine, GF 109203X acted as a competitive inhibitor with ATP, and did not affect phorbol ester binding. This compound inhibited PKC-mediated phosphorylation of P47 in human platelets (IC_{50} = 0.19 μM), but not MLCK activity, whereas staurosporine inhibited both. A number of other bisindolylmaleimides have also been described with similar specificities.

Aminoalkyl bisindolylmaleimides have also been found to be potent and selective PKC inhibitors in vitro (Davis et al., 1989,1992). One of these compounds, Ro 31-8220 (3-[1–3-amidinothio propyl-1H-indol-3-yl]-3-[1-methyl-1H-indol-3-yl]maleimide), inhibits PKC by a competitive mechanism with respect to ATP (K_i = 2.8 nM), as do K-252a and staurosporine (K_i = 140 μM). These compounds interact with the catalytic domain of PKC since they also inhibit the catalytic fragment of PKC (Wilkinson et al., 1993). Ro 31-8220 exhib-

Table 7

Inhibition of PKC Isoenzymes by Isoform-Selective PKC Inhibitors

Inhibitor	Rat brain PKCαβγ	cPKC				nPKC		aPKC	
		PKCα	PKβ1	PKCβ2	PKCγ	PKCε	PKCδ	PKCζ	PKCη
Staurosporine[a]	22(7)[d]	28(1.5)[b]	13	11	32	25(9)[d]	(9)[d]	(220)[b](16)[d]	—
UCN-01 (7-hydroxy-staurosporine)[b]	—	0.44	1.7	—	0.94	25	20	3800	—
Ro 31-7208[a]	323	160	310	280	377	330	—	—	—
Ro 31-7549[a]	158	53	195	163	213	175	—	—	—
Ro 31-8220[a]	23	5	24	14	27	24	—	—	—
Ro 31-8425[a]	15	8	8	14	13	39	—	—	—
Ro 32-0432[a]	21	9	28	31	37	108	—	—	—
Gö 6976[c]	7.9	2.3	6.2	—	—	No inhibition	No inhibition	No inhibition	—
GF 109203X/Gö 6850[c]	31(10)[e]	8.4(20)[e]	18(17)[e]	(16)[e]	(20)[e]	132	210	5800	—
LY 333531[f]	—	360	4.7	5.9	300	600	250	$>10^5$	52
Staurosporine[f]	—	45	23	19	110	18	28	$>1.5 \times 10^3$	5

IC_{50} values (in nM) are taken from [a]Wilkinson et al. (1993), [b]Mizuno et al. (1995), [c]Martiny-Baron et al. (1993), [d]Gschwendt et al. (1992), [e]Toullec et al. (1991), or [f]Ishii et al. (1996).

ited a 150-fold selectivity for PKC over the other protein-serine/threonine kinases tested (Table 6), and inhibited PKC-mediated phosphorylation of P47 in human platelets (IC_{50} = 700 nM) and downregulation of CD3 in human T-cells (IC_{50} = 500 nM). Ro 31-8220 was 22 times more potent as an inhibitor of membrane compared to cytosolic derived PKC, whereas GF 109203X was equipotent (Budworth and Gescher, 1995). These differences in potency of Ro 31-8220 and GF 109203X, along with the opposite preference of calphostin C (*see* Section 2.4.2.), must be considered in the interpretation of their pharmacological effects in intact cells. Slight selectivity for PKCα over other isoforms has been reported for several bisindolylmaleimides (Table 7; Wilkinson et al., 1993). The macrocyclic bisindolylmaleimide LY333531 showed selectivity toward inhibition of PKCβ by a competitive mechanism vs ATP both in vitro (Table 7) and in diabetic rats in vivo (Ishii et al., 1996). A derivative with a conformationally restricted side chain, Ro 31-8425, exhibited improved selectivity for PKC (Table 6), and inhibited PKC-mediated P47 phosphorylation and CD3 downregulation with IC_{50} values of 460 and 600 nM, respectively (Muid et al., 1991).

2.4.5.1.1. USE OF BISINDOLYLMALEIMIDES

GF 109203X (bisindolylmaleimide I), several other bisindolylmaleimides and Ro 31–8220 are available commercially; other aminoalkyl bisindolylmaleimides are not commercially available. These compounds are prepared as stock solutions in DMSO. Although they inhibit PKC with IC_{50} values in the nanomolar range, higher concentrations (low micromolar) are required to inhibit PKC-mediated processes in intact cells (*see* Section 1.). This is probably a result of the higher ATP concentrations (millimolar) in intact cells compared to in vitro assays (micromolar), the higher PKC concentrations in intact cells (micromolar) compared to in vitro assays (nanomolar), and membrane permeability. Ro 31–8220 and Ro 31-8425 have been shown to inhibit phorbol ester-induced P47 phosphorylation in platelets and CD3 downregulation in T-cells with submicromolar IC_{50} values (Nixon et al., 1992). Preincubation for 10 min with Ro 31-8220 has been used to demonstrate a role for PKC in phorbol ester-induced neutrophil function (IC_{50} = 1 μ*M*; Keller and Niggli, 1993). Preincubation of cerebrocortical synaptosomes with 10 μ*M* Ro 31-8220 inhibited 4-aminopyridine-evoked glutamate release, MARCKS phosphorylation, and $[Ca^{2+}]_i$ increase (Coffey et al., 1993). Ro 31-8425 has been used in several studies of immune cell function to demonstrate a role for PKC in the regulation of extracellular signal-regulated kinase-2 (ERK2) in T-cells (pretreatment with 100 μ*M* for 10 min; Izquierdo et al., 1994) and to rule out PKC involvement in T-cell-receptor-stimulated Ca^{2+} entry into Jurkat E6 cells (not inhibited by 10 μ*M* Ro 31-8425; Conroy et al., 1994) or in C5a-induced Mac-1 expression in human neutrophils (Sullivan et al., 1994).

2.4.5.2. INDOLECARBAZOLES

One of the most potent and selective representatives of these staurosporine derivatives is Gö 6976 (12-[2-cyanoethyl]-6,7,12,13-tetrahydro-13-methyl-5-oxo-5*H*-indolo[2,3-a]pyrrolo[3,4-c]carbazole), which is highly selective for PKC (Table 6; Martiny-Baron et al., 1993). Gö 6976 inhibits PKC competitively with respect to ATP (K_i = 2.8 n*M*), as do other staurosporine-related inhibitors. Of particular interest, Gö 6976 selectively inhibits the Ca^{2+}-dependent cPKC isoforms (Table 7; Gschwendt et al., 1995). Gö 6850, also known as GF 109205X, showed a less marked selectivity for cPKC, and also inhibited nPKC (Wenzel-Seifert et al., 1994). The 7-hydroxy derivative of staurosporine, UCN-01, also showed a moderate selectivity for cPKC vs nPKC or aPKC isozymes (Mizuno et al., 1995).

2.4.5.2.1. USE OF INDOLECARBAZOLES

Gö 6976 and Gö 6850 are prepared as stock solutions (1 m*M*) in DMSO. By comparing the effects of 1 μ*M* Gö 6796 (preincubated for 3 min before stimuli), which inhibits cPKC, but not nPKC isoforms, with the effects of 1 μ*M* Gö 6850 (GF 109203X), which inhibits both, evidence was obtained for the specific involvement of nPKC in various human neutrophil functions (Wenzel-Seifert et al., 1994). Preincubation of Swiss 3T3 cells with Gö 6976 for 5 min inhibited phorbol ester-stimulated phosphorylation of the PKC-specific substrate MARCKS (IC_{50} = 3 μM); Gö 6983 (not commerically available) was more cell-permeant than Gö 6976 with an IC_{50} = 0.3 μM (Gschwendt et al., 1992). Certain of the indolecarbazoles and bisindolylmaleimides, like staurosporine, have isozyme selective effects on PKC translocation, which may be involved in some of their effects (Courage et al., 1995).

2.4.5.3. NPC 15437

NPC 15437 (2,6-diamino-N-([1-1-oxotridecyl-2-piperidinyl]methyl)hexanamide) is a novel selective inhibitor of PKC with respect to the few protein kinases tested (Table 8; Sullivan et al., 1992). NPC 15437 is a competitive inhibitor of the activation of PKCα by phorbol ester (K_i = 5 μM) and phosphatidylserine (K_i = 12 μM), is mixed with respect to activation by Ca^{2+}, and does not inhibit the catalytic fragment of PKC. Mutant PKC constructs were used to show that inhibition by NPC 15437 required residues 12–42 of the amino-terminal region of the C1 regulatory domain, which contains the pseudosubstrate binding domain and part of the first cysteine-rich repeat sequence (Sullivan et al., 1991).

2.4.5.3.1. USE OF NPC 15437 IN INTACT CELLS

NPC 15437 (NOVA Pharmaceuticals, Baltimore, MD) dose-dependently antagonized phorbol ester-induced phosphorylation of the 47-kDa protein pleckstrin in intact human platelets (IC_{50} = 34 μM) (Sullivan et al., 1992). Coadministration of 20 μM NPC 15437 prevented phorbol ester-induced c-*Fos* induction in GTI-7 cells, a mouse hypothalamic GnRH neuronal cell line (Wetsel et al., 1993). NPC 15437 (40 μM) has also been shown to inhibit phorbol ester-induced upregulation and hyperphosphorylation of CD34 in myeloid cells (Fackler et al., 1992). Studies have employed NPC 15437 in vivo to implicate PKC in retention of learned avoidance in rats (intracranial injection) (Walker and Gold, 1994), and on acquisition and memory retention of a Y-maze avoidance task in mice (ip injection) (Mathis et al., 1992).

Table 8
Inhibition of PKC by NPC 15437[a]

Inhibitor	PKC	PKA	MLCK
NPC 15437	19	>>300	>>300
Sphingosine	57	98	145
Staurosporine	0.08	0.17	0.11
H-7	60	167	184

[a]IC_{50} values (in μM) taken from Sullivan et al. (1992). Abbreviations as in Table 1.

2.5. Cyclin-Dependent Kinase Inhibitors

Members of the cyclin-dependent kinase (CDK) family regulate transitions between phases of the cell cycle (Dorée and Galas, 1994; Morgan, 1995). These proline-directed protein-serine/threonine kinases consist of a catalytic subunit and a regulatory subunit (cyclin); enzymatic activity is also regulated by subunit phosphorylation and by a number of small protein inhibitors that interact with cyclins, CDKs, or their complexes. Six CDKs have been identified in humans, namely CDK1 (also known as cdc2) and CDK2–6; all are regulated by transient association with specific members of the family of cyclins, which have been identified except for that associated with CDK3. CDK-activating kinase (CAK; also known as CDK7) phosphorylates and activates CDK1, CDK2, and possibly other CDKs. All transitions (or checkpoints) in the cell cycle are controlled by changes in the activity of CDKs.

A diverse group of proteins have been identified that bind and inactivate CDK-cyclin complexes, and many function as tumor suppressers (Morgan, 1995). Two distinct families of CDK inhibitors (CKIs), which differ in their CDK specificities, have been identified in mammalian cells: the p21/p27 family and the p16/p18 family (Aprelikova et al., 1995). Knowledge of CKI inhibitory mechanisms is incomplete, and these proteins have yet to become useful analytical reagents. However, a number of small molecule CDK inhibitors have recently been identified (Table 9). N^6-(Δ^2-isopentenyl) adenine and N^6-dimethylaminopurine are non-

Table 9
IC$_{50}$ Values for Cyclin-Dependent Kinase Inhibitors

Inhibitor	cdc2/CDK1	MAPK	PKC	PKA	CKII	CKI	EGF-R
K252a[a]	1.8	0.20	1.2	1.3	9.0	59	21
Staurosporine[a]	0.0069	0.0069	0.0088	1.1	15	50	0.43
Butyrolactone I[a]	0.68	94	160	260	240	>590	>590
L86-8275[b]	0.3	—	6	145	—	—	25
Olomoucine[c]	7	30	>1000	>2000	>2000	—	440

IC$_{50}$ values (in µM) taken from [a]Kitagawa et al. (1993), [b]Losiewicz et al. (1994), and [c]Vesely et al. (1994). Abbreviations: cdc2/CDK1, mouse cyclin B-cdc2 kinase; MAPK, human 42-kDa-MAP kinase; PKC, mouse protein kinase C; PKA, catalytic subunit of bovine cAMP-dependent kinase; CKII and CKI, porcine casein kinase II and I; EGF-R, human EGF receptor tyrosine kinase.

specific compounds that inhibit CDKs and many other protein kinases with IC_{50} values of 50–100 μM (Vesely et al., 1994). The nonspecific protein kinase inhibitors staurosporine (IC_{50} = 4–5 nM; Gadbois et al., 1992) and K252a (Kitagawa et al., 1993) are also potent inhibitors of CDK1. Butyrolactone I (α-oxo-β-[p-hydroxy-m-3,3-dimethylallyl-benzyl]-γ-methoxycarbonyl-γ-butyrolactone) is a selective inhibitor of CDK1 and CDK2 that is competitive with ATP (K_i = 0.36 μM) and not protein substrate (Kitagawa et al., 1993). The flavonoid L86–8275 ([–]cis-5,7-dihydroxy-2-[2-chlorophenyl]-8(4-3-hydroxy-1-methyl-piperidinyl]-4H-benzopyran-4-one) inhibited cdc2/CDK1 (IC_{50} = 0.5 μM) by a competitive mechanism with respect to ATP (K_i = 0.04 μM) (Losiewicz et al., 1994); its effect on related CDKs has not been reported. The most extensively characterized CDK inhibitor is the purine analog olomoucine. Olomoucine (2-[2-hydroxyethylamino]-6-benzylamino-9-methylpurine) is a specific inhibitor of CDK1, CDK2, CDK5, and ERK1/MAP-kinase protein kinases (IC_{50} values of 3–25 μM), but not of CDK4 or CDK6, among 35 kinases tested (Table 9; Vesely et al., 1994). Kinetic analysis of CDK1 inhibition revealed competitive inhibition with respect to ATP and noncompetive inhibition with respect to protein substrate.

2.5.1. Use of CDK Inhibitors

Butyrolactone I inhibited histone H1 phosphorylation in nuclear extracts from FM3A cells (IC_{50} = 2.6 μM) and retinoblastoma gene product (pRB) phosphorylation in both nuclear extracts (IC_{50} = 3.9 μM) and intact human IMR32 neuroblastoma cells (Kitagawa et al., 1993). L86–8275 inhibited the S/G2 transition in synchronized MDA-468 breast carcinoma cells (IC_{50} ≈ 125 nM), and prevented the associated increase in CDK1-mediated histone H1 phosphorylation through inhibition of CDK1 (Worland et al., 1993). Olomoucine and the inactive control compound iso-olomoucine are prepared as 10-mM stock solutions in DMSO. Olomoucine inhibited DNA synthesis and M-phase promoting factor activity in *Xenopus* egg extracts at 10–15 μM in vitro and inhibited the G2/M transition in starfish oocytes in vivo (IC$_{50}$ = 30 μM) (Veselý et al., 1994). Although CDKs have been identified in neuronal tissues (Lew and Wang, 1995), limited information on the use of these inhibitors in neuronal tissues is not available. The CDK inhibitors flavopiridol (0.3–3 μM) and olomoucine (50–200 μM) suppressed the death and inhibited proliferation of cultured postmitotic differentiated PC12 cells and sympathetic neurons (Park et al., 1996).

2.6. Mitogen-Activated Protein Kinase (MAPK) and MAPK Kinase (MEK) Inhibitors

MAPKs are ubiquitous protein serine/threonine kinases involved in protein kinase cascades that regulate the cell cycle, cell growth, and differentiation (Cobb and Goldsmith, 1995). MAPK cascades are activated by extracellular signals, such as growth factors, that activate Raf protein kinase activity through Ras to phosphorylate and activate MAPK kinases, also called MAP/ERK kinases (MEK). The dual specificity kinases MEK1 and MEK2 are dedicated kinases that phosphorylate only MAPKs (extracellular signal-regulated protein kinases 1 and 2 [ERK1 and ERK2]) on a threonyl and a tyrosyl residue separated by a single amino acid. Phosphorylated and active MAPKs then translocate from the cytosol to the nucleus and relay the signal downstream by phosphorylating other protein kinases and regulatory proteins, including transcription factors.

PD 98059 (2-[2'-amino-3'-methoxyphenyl]-oxanaphthalen-4-one) has been identified as a noncompetitive cell-permeant inhibitor of MAPK kinase (MEK) without significant activity on MAPKs (Dudley et al., 1995). PD 98059 was a selective inhibitor of dephosphorylated MEK1 (IC$_{50}$ = 2–7 μM) vs MEK2 (IC$_{50}$ = 50 μM) and did not inhibit 18 other protein serine/threonine kinases in vitro, including PKA, PKCα, and MLCK (Alessi et al., 1995). Olomoucine (*see* Section 2.5.) inhibited MAPK (IC$_{50}$ = 25 μM) at fourfold lower potency than CDK2 (IC$_{50}$ = 7 μM).

2.6.1. Use of PD 98059

PD 98059 is prepared as a 50 mM stock solution in DMSO, stored in aliquots at –80°C, and diluted into aqueous buffers at <100 μM immediately prior to use. Pretreatment of PC12 cells with PD 98059 for 30 min blocked NGF-stimulated MAPK activity (IC$_{50}$ = 2 μM; maximal effect at 10–100 μM), phosphorylation of MAPK on tyrosine, and neurite formation without altering cell viability (Pang et al., 1995). Preincubation of Swiss 3T3 cells for 90 min with 50 μM PD 98059 inhibited the activation of MAPK by EGF (Alessi et al., 1995).

2.7. Protein-Tyrosine Kinase Inhibitors

Protein phosphorylation on tyrosine residues, catalyzed by protein-tyrosine kinases (PTKs), regulates cell growth, proliferation, and differentiation, the cell cycle, cytoskeletal interactions, and immune cell signaling (Fantl et al., 1993; *see also* Chapter 6). The receptor tyrosine kinases (RTKs) participate in transmembrane signaling by growth factors and other extracellular signals, whereas intracellular PTKs mediate signaling within the cell, including the nucleus. Although phosphorylation on tyrosine residues constitutes only 0.01% of total cellular phosphorylation, it is prominent in the mechanisms of many growth factors and hormones, such as insulin, and in the pathophysiology of cancer and other proliferative diseases. It is also a promising target for new drug development (Powis, 1991; Levitzki and Gazit, 1995).

Signaling through PTKs involves multiple protein–protein interactions. For example, ligand binding to RTKs, which consist of an extracellular domain linked by a single transmembrane α-helix to a cytosolic domain containing the PTK catalytic activity, causes most RTKs to dimerize and trans-autophosphorylate. The phosphotyrosine residues in the activated RTKs transduce their hormone signals to intracellular signaling molecules by promoting protein associations with adapter proteins through Src homology 2 (SH2) domains, which bind specific phosphotyrosine-containing sequences, to form a signaling complex, as exemplified by the Ras-MAP kinase cascade (Pawson, 1995). A number of approaches have been considered to intercept this complex signaling pathway at various steps, including inhibition of growth factor binding to RTKs, blockade of RTK interactions with downstream targets through SH2 domains and inhibition of PTK activity (Levitzki and Gazit, 1995). This section will consider direct inhibition of PTK activity, which has been the most productive target so far. A peptide derived from the noncatalytic domain of pp60$^{v\text{-}src}$ (residues 137–157) has been described that inhibits pp60$^{v\text{-}src}$ and the EGF receptor PTK in vitro (Sato et al., 1990). In addition, a number of cell-permeable low-mol-wt inhibitors have been identified. Although a high degree of homology exists among the catalytic domains of different PTKs, some compounds exhibit selectivity between different enzymes, which ultimately may lead to the development of more specific PTK inhibitors as more structural data become available.

2.7.1. Natural PTK Inhibitors

The efficacy of PTK inhibitors depends on their ability to inhibit the initial kinase autophosphorylation event and the phosphorylation of downstream substrates; these potencies may differ as a result of autophosphorylation-induced conformational changes in the PTK (Levitzki and Gazit, 1995). Preferably, PTK inhibitors should be competitive with respect to phosphate acceptor and potent inhibitors of autophosphorylation, which would lead to complete proximal blockade of the signaling pathway.

A number of PTK inhibitors have been isolated from microbial extracts that exhibit broad specificity with potencies in the micromolar range. The flavonoid compound quercetin was identified as an inhibitor of both protein-serine/threonine kinases and PTKs (Table 10). The isoflavonoid genistein (5,7-dihydroxy-3[4-hydroxyphenyl]-4*H*-1-benzopyran-4-one; 4',5,7-trihydroxyisoflavone) was isolated from *Pseudomonas* as an inhibitor of the EGF receptor PTK (Akiyama and Ogawara, 1991). It inhibits multiple PTKs in vitro (Table 10; Akiyama et al., 1987) by a competitive mecha-

Table 10
IC$_{50}$ Values for Natural Protein-Tyrosine Kinase Inhibitors

Inhibitor	EGF-R	pp60[v-src]	PKC	PKA	PKG	CaMKII
Quercetin[a]	26	26	83	>330	—	—
Genistein[a]	22 (2.6)[g]	26	>370	>370	—	—
Genistein[b]	—	18	185	>200	—	>200
Lavendustin A[c]	0.011	>100	>260	>260	—	—
Lavendustin A[b]	—	0.5	>100	>100	—	>200
Herbimycin A[d]	—	0.9	—	—	—	—
Erbstatin[e]	3.4	—	20	0.8	4.5	—
Methyl-2,5-dihydroxycinnamate (MDC)[f]	0.8	—	—	—	—	—

IC$_{50}$ values (in μM) taken from [a]Akiyama et al. (1987), [b]O'Dell et al. (1991), [c]Onoda et al. (1989), [d]Uehara et al. (1989), [e]Bishop et al. (1990), and [f]Umezawa et al. (1990). [g]Value (in parentheses) for inhibition of EGF-R autophosphorylation. Abbreviations: EGF-R, EGF receptor tyrosine kinase; otherwise as for Table 1.

nism with respect to ATP (K_i = 14 μM) and a noncompetitive mechanism with respect to phosphate acceptor. Major limitations of genistein are its low membrane permeability and its ability to inhibit the activity of enzymes other than PTKs, including β-galactosidase, protein-histidine kinase, DNA topoisomerases I and II, and S6 kinase, and to block Ca^{2+} channel currents and the GLUT1 hexose transporter (Vera et al., 1996). Daidzein (4',7-dihydroxyisoflavone; IC_{50} > 400 μM for EGF receptor) or genistin are available for use as negative control compound for genistein.

Lavendustin A (5-amino-[N-2,5-dihydroxybenzyl-N-2-hydroxybenzyl]-salicylic acid) is a potent inhibitor of PTKs with good selectivity, but poor membrane permeability (Table 10; Onoda et al., 1989). It is a slow and tight binding inhibitor of the EGF receptor by a mixed competitive mechanism with respect to ATP and substrate (Hsu et al., 1991). In contrast, the pp60[F527] PTK was inhibited by a noncompetitive mechanism with respect to ATP and uncompetitive with respect to phosphate acceptor (Agbotounou et al., 1994). Lavendustin B is available as a negative control compound (IC_{50} = 1.3 μM for EGF receptor). The benzoquinoid PTK inhibitors lavendustin A and methyl-2,5-dihydroxycinnamate have also been shown to have a direct antimuscarinic action unrelated to effects on tyrosine phosphorylation, thus limiting their utility in intact cells (Otero and Sweitzer, 1993).

Herbimycin A is a benzenoid ansamycin antibiotic isolated from *Streptomyces*, which was able to reverse Rous sarcoma virus (RSV) transformation by inhibiting pp60[v-src] PTK activity (IC_{50} = 12 μM; Uehara et al., 1989; Uehara and Fukazawa, 1991). It irreversibly inhibited pp60[v-src] by binding to the kinase through a sulfhydryl group, which targets the enzyme for degradation. In addition to pp60[v-src], herbimycin A also inhibited other cytoplasmic PTKs, including pp120[v-abl], pp130[v-fps], and pp210[bcr-abl]. It does not appear to affect PKC or PKC-dependent functions, although this has not been confirmed in vitro. Caution is required in the use of herbimycin A as a PTK inhibitor, since it has also been shown to modify NF-κB directly (Mahon and O'Neill, 1995).

Erbstatin was isolated from *Streptomyces* as an inhibitor of the EGF receptor PTK (Umezawa and Imoto, 1991). Erbstatin is too unstable for reliable experimentation; however, a stable analog, methyl-2,5-dihydroxycinnamate (MDC; erbstatin analog), is available with greater potency (Umezawa et al., 1990,1992). Erbstatin inhibits PTK by a mechanism competitive with both ATP and phosphate acceptor (Posner et al., 1993). The selectivity of erbstatin as a PTK inhibitor is controversial (Tables 10 and 11); it has been reported to be a potent inhibitor of PKC (Bishop et al., 1990).

A report by Barret et al. (1993) casts doubt on the specificities of most PTK inhibitors (Table 11). The reason(s) for the differences between their data on the potency and specificity of various PTK inhibitors and the data of others is unclear, but may include different substrates used, differences in enzyme purity and source, and/or differences in ATP concentrations. These discrepancies stress the need for caution in interpreting results obtained with these compounds.

2.7.1.1. USE OF NATURAL PTK INHIBITORS

Natural PTK inhibitors have been shown to have antiproliferative effects in intact cells (Barret et al., 1993). Genistein and lavendustin A, prepared as stock solutions in DMSO, have been used in a number of studies of PTK action in neurons. Genistein is active with intact cells at concentrations of 40–120 μM (Akiyama et al., 1987). It inhibited EGF-stimulated increases in total cellular phosphotyrosine content and in EGF receptor phosphorylation in A431 cells (IC_{50} = 110 μM; Akiyama et al., 1987), and EGF-, PDGF-, or T-cell receptor-stimulated phosphatidylinositol tumor (Akiyama and Ogawara, 1991). Antiproliferative effects of genistein were observed on RSV-transformed rat 3Y1 cells (IC_{50} = 26 μM) and on *ras*-transformed, but not normal, NIH 3T3 cells (at 37 μM). Genistein also inhibited the mitogenic effect on NIH 3T3 cells of EGF (IC_{50} = 12 μM), insulin (IC_{50} = 20 μM) and thrombin (IC_{50} = 20 μM).

In nervous tissue, lavendustin A (5–10 μM) or genistein (100 μM), but not genistin, blocked induction of long-term

Table 11
IC$_{50}$ Values of Various Inhibitors of Protein-Tyrosine Kinases[a]

	HPK40[b]	EGF receptor	pp60[c-src]	p56[lck]	PKA	PKC
Staurosporine	0.020	0.010	0.090	0.200	0.080	0.008
Suramin	400	70	200	200	10	60
Erbstatin	3	>100	70	900	>1000	>1000
Genistein	>100	300	>1000	>1000	300	>1000
Quercetin	200	50	600	>1000	>1000	>1000
H-7	>1000	>1000	>1000	>1000	>1000	80
H-8	>1000	>1000	200	800	>1000	100
H-9	>1000	>1000	>1000	>1000	>1000	200
HA1004	>1000	>1000	>1000	>1000	>1000	10
PKI[b]	>10	>10	>10	>10	0.001	>10
Cibacron blue	2.5	100	300	200	6	100
Tyrphostin 46	>100	20	1000	>1000	>1000	>1000
Tyrphostin 50	>100	>100	>1000	>1000	>1000	>1000
Tyrphostin 9	>100	>100	>1000	>1000	>1000	>1000

[a]IC$_{50}$ values (in μM) taken from Barret et al. (1993), with permission.
[b]HPK40 is a PTK purified from HL60 cells.

164

potentiation in guinea pig CA1 hippocampal neurons (O'Dell et al., 1991), and genistein (100–200 µM), but not genistin or diadzein, increased endogenous dopamine release from mouse striatal slices (Bare et al., 1995). Because of its low tissue permeability, both studies employed 30-min preincubations with relatively high concentrations of inhibitors. In contrast, preincubation for 20 min of cultured striatal neurons with 0.3 µM lavendustin A or 3 µM genistein inhibited NMDA-induced cGMP production (Rodriquez et al., 1994). Genistein (20 µM) or lavendustin A (10 µM), but not genistin or lavendustin B, inhibited tyrosine phosphorylation detected by immunocytochemistry during synapse formation between identified leech neurons (Catarsi et al., 1995). Genistein (37–300 µM) inhibited tyrosine kinase activity, phosphotyrosine content, and protein synthesis in cultured cortical neurons (Hu et al., 1993). Injection into the gerbil hippocampus of genistein or lavendustin A, but not genistin, blocked ischemia-induced delayed neuronal death when administered 30 min prior to ischemia (Kindy, 1993). Genistein or lavendustin A reduced phosphotyrosine content and potentiated substrate-induced neurite growth in cultured neurons (Bixby and Jhabvala, 1992), whereas genistein, lavendustin A, herbimycin A, or MDC inhibited filopodia formation after axotomy in cultured *Aplysia* neurons (Goldberg and Wu, 1995). Genistein (37-370 µM), herbimycin A (10 µM), or tyrphostin RG-50864, but not lavendustin A (75 µM) or daidzein (390 µM), induced neurite outgrowth in PC12 cells (Miller et al., 1993). Genistein (100 µM) or lavendustin A (10 µM) inhibited, and pp60[c-src] potentiated, NMDA receptor-mediated whole-cell currents in spinal dorsal horn neurons (Wang and Salter, 1994). Genistein (20–200 µM; IC_{50} = 30 µM) or herbimycin A (1–15 µM), but not daidzein, inhibited L-type Ca^{2+}-channel activity in pituitary GH_3 cells (Cataldi et al., 1996). Intracerebroventricular injection of lavendustin A or herbimycin A inhibited the induction of long-term potentiation in rat dentate gyrus in vivo (Abe and Saito, 1993). Lavendustin A (1 µM) or herbimycin A (1 µM), but not staurosporine (0.1

μ*M*) blocked basic fibroblast growth factor-induced acceleration of axonal branch formation in cultured rat hippocampal neurons (Aoyagi et al., 1994).

Herbimycin A, prepared as a stock solution in DMSO, is readily inactivated by sulfhydryl compounds, which should be excluded from buffers (Uehara et al., 1989). Herbimycin A (0.17–1.7 μ*M*) increased phosphotyrosine content and reversibly induced reversion to normal cell morphology in fibroblasts transformed by the PTK oncogenes *src, yes, fbs, ros, abl,* and *erbB,* but not by *ras, myc,* or *raf.* It also induced differentiation of the human leukemia cell line K562 and of mouse erythroleukemia and embryonal carcinoma (F9) cells. Herbimycin A has been used in several studies to analyze the role of PTKs in neuronal systems (*see* preceding paragraph). In addition, herbimycin A (1.7 μ*M*) completely inhibited PDGF-induced Ca^{2+} mobilization and the inhibitory effect of neuropeptide Y on angiotensin II- or bradykinin-induced Ca^{2+} mobilization in human neuroblastoma cells (Shigeri and Fujimoto, 1994). An interesting study by Doherty et al. (1994) showed that herbimycin A (0.5 μ*M*), but not lavendustin A (20 μ*M*) or MDC (52 μ*M*), indirectly stimulated PTK activity and neurite outgrowth in cultured cerebellar granule cells, possibly by inhibiting a PTK that tonically suppresses the activity of a second PTK, which is itself inhibited by lavendustin A or MDC.

The erbstatin analog MDC, prepared as a stock solution in DMSO, is about four times more stable than erbstatin in calf serum. As a rule, use of MDC requires long incubation time (hours) and/or relatively high concentrations (30–500 μ*M*). It inhibited EGF receptor autophosphorylation in A431 cells at 130–520 μ*M*, and delayed S phase induction by EGF in rat kidney cells at 64 μ*M* (Umezawa et al., 1990). MDC inhibited EGF-stimulated growth of NIH 3T3 cells (IC_{50} = 2.6 μ*M*; Umezawa et al., 1992). However, 5 μ*M* MDC or 100 μ*M* genistein effectively inhibited ADP- and thapsigargin-evoked Ca^{2+} entry and tyrosine phosphorylation in intact human platelets (Sargeant et al., 1993). Comparable potency was observed for inhibition of PDGF-induced chemotaxis and tyrosine phosphorylation in rat aortic smooth muscle cells by MDC (IC_{50} = 5 μM) and genistein (IC_{50} = 150 μM) (Shimokado et al., 1994). Several reports have described the use of MDC in neurons *(see above)*.

2.7.2. Tyrphostins

A vigorous effort has been made to synthesize potent and selective PTK inhibitors based on detailed structure–activity relationship studies (Levitzki and Gazit, 1995). Elements of the natural PTK inhibitors erbstatin and lavendustin A and the substrate tyrosine have been incorporated to synthesize hundreds of hydroxylated benzylidenemalononitrile compounds, named tyrphostins. Subsequently, bisubstrate inhibitors based on the transition state of tyrosyl phosphorylation by PTK have been designed and synthesized that contain a quinoline moiety as the ATP analog. Different tyrphostins exhibit different modes of inhibition of PTK with respect to ATP and phospho-acceptor substrate (Posner et al., 1993), although they were initially designed as phospho-acceptor substrate inhibitors. Irrespective of their mode of inhibition, tyrphostins, like most other PTK inhibitors, are 10^2–10^5 less potent against protein-serine/threonine kinases, although detailed data are not available for most tyrphostins. At least one study suggested that selectivity is not this great for tyrphostin AG213, which exhibited only 7.5-fold selectivity for pp60[c-src] over PKC (O'Dell et al., 1991). A number of tyrphostins preferentially inhibit the EGF receptor PTK over the insulin receptor PTK, although multiple PTKs are usually inhibited (Table 12). The first-generation tyrphostins inhibited the EGF receptor PTK in vitro in the 0.3–1 μM range. Subsequent conformationally restricted and alkyl-aryl substituted derivatives exhibited differences in PTK selectivity, the most striking being the selectivity of tyrphostin AG1478 for the EGF receptor PTK and of tyrphostin AG1296 for the PDGF receptor PTK (Table 12). A number of tyrphostins exhibit good inhibitory activity in intact cells, and are being used as starting points for drug development in the treatment of various proliferative disorders.

Table 12
Selectivity of Protein-Tyrosine Kinase Inhibitors In Vitro[a]

Tyrphostin	EGFR[b]	HER2–Neu	PDGFR	Trk	p210[Bcr-Abl]	InsR
AG18 (A23; RG 50810)	35	—	25	>100	75	4000
AG82 (A25)	3	—	—	>100	3.6	—
AG213 (A47; RG 50864)	0.8	—	3	>100	6	640
AG494 (B48)	0.7	42	6	—	75	>100
AG555 (B46)	0.7	35	—	—	—	>100
AG825	19	0.35	40	—	75	>100
AG879	>500	1.0	>100	10	—	—
AG1112	18.5	—	—	—	0.8	—
AG1296	>100	>100	0.5	—	>50	—
AG1478	0.003	>100	>100	—	>50	—
PD 153035[c]	0.029	—	>50	—	—	>50
CGP 53716[d]	>100	—	0.1	—	—	—

[a]Data for IC$_{50}$ values (μM) modified from Levitzki and Gazit (1995) except as noted.
[b]Abbreviations: EGFR, EGF receptor; HER2-Neu, ErbB2/neu receptor kinase (HER2); PDGFR, PDGF receptor; InsR, insulin receptor.
[c]Data from Fry et al. (1994).
[d]Data from Buchdunger et al. (1995).

Although there are extensive data on the PTK inhibitory activity of these compounds, they may also have significant additional biological activities. Tyrphostin A9 is a potent uncoupler of oxidative phosphorylation. Several widely used tyrphostins have recently been shown to inhibit the GTPase activity of transducin and the enzymatic activities of other GTP-utilizing enzymes, including guanylyl cyclase and fructose-6-phosphate kinase, which complicates the interpretation of results obtained in intact cells (Wolbring et al., 1994). Tyrphostin 23 or 1 (an inactive analog) and genistein, but not daidzein, have been described as blockers of Ca^{2+} channels, with potencies comparable to their PTK inhibitory potencies (Wijetunge et al., 1992). Inhibition of topoisomerase I and/or II by tyrphostins (AG786, AG555, AG18, and AG213), as well as by erbstatin or genistein, may also play a role in their effects on whole cells, and provides an additional mechanism for their antiproliferative effects (Aflalo et al., 1994; Markovits et al., 1994). Tyrphostins, genistein, and erbstatin have been shown to increase cGMP in rat pinealocytes by inhibition of its metabolism, which may lead to complicating effects on this pathway (Ogiwara et al., 1995). Inactive tyrphostin analogs are available as control compounds, but topoisomerase inhibition correlates with PTK inhibition in vitro.

2.7.2.1. USE OF TYRPHOSTINS

Tyrphostins are prepared as stock solutions in DMSO or ethanol at 1000X concentrations. The nomenclature for tyrphostins is complicated by the use of identical sequential numbers for different chemical structures in two original publications, such that each compound can have multiple names (Gazit et al., 1989, 1991). Tyrphostin RG 50864 (100 μM) inhibited EGF-induced phosphorylation of phospholipase Cγ in HER 114 cells (Margolis et al., 1989). Several analogs exhibit antiproliferative effects on cultured cells at micromolar concentrations (Levitzki and Gazit, 1995). For example, 5 μM AG490, but not 11 other tyrphostins, selectively blocked leukemia cell growth in vitro and in vivo by inducing programmed cell death, with no deleterious effects on normal hematopoiesis; this effect correlated with specific inhibition of constitutively active Jak-2 PTK by AG490 (Meydan et al., 1996). Preincubation for 15 min with the selective tyrphostin AG1478 (0.25 μM) suppressed transactivation of the EGF receptor by G-protein-coupled receptors in Rat-1 cells (Daub et al., 1996). Selective effects of tyrphostins in intact cells have been observed for 2.5 μM AG1478, which blocked EGF receptor autophosphorylation in HER 14 keratinocytes without affecting the PDGF receptor, whereas 30 μM AG879, a pp140[c-trk] inhibitor, did not (Osherov and Levitzki, 1994), and for AG1296, which reversibly blocked PDGF receptor autophosphorylation in Swiss 3T3 cells (IC$_{50}$ = 0.3–0.5 μM) without affecting the EGF receptor (Kovalenko et al., 1995). Genistein (100 μM), tyrphostin B42 (10 μM), or B44 (30 μM), but not daidzein (100 μM), inhibited muscimol-stimulated $^{36}Cl^{-}$ uptake in mouse brain microsacs (Valenzuela et al., 1995). PTK specificity of different PTK inhibitors were suggested by the observation that tyrphostins 23 or 25 (150 μM) or MDC (77 μM), but not genistein, lavendustin A, or tyrphostins 34 or 37, inhibited cell adhesion molecule-stimulated neurite outgrowth in rat cerebellar neurons (Williams et al., 1994). Genistein or tyrphostin 25 (10 μM) inhibited inducible NO synthase expression in cultured glial cells (Feinstein et al., 1994). Pretreatment of PC12 cells for 90 min with tyrphostin AG879 inhibited NGF-dependent pp140[c-trk] tyrosine phosphorylation and neurite outgrowth (IC$_{50}$ = 10 μM), but not EGF or PDGF receptor phosphorylation (Ohmichi et al., 1993). Tyrphostin RG 50864 (IC$_{50}$ = 50 μM), but not the inactive analog RG 50862, inhibited acetylcholine receptor clustering in cultured *Xenopus* myocytes (Peng et al., 1993).

2.7.3. Selective PTK Inhibitors: PD 153035 and CGP 53716

PD 153035 is an extremely potent (K_i = 5.2 pM) tricyclic quinazoline analog ATP site inhibitor of the EGF receptor kinase. For comparison, PD 153035 inhibited the EGF receptor in vitro with an IC$_{50}$ value of 29 pM, which was 4–5 orders of magnitude less than those observed for erbstatin (370 nM) and genistein (1200 nM) under similar conditions (Fry et al., 1994). Inhibition of the EGF receptor was selective, since minimal inhibition of recombinant PDGF receptor, FGF receptor CSF-1 receptor, insulin receptor, or pp60$^{v\text{-}src}$ PTK activity was observed at concentrations up to 50 µM (Table 12). The potency and selectivity of PD 153035 was also evident in intact Swiss 3T3 fibroblasts and A-431 epidermoid carcinoma cells.

CGP 53716 is a 2-phenylaminopyrimidine compound identified as a selective inhibitor of the PDGF receptor PTK (Buchdunger et al., 1995). This compound inhibited autophosphorylation of the immunoprecipitated PDGF receptor with an IC$_{50}$ value of 0.1 µM, which was three orders of magnitude less than its IC$_{50}$ for five other PTKs, including the EGF receptor (Table 12). However, the v-Abl PTK was inhibited with an IC$_{50}$ value of 0.6 µM. A number of protein serine/threonine kinases showed little or no inhibition. Selective inhibition of PDGF receptor autophosphorylation and PDGF-mediated tyrosine phosphorylation and c-*fos* mRNA induction was demonstrated in intact BALB/c 3T3 cells.

These compounds are just two examples of a number of potent and relatively selective PTK under development as antiproliferative agents. Their selectivity promises to make them useful analytical reagents in the study of neuronal protein tyrosine phosphorylation.

3. Protein Phosphatase Inhibitors

3.1. Protein-Serine/Threonine Phosphatase Inhibitors

Protein-serine/threonine phosphatases dephosphorylate phosphoserine and phosphothreonine residues in proteins. Based on their substrate specificities and regulator sensitivities, they are classified as type 1 phosphatase (protein phosphatase-1; PP1), which is inhibited by inhibitor-1 (I-1) and inhibitor-2 (I-2) and dephosphorylates the β-subunit of phosphorylase kinase, and as type 2 phosphatases (PP2A, PP2B, and PP2C), which are insensitive to I-1 and I-2, and dephosphorylate the α-subunit of phosphorylase kinase (Cohen, 1991). Type 2 phosphatases are further classified by their divalent cation sensitivity; PP2A is independent, PP2B requires Ca^{2+}/calmodulin, and PP2C requires Mg^{2+}. Molecular cloning revealed that the catalytic subunits of PP1, PP2A, and PP2B are closely related compared to PP2C, and that multiple isoforms of each enzyme exist (Wera and Hemmings, 1995). Many novel protein phosphatases have been identified recently that do not fit into the above classifications, but are closely related to the existing classes. Although purified PP1 and PP2A display broad and overlapping substrate specificities, the holoenzyme forms exhibit altered specificities in part because of influences from associated regulatory subunits. Numerous aspects of neuronal function are regulated by protein phosphatases, including neurotransmitter release, neurotransmitter receptors, and ion channels (Nairn and Shenolikar, 1992). The identification of a number of potent, low-mol-wt microbial toxin protein phosphatase inhibitors has advanced the study of protein phosphatase function (Fig. 1). In cell-free extracts, several nonspecific inorganic compounds inhibit protein phosphatase activity, including sodium fluoride (50 mM), sodium pyrophosphate (5 mM), EDTA (1–2 mM; chelates Mg^{2+} and blocks protein kinases and PP2C), and EGTA (0.1–1 mM; chelates Ca^{2+} and blocks PP2B as well as other Ca^{2+}-dependent enzymes).

3.1.1. Okadaic Acid

Okadaic acid (Fig. 1) is a complex polyether fatty acid produced by marine dinoflagellates, and is a potent tumor promoter and a causative agent of diarrhetic shellfish poisoning. It is the best-studied and most widely used phosphatase inhibitor because of its membrane permeability and high potency. Okadaic acid is a specific inhibitor of PP2A and PP1, which are structurally related; PP2B, which is also related, is far less sensitive and PP2C, which is unrelated, is unaffected (Table 13). Protein kinases, protein-tyrosine phosphatases, and alkaline phosphatases are also insensitive. In cell-free systems, 1–2 nM okadaic acid usually inhibits PP2A completely ($IC_{50} \approx 0.1$ nM), whereas PP1 is unaffected by 1 nM okadaic acid, but is completely inhib-

PP1-PP2A Inhibitors

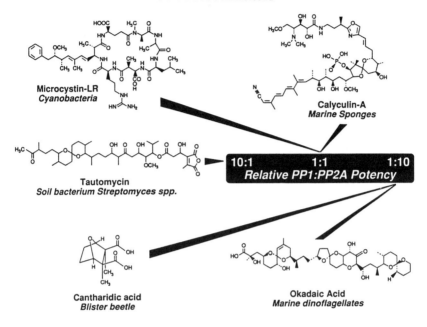

Fig. 1. Chemical structures of various toxin inhibitors of protein-serine/threonine phosphatases. Biological sources of the toxins are indicated in italics below the inhibitor names. No potent inhibitors of PP2C, the fourth major class of protein-serine/threonine phosphatases, have been identified as yet. From Armstrong (1995), with permission.

ited at 5 μ*M* (IC$_{50}$ ≈ 10–15 n*M*). This differential selectivity can be used to quantify PP2A and PP1 individually in dilute cell extracts (Cohen et al., 1990).

Table 13

Properties of Protein-Serine/Threonine Phosphatase Inhibitors[a]

Inhibitor	Chemical nature	Potency	Comments
Protein phosphatase-1 and -2A Okadaic acid, acanthifolcin and dinophysistoxin	Polyether carboxylic acids	PP2A >PP1 >>>PP2B	Cell-permeable; 1 nM specifically inhibits PP2A in dilute assays
Tautomycin	Polyketide	PP1 >PP2A >>>PP2B	Cell-permeable; used in parallel with okadaic acid, distinguishes role of PP1 and PP2A
Microcystins, nodularins, and mutoporin	Cyclic peptides	PP2A ~ PP1 >>>PP2B	Do not enter some mammalian cells. Microcystin has a covalent interaction with PP1 and PP2A, and is used as an affinity-purification ligand
Calyculin A	Phosphorylated polyketide	PP2A ~ PP1 >>>PP2B	Cell-permeable
Cantharidin	Terpeniod	PP2A >PP1 >>>PP2B	Cell-permeable; requires relatively high concentrations, but is inexpensive

Inhibitor-1 and -2; DARPP-32	Heat-stable proteins	Inhibit PP1 only	Not cell-permeable but useful in cell extracts to identify role of PP1. Inhibitor-1 and DARPP-32 are activated by PKA and inactivated by PP2A and PP2B
Protein phosphatase2B Cyclosporin A	Cyclic undecapepeptide	Binds to cyclophilin in cell; complex inhibits PP2B with nanomolar affinity	Immunosuppressive drug
FK-506	Macrocyclic lactone	Binds to FKBP in cell; complex inhibits PP2B with nanomolar affinity	Immunosuppressive drug

[a]Modified from MacKintosh and MacKintosh (1994), with permission.

3.1.1.1. Use of Okadaic Acid

Okadaic acid is prepared at 5 mM as a stock solution in DMSO and stored at −20°C. Owing to potential problems with stability of dissolved okadaic acid, several salt forms of okadaic acid are available with improved stability after dissolution. 1–Norokadaone and methyl okadaate are available as negative control compounds. PP2A and PP1 can be quantified (Cohen et al., 1990; Cohen et al., 1989) in dilute cell extracts by comparing the dephosphorylation of an exogenous substrate for both enzymes (usually [^{32}P]phosphorylase a) in the absence of okadaic acid (total activity), in the presence of 1–2 nM okadaic acid (to block PP2A; total activity – inhibited activity = PP2A activity), in the presence of 5 µM okadaic acid (to block PP1 and PP2A; activity at 1 nM okadaic acid −activity at 5 µM okadaic acid = PP1 activity). PP2B and PP2C activities are negligible because of the exclusion of Ca^{2+} and Mg^{2+} in the assays. Alternatively, 0.2 µM I-2 or phospho-I-1 can be used to determine specifically PP1 activity (*see* Section 3.1.3.). The assay is sensitive to enzyme concentration, since the amount of okadaic acid needed to inhibit PP2A completely is similar to the amount of PP2A in the assay. This technique has been used to determine the subcellular distributions of PP1 and PP2A in rat forebrain (Sim et al., 1994) and in rabbit brain (Cohen et al., 1989). Selective inhibition of PP2A in synaptic plasma membranes by 2 nM okadaic acid and inhibition of PP1 as well by 1 µM okadaic acid (confirmed by sensitivity to I-2) were demonstrated using both endogenous and exogenous substrates (Han and Dokas, 1991). PP2C can be measured using a substrate for this enzyme (usually ^{32}P-casein) in the presence of 5 µM okadaic acid (to block PP1 and PP2A) and 0.1 mM EGTA (to block PP2B), and in the absence or presence of 20 mM Mg^{2+}.

In intact cell studies, both PP1 and PP2A are usually present at concentrations of 0.1–1 µM; thus, at least 1 µM okadaic acid is required to inhibit PP2A completely, and selective inhibition of PP1 and PP2A is therefore not possible (Hardie et al., 1991). Since increasing protein phosphorylation by inhibiting protein phosphatases with okadaic acid is equivalent to activating multiple protein kinases, treatment of intact cells with this inhibitor has multiple physiological effects, both inhibitory and stimulatory, such as stimulation of glycogenolysis and gluconeogenesis in hepatocytes, or of glucose transport and lipolysis in adipocytes. In short-term incubations (15 min), hepatocyte viability is not affected; given the central regulatory importance of protein phosphorylation in cell function, it is anticipated that longer incubations may result in secondary effects. As with protein kinase inhibitors, specificity of protein phosphatase inhibitors is improved with the use of negative control compounds, several inhibitors with differing protein phosphatase specificity, and/or intracellular application of purified protein phosphatases.

Okadaic acid has been employed in a number of studies of protein phosphatase function in neuronal systems. Okadaic acid (1 μM) increased total protein phosphorylation and phosphorylation of tyrosine hydroxylase (Haavik et al., 1989), and increased Ca^{2+}-independent norepinephrine release (Wu and Wagner, 1991) in bovine adrenal chromaffin cells. Extracellular application of 1 μM okadaic acid enhanced neurotransmitter release at frog and lobster neuromuscular junctions (Abdul-Ghani et al., 1991). In synaptosomes, 1 μM okadaic acid (preincubated for 5 min) increased basal phosphorylation of P96 and P75 (Sihra et al., 1992); 0.1 μM okadaic acid or 0.1 μM calyculin A (preincubated for 15 min), but not 1-norokadaone, reduced KCl-evoked, but not ionomycin-induced, increases in $[Ca^{2+}]_i$ and glutamate release (Nichols et al., 1996); 1 μM okadaic acid (preincubated for 5 min) increased basal release of glutamate, aspartate, and GABA, and increased basal protein phosphorylation (Sim et al., 1993); and 2 μM okadaic acid (preincubated for 5 min) potentiated Ca^{2+}-dependent phosphorylation of multiple phosphoproteins in lysed synaptic membranes (Hemmings and Adamo, 1996). Okadaic acid (1–20 μM) induced τ phosphorylation in brain slices, whereas FK-520 (1–10 μM), a PP2B inhibitor, did not; however, 5 μM FK-520 plus 0.1 μM okadaic acid mimicked the effect of 10 μM okadaic acid. These results suggested that 0.1 μM okadaic acid indirectly activated a MAP kinase to phosphorylate τ, which was dephosphorylated by PP2B (Garver et al., 1995). Exposure of cultured rat sensory neurons to 0.5–1 μM okadaic acid for 30 min evoked Ca^{2+}-dependent release of substance P and calcitonin gene-related peptide (Hingtgen and Vasko, 1994). Okadaic acid (0.6 μM) or calyculin A (0.5 μM) inhibited lactacystin-induced differentiation of Neuro 2a cells (Tanaka et al., 1995). Okadaic acid (IC_{50} = 7 nM) inhibited NGF-induced neurite outgrowth in PC12 cells at very low concentrations (Chiou and Westhead, 1992).

Okadaic acid and other phosphatase inhibitors have been especially valuable in identifying ion channels that are regulated by protein phosphorylation (Armstrong, 1995). These studies have implicated protein phosphatases in the regulation of Ca^{2+}, K^+, Na^+, Cl^-, and glutamate-activated channels. Much of the work on Ca^{2+} channels has involved cardiac tissue, but a number of studies have involved neuronal tissues. Okadaic acid (0.1 μM) occluded inhibition of ω-conotoxin GVIA-sensitive Ca^{2+} currents by D_2 receptors expressed in NG 108–15 cells (Brown and Seabrook, 1995). Intracellular injection into snail neurons of okadaic acid or microcystin-LR, but not cyclosporin A, mimicked and occluded the enhancement of Ca^{2+} channels by muscarinic agents or serotonin (Golowash et al., 1995). Application of 0.5 μM okadaic acid increased kainate currents in perforated patches from hippocampal neurons (Wang et al., 1991).

3.1.2. Cantharidin, Tautomycin, Microcystins, Nodularin, and Calyculin A

These compounds, like okadaic acid, are active as tumor promoters. Cantharidin is a natural insecticide from the blister beetle that inhibits PP2A (IC_{50} = 40 nM) at 10-fold greater potency than PP1 (IC_{50} = 470 nM), and is inactive against PP2B and PP2C at concentrations up to 30 μM (Li et al., 1993). Cantharidin, like okadaic acid, inhibited both the purified catalytic subunits of PP1 and PP2A and their native holoenzyme forms present in a PC12 cell extract (Honkanen, 1993). Cantharidin, and its less potent derivative endothall, are cell-permeant and economical inhibitors that have yet to find widespread use.

Tautomycin, the microcystins, and calyculin A all inhibit PP1 with greater potency than okadaic acid. These differences in relative potencies can be exploited in parallel experiments using two or more inhibitors at different concentrations to tease out the involvement of a specific protein phosphatase in the control of a particular cellular process; a similar approach has been demonstrated for the isoquinolinesulfon-amide protein kinase inhibitors (Section 2.1.2.). Tautomycin is a potent inhibitor of PP1 ($IC_{50} \approx 0.2$ nM) and of PP2A ($IC_{50} \approx 1$ nM), compared to PP2B ($IC_{50} \approx 100$ μM) and PP2C (not inhibited); it also does not affect PKA, PKC, or several other protein kinases (MacKintosh and Klumpp, 1990). The microcystins are cyclic cyanobacterial heptapeptide toxins that are potent environmental hepatotoxins. Microcystin-LR inhibited both PP1 and PP2A with IC_{50} values of ~0.1 nM, compared to 10 and 0.1 nM, respectively, for okadaic acid under similar in vitro conditions; PP2B was inhibited with 1000-fold lower potency ($IC_{50} = 0.2$ μM), whereas PP2C, PKA, PKC, CaMKII, and five other protein kinases were not inhibited (MacKintosh et al., 1990). The results of an independent study indicated that microcystin-LR was more potent as an inhibitor of PP2A ($IC_{50} = 0.04$ nM) than of PP1 ($IC_{50} = 1.7$ nM) (Honkanen et al., 1990). Microcystin-RR is an analog of microcystin-LR with reduced toxicity that also inhibited PP2A ($IC_{50} = 1.4$ μM); microcystin-YR is a tyrosine-containing analog of microcystin-LR suitable for radioiodination. Nodularin is a cyclic pentapeptide cyanobacterial hepato-toxin that inhibits PP2A ($IC_{50} = 0.09$ nM); it is similar to microcystin-LR, but has improved water solubility (Matsushima et al., 1990). These peptide inhibitors do not readily cross cell membranes, and are useful primarily in broken cell preparations or for microinjection studies.

Calyculin A, a cytotoxic compound isolated from a marine sponge, is a phosphorylated polyketide that is a potent inhibitor of both PP1 ($IC_{50} = 2$ nM) and PP2A ($IC_{50} = 0.5$–1 nM); thus, although okadaic acid inhibits PP2A with

100-fold greater potency than PP1 (IC_{50} = 500 nM), under similar conditions, calyculin A is equipotent (Ishihara et al., 1989). The relative potency of these two toxins can indicate whether an effect is caused by the inhibition of PP1 or PP2A. Calyculin A is inactive against acid and alkaline phosphatase; however, it has not been tested against PP2B or PP2C. Calyculin A appears to have good cell-permeability because of its hydrophobicity, but is light-sensitive. An inhibitory effect of calyculin A on Ca^{2+} channels independent of phosphatase inhibition has been reported that could affect the interpretation of studies in whole-cells (Gutierrez et al., 1994).

3.1.2.1. USE OF CANTHARIDIN, TAUTOMYCIN, MICROCYSTINS, AND CALYCULIN A

Cantharidin derivatives (soluble in DMSO) have received little attention in the neuroscience literature, in part because of their low potencies. Cantharidin was found to exert a positive inotropic effect in isolated guinea pig papillary muscle (1–100 µM), associated with increases in L-type Ca^{2+} channel currents and regulatory protein phosphorylation (Neumann et al., 1995).

Tautomycin (soluble in DMSO, ethanol, and methanol) has been shown to inhibit Ca^{2+} channel current in isolated smooth muscle cells at 1–100 µM (Groschner et al., 1995). The relative potencies of calyculin A > tautomycin > okadaic acid in inhibiting PP1 correlated with their ability to inhibit thrombin-induced Ca^{2+} influx in intact human platelets (Murata et al., 1993). Similarly, the sensitivity of smooth muscle heavy meromyosin phosphatase to microcystin-LR > tautomycin >> okadaic acid was used to implicate PP1 (Gong et al., 1992).

Microcystin-LR (soluble in H_2O, DMSO, ethanol, and methanol) has been used in a number of neuronal systems. Intracellular application of CaMKII or microcystin-LR enhanced AMPA and NMDA responses in isolated rat dorsal horn neurons (Kolaj et al., 1994), and reduced desensitization of the NMDA receptor in outside-out patches (10 µM, but not 50 nM by intracellular perfusion; Tong and Jahr, 1994).

Microcystin-LR, okadaic acid, or calyculin A blocked an endogenous phosphatase that modulates the cGMP sensitivity of ion channels in excised retinal rod patches (Gordon et al., 1992). The induction of long-term depression was blocked by extracellular application of 1 μM okadaic acid or 1 μM calyculin A, or intracellular microcystin-LR (10 μM in pipet), but not by 1 μM 1–norokadaone in rat hippocampal slices (Mulkey et al., 1993). In whole patch recordings of locus coeruleus neurons, 50 nM or 4 μM microcystin-LR, but not 1 μM okadaic acid, decreased the rate of recovery from desensitization of μ opioid receptor-induced currents (Osborne and Williams, 1995). Intracellular application of microcystin-LR prolonged serotonin- and cAMP-induced inward currents in *Aplysia* sensory neurons (Ichinose et al., 1990).

Calyculin A (soluble in DMSO and ethanol) has found numerous applications as an equipotent cell-permeant inhibitor of both PP1 and PP2A. Bath application of 0.5–1 μM calyculin A enhanced AMPA-, but not NMDA-mediated synaptic transmission in rat hippocampal slices, an effect that could be prevented by KN-62 (Figurov et al., 1993); a similar study confirmed that bath application of 0.3 μM calyculin A enhanced AMPA-mediated responses, as did intracellular application of microcystin-LR or okadaic acid (Wyllie and Nicoll, 1994). Using a phosphorylation state-specific antibody, the phosphorylation of AMPA receptors was enhanced by calyculin A in rat cerebellar slices (Nakazawa et al., 1995). In cultured hippocampal neurons, calyculin A (20 nM) or okadaic acid (10 nM) enhanced NMDA currents, whereas purified PP1 or PP2A reduced channel opening (Wang et al., 1994b). Differential sensitivity to calyculin A vs okadaic acid was used to implicate PP1 or PP2A in two distinct effects on guinea pig L-type Ca^{2+} channels (Wiechen et al., 1995). Inhibition of acetylcholine-mediated reductions in isoproterenol-induced L-type Ca^{2+} currents in whole cells by injected okadaic acid (3–9 μM) or cantharidin (3 μM), but not calyculin A (1 μM), cyclosporin A (10 μM), or I-2, supported a role for PP2A in this effect (Herzig et al., 1995). Calyculin A and okadaic acid inhibited equipotently Ca^{2+}- and KCl-evoked [³H]acetylcholine release from rat hippocampal synaptosomes, whereas orthovanadate had no effect (Vickroy et al., 1995). Calyculin A caused a rapid retraction of neurites in NGF-treated PC12 cells (Reber and Bouron, 1995).

3.1.3. Inhibitor-1 (I-1), DARPP-32, and Inhibitor-2 (I-2)

PP1 activity is highly regulated in vivo by a number of thermostable modulatory proteins, including inhibitor-1 (I-1), dopamine- and cAMP-regulated phosphoprotein, M_r = 32,000 (DARPP-32), and inhibitor-2 (I-2) (MacKintosh and MacKintosh, 1994). Phosphorylation of I-1 (165 amino acids; M_r = 26,000; mol wt = 18,640) at Thr^{35} by PKA converts it to a specific and potent (K_i = 1.6 nM) inhibitor of PP1 with a complex mixed competitive/noncompetitive mechanism of inhibition (Foulkes et al., 1983). DARPP-32 (202 amino acids; M_r = 32,000; mol wt = 22,591) is similar to I-1 in its NH_2-terminal domain; phosphorylation of DARPP-32 at Thr^{34} by PKA converts it to a specific and potent (IC_{50} = 1 nM) inhibitor of PP1 with similar inhibitory kinetics to I-1 (Hemmings et al., 1984). Both inhibitors are efficiently dephosphorylated by PP2A or PP2B. Structure–activity studies have defined the inhibitory domain of DARPP-32 as residues 6–38 (Hemmings et al., 1990; Desdouits et al., 1995), and of I-1 as residues 9–54 (Aitken and Cohen, 1982). I-2 (203 amino acids; M_r = 32,000; mol wt = 22,835) forms an inactive complex with the catalytic subunit of PP1 (known as Fc or the ATP-Mg^{2+}-dependent protein phosphatase); phosphorylation of I-2 by glycogen synthase kinase-3 (Fa) leads to activation of the complex. Dephoso-I-2 inhibits PP1 (K_i = 3.1 nM) by a competitive mechanism (Foulkes et al., 1983); attempts to identify active peptide fragments have been unsuccessful.

3.1.3.1. Use of I-1, DARPP-32, and I-2

Only I-2 is commercially available (recombinant rabbit; Calbiochem). I-1, DARPP-32, and I-2 can be purified from rabbit skeletal muscle (I-1 and I-2; Cohen et al., 1990) or brain (DARPP-32; Hemmings et al., 1995), or I-1 and DARPP-32 peptides can be custom-synthesized. The use of I-2 in cell extracts to identify PP1 has been discussed (Section 3.1.1.1.).

Although the catalytic subunit is inhibited almost instantaneously by I-2, native holoenzyme forms of PP1 require preincubation with higher concentrations (0.2 μM) of I-2 for 15–20 min for effective inhibition (Gong et al., 1992); a similar phenomenon has been observed with I-1. The thermostable PP1 inhibitors have been employed in intact cell studies owing to their excellent potency and specificity. Thiophosphorylated I-1 (I-1 phosphorylated by PKA using ATPγS to reduce dephosphorylation), I-2, or 10 μM microcystin-LR reduced the amplitude of outward currents elicited by FMRFamide in *Aplysia* sensory neurons (Endo et al., 1995). Intracellular application of thiophosphorylated I-1 blocked long-term depression in rat hippocampal slices (Mulkey et al., 1994). Phospho-DARPP-32(8–38) peptide mimicked the inhibitory effect of dopamine on Na^+,K^+-ATPase activity in renal tubule cells (10 μM; Aperia et al., 1991), and intracellular dialysis with 100 μM DARPP-32(8–38) antagonized dopamine-induced reductions in N- and P-type Ca^{2+} currents in dissociated striatal neurons (Surmeier et al., 1995). A phospho-I-1 peptide fragment (5 μM), but not 2 μM okadaic acid, mimicked forskolin activation of Ca^{2+} currents in cultured dorsal root ganglion (DRG) neurons (Dolphin, 1992). In planar lipid bilayers, 80 nM thiophosphorylated I-1 or 0.25 μM microcystin-LR enhanced the modulation of type-2 Ca^{2+}-dependent K^+ channels by ATP, an effect that was inhibited by 0.2 μM PKC(19–36) (Reinhart and Levitan, 1995). The use of I-1, I-2, or DARPP-32 peptides has been useful in demonstrating a role for PP1 in the regulation of each of these ion channel proteins.

3.1.4. PP-2B Inhibitors: Cyclosporin A, FK-506, Autoinhibitory Peptide, and Pyrethroids

PP2B (also known as calcineurin; CaN) is a Ca^{2+}/calmodulin-dependent protein phosphatase that consists of a 61 kDa catalytic (A) subunit and a 19-kDa (B) subunit (Klee et al., 1988). In contrast to PP1 and PP2A, PP2B exhibits a relatively selective substrate specificity, which includes a number of regulatory proteins (e.g., I-1, DARPP-32, the RII subunit of PKA). The catalytic domain of PP2B shows about 40% sequence identity to the catalytic domains of PP1 and PP2A; it is also inhibited by okadaic acid, but at much higher concentrations (IC_{50} = 5 μM). The B-subunit is structurally related to calmodulin and binds Ca^{2+}. In addition to the catalytic core and binding sites for calmodulin and the B-subunit, the catalytic A-subunit contains an intrasteric autoinhibitory regulatory domain (Hashimoto et al., 1990). Deletion of this domain results in a constitutively active phosphatase, and a peptide based on this domain [CaN A(467–492)] inhibits PP2B activity competitively (K_i = 4 μM; Parsons et al., 1994), and not PP1, PP2A, or CaMKII (Hashimoto et al., 1990). PP2B is also inhibited by the immunosuppressive drugs cyclosporin A and FK-506 or its ethyl derivative FK-520 in association with specific endogenous "immunophilin" binding proteins (Liu et al., 1991). The complexes cyclophilin-cyclosporin A and FK-506-FKBP (FK-506 binding protein; FKBP-12) bind to and inhibit bovine brain and T-cell PP2B, the critical target for immunosuppression, in the presence of Ca^{2+}/calmodulin. FK-506 is about 10-fold more potent than cyclosporin A in inhibiting PP2B following a 30 min preincubation (Antoni et al., 1993). Rapamycin is an analog of FK-506 that does not inhibit PP2B. The type II pyrethroid insecticides (cypermethrin, deltamethrin, and fenvalerate) are potent inhibitors of PP2B (IC_{50} values of 10^{-9}–$10^{-10}M$). However, these compounds also have potent activating effects on voltage-dependent Na^+ channels (Enan and Matsumura, 1992).

3.1.4.1. USE OF PP2B INHIBITORS

The immunosuppressant drugs cyclosporin A (available from Sandoz Pharmaceuticals, East Hanover, NJ), FK-506 (available from Abbott [Chicago, IL] and Fugisawa Pharmaceutical [Osaka, Japan]), and FK-520 (available from Merck, Piscataway, NJ) are prepared as stock solutions in γ-cyclodextrin, ethanol, or DMSO. FK-506 ($10^{-8}M$) or CaN A(467–491) (10 μM) inhibited α-adrenergic stimulation of Na^+,K^+-ATPase activity in renal tubule cells (Aperia et al.,

1992). The autoinhibitory peptide of PP2B (300 μ*M*) reduced glycine-insensitive desensitization of NMDA receptors when injected into rat hippocampal neurons (Tong and Jahr, 1994). FK-506 (1 μ*M*) facilitated the induction of long-term potentiation (Funauchi et al., 1994) and blocked long-term depression (Torii et al., 1995) in rat visual cortex. Cyclosprin A (1–20 μ*M*) or FK-506 (0.1–1 μ*M*), but not rapamycin (5 μ*M*), increased the rate of glutamate release by a presynaptic mechanism in rat cortical neurons (Victor et al., 1995) and enhanced Ca^{2+}-dependent ACTH release from AtT-20 pituitary cells (Antoni et al., 1993). Using dissociated hippocampal neurons, 10 μ*M* okadaic acid prolonged NMDA channel openings only when Ca^{2+} entry was permitted, an effect that was mimicked by 10–50 nM FK 506, whereas PP2B had the opposite effect; these results implicated PP2B in regulating NMDA receptors (Lieberman and Mody, 1994). Cyclosporin A (1 μ*M*) or FK-520 (1 μ*M*) inhibited depolarization-evoked dynamin I dephosphorylation in rat cortical synaptosomes (Liu et al., 1994; Nichols et al., 1994), and enhanced 4-aminopyridine-evoked glutamate release (Nichols et al., 1994). FK-520 (5 μ*M*) or cypermethrin (0.1 μ*M*) affected τ phosphorylation in rat brain slices (Garver et al., 1995). Extracelluar cyclosporin A (250 μ*M*) or FK-506 (50–100 μ*M*), but not 50 μ*M* rapamycin, inhibited long-term depression in rat hippocampal slices; intracellular application with pipets containing 10–50 μ*M* FK-506 or 250–500 μ*M* CaN A(467–492) was also effective (Mulkey et al., 1994). Cyclosporin A (1 μ*M*) and FK-506 (1 μ*M*) delayed neuritogenesis and inhibited neurite extension in cultured chick dorsal root ganglion neurons (Chang et al., 1995).

3.2. Protein-Tyrosine Phosphatase Inhibitors

Unlike the protein tyrosine kinases and protein-serine threonine kinases, the protein-tyrosine phosphatases (PTPases) and protein-serine/threonine phosphatases lack structural and sequence similarities. The PTPases consist of over 40 enzymes subcategorized into three groups, receptor-like PTPases, intracellular PTPases, and dual-specificity PTPases, all of which contain an active site signature motif containing an essential cysteinyl residue (Stone and Dixon, 1994). They were originally distinguished from the protein-serine/threonine phosphatases by their sensitivity to inhibition by vanadate and Zn^{2+} rather than F^- and EDTA. Some receptor PTPases were subsequently found to be resistant to Zn^{2+}, and a PTPase sensitive to F^- has also been reported. Although there is no specific inhibitor, vanadate (HVO_4^{2-}) has been widely used to inhibit PTPases (Gordon, 1991). Molybdate is also effective as a PTPase inhibitor. Higher concentrations of vanadate can inhibit protein-serine/threonine phosphatases. Both vanadate and molybdate have pleiotropic effects on intact cells and must be used judiciously.

3.2.1. Vanadate

Stock solutions of sodium orthovanadate (Na_3VO_4 in H_2O; multiple vendors) are prepared by adjusting a $1M$ solution to pH 10.0, boiling until translucent to depolymerize, and readjusting to pH 10.0. For inhibition of tyrosine dephosphorylation during tissue homogenization, the lysis buffer should contain 1 mM vanadate, which can be reduced in subsequent assays. For intact cells in culture, toxic effects become evident beyond 6-h incubations with 0.1 mM vanadate. Because of the diverse effects of this compound (Gordon, 1991), the results obtained in intact cells may not be explained solely by PTPase inhibition. Nevertheless, intracellular application of sodium orthovanadate (0.1 μM) in cultured dorsal horn neurons potentiated NMDA currents, in contrast to genistein (0.1 μM), which was inhibitory (Abdul-Ghani et al., 1991). Superfusion of pituitary GH_3 cells for 15 min with 100 μM vanadate abolished the inhibition of L-type Ca^{2+} channels observed with exposure to 200 μM genistein (Cataldi et al., 1996).

The addition of hydrogen peroxide to vanadate results in the generation of pervanadate (Evans et al., 1994). A stock solution (50 mM) is prepared by mixing an equimolar amount of 30% hydrogen peroxide with 1 mL of 0.5M sodium vanadate. This solution is incubated at room temperature for 20 min and then diluted to 10 mL with 100 mM Tris, 100 mM HEPES, pH 7.4. Residual peroxide is removed by adding catalase to 2 µg/mL and the solution incubated at room temperature for 30 min. This solution is stable for 2 h and exhibits limited toxicity on cultured YT cells (Evans et al., 1994). Although both vanadate and pervanadate (25–100 µM) inhibit PTPases, their effects on cellular activation differ for reasons that remain unclear.

3.2.2. Phenylarsine Oxide and Diamide

Phenylarsine oxide (PAO) is a membrane-permeable trivalent arsenical compound that has multiple biochemical effects that are thought to be mediated by covalent binding to vicinal sulfhydryl groups. Recently, PAO was found to be a potent inhibitor of PTPases (Garcia-Morales et al., 1990; Liao et al., 1991) in low micromolar concentrations (1–25 µM). However, PAO is also a potent inhibitor of glucose transport, receptor endocytosis, and NADPH oxidase (Le Cabec et al., 1995), factors that must be considered when interpreting results obtained in intact cells.

Diamide (azodicarboxylic acid bis[dimethylamide]) also inhibits PTPases by interacting with vicinal sulfhydryl groups of the enzyme (Monteiro et al., 1991). Diamide (0.1–0.5 mM) did not affect the tyrosine kinase activity associated with the EGF receptor. The inhibition of PTPase activity by pervanadate, PAO, or diamide can be reversed by sulfhydryl reducing agents, such as 2-mercaptoethanol or dithiothreitol.

Acknowledgments

Thanks are given to Hans-G. Genieser (BIOLOG Life Science Institute, Bremen, FRG), Angus C. Nairn (Rockefeller University, New York), and D. James Surmeier (University of Tennessee at Memphis) for helpful comments, as well as to Lisa Cabrera and Marife Enriquez for excellent secretarial assistance. The National Institutes of Health is acknowledged for support of work in the author's laboratory.

Appendix I

Sources of Protein Kinase Inhibitors

Alamone Labs, Jerusalem, Israel (800) 618–1644
   Chelerythrine
Bachem California, Torrance, CA (310) 539–4171

[Ala^{286}]CaMKII(281–302)	PKC(19–31)
CaMKII(281–309)	PKC(19–36)
CaMKII(290–309)	PKI(5–22)amide
MLCK inhibitor peptide	PKI(5–24)
Myristoyl-PKC(20–28)	PKI-tide
PKC(20–28)	p60$^{v\text{-}src}$(137-157)

Boehringer Mannheim Corp., Indianapolis, IN
   (800) 428–5433

GF 109203X	PKC(19–31)
Staurosporine	Tyrphostin 47 (RG 50864; AG 213)

Biolog Life Science Institute, La Jolla, CA (619) 457-1573

(Rp)-cAMPS	(Rp)-cGMPS
(Rp)-8-Br-cAMPS	(Rp)-8-Br-cGMPS
(Rp)-8-Cl-cAMPS	(Rp)-8-pCPT-cGMPS
(Rp)-8-CPT-cAMPS	(Rp)-8-Br-PET-cGMPS
(Rp)-8-PIP-cAMPS	

BIOMOL, Plymouth Meeting, PA (800) 942–0430

H-7	Erbstatin analog (methyl
H-8	2,5-dihydroxycinnamate)
H-9	Genistein
H-89	Lavendustin A
HA1004	Tyrphostin 1
ML-7	Tyrphostin 23 (RG 50810;
ML-9	RG 50858; AG18)
Calphostin C	Tyrphostin 25
KT5720	Tyrphostin 47 (RG 50864;
KT5823	AG213)
KT5926	Tyrphostin 51

Calbiochem-Novabiochem International, La Jolla, CA
   (800) 854-3417

A3	Genistein

Bisindolylmaleimide I
Bisindolylmaleimide II
Bisindolylmaleimide III
Bisindolylmaleimide IV
Bisindolylmaleimide V
Calphostin C
Chelerythrine
Gö 6976
H-7
Iso-H-7
H-8
H-9
H-89
HA-100
HA1004
HA1077
K-252a
K-252b
K-252c
KN-62
KN-93
KT5720
KT5823
KT5926
ML-7
ML-9
CaMKII(290–309)
PKI(6–25)
PKC(19–36)
Ro 31–8220
Sphingosine
Staurosporine
Daidzein
Erbstatin analog (methyl
   2,5-dihydroxycinnamate)

Herbimycin A
Lavendustin A
Lavendustin B
Wortmannin
Tyrphostin A1
Tyrphostin A8 (AG10)
Tyrphostin A9 (AG17)
Tyrphostin A23 (RG 50810
   RG 50858; AG18)
Tyrphostin A25 (AG82)
Tyrphostin A46 (AG99)
Tyrphostin A47 (RG 50864;
   AG213)
Tyrphostin A48
Tyrphostin A51
Tyrphostin A63
Tyrphostin AG126
Tyrphostin AG370
Tyrphostin AG879
Tyrphostin AG1288
Tyrphostin AG1295
Tyrphostin AG1296
Tyrphostin AG1478
Tyrphostin B42 (AG490)
Tyrphostin B44
Tyrphostin B46 (AG555)
Tyrphostin B48 (AG494)
Tyrphostin B50
Tyrphostin B56 (AG556)
Tyrphostin RG 13022
Tyrphostin RG 14620
Bistyrphostin
Tyrosine-specific protein
   kinase Inhibitor [pp60$^{v\text{-}src}$
   (137–157)]

Gibco BRL, Gaithersburg, MD (301) 840–8000

Genistein
Herbimycin A

[Glu^{27}]PKC(19–36)
PKC(19–36)

Lavendustin A   PKI(6–22)amide
Methyl 2,5-dihydroxy-   Tyrphostin
   cinnamate
LC Laboratories, Woburn, MA (617) 938–1700
   Tyrphostin A1   Lavendustin A
   Tyrphostin A3   Lavendustin B
   Tyrphostin A9 (RG 50872;AG17)   Calphostin C
   Tyrphostin A10 (AG126)   Chelerythrine
   Tyrphostin A23 (RG 50858;AG18)   H-7
   Tyrphostin A25 (RG 50875;AG82)   Iso-H-7
   Tyrphostin A46 (B40;AG99)   H-8
   Tyrphostin A47 (RG 50864;AG213)   H-9
   Tyrphostin A48   H-89
   Tyrphostin A51(AG183)   HA1004
   Tyrphostin A63   K-252a
   Tyrphostin B7 (AG370)   K-252b
   Tyrphostin B42 (AG490)   KT5720
   Tyrphostin B44   KT5823
   Tyrphostin B46(AG555)   KT5926
   Tyrphostin B48(AG494)   ML-9
   Tyrphostin B50   PKC(19–31)
   Tyrphostin B52(AG698)   PKC(19–36)
   Tyrphostin B56(AG556)   PKI(5–24)
   Tyrphostin B66(AG528)   PKI(5–24)amide
   Tyrphostin Bistyrphostin(AG537)   PKI(5–22)amide
   Herbimycin A   PKI(14–24)amide
   Genistein   (Rp)cAMPS
   Daidzein   Staurosporine
                                      Wortmannin
Peninsula Laboratories, Belmont, CA (800) 922–1516
   PKI(5–24)   PKI(14–24)
   PKI(5–24)amide   Myosin kinase inhibiting
   PKI(5–22)amide   peptide
Promega, Madison, WI (800) 356-9526
   Myristoyl PKC (19–31)   PKI(5–24)amide
   Olomoucine
Research Biochemicals International, Natick, MA
   (800) 736–3690

Tyrphostin AG 1295
Tyrphostin AG 1478
Calphostin C
Chelerythrine
Erbstatin analog
  (methyl 2,5-dihydroxy
  cinnamate)
Genistein
GF 109203X
H-7

H-8
H-9
HA-1004
HA-1077
Herbimycin A
Lavendustin A
NPC-15437
Staurosporine
PKC(19–36)

Seikagaku America, Ijamsville, MD (800) 237-4512

HA1004
H-7
H-8
H-9
H-85
H-88
H-89

KN-04
KN-62
KN-92
KN-93
ML-7
ML-9

Sigma Chemical, St. Louis, MO (800) 325–3010

Calphostin C
CaMKII(290–307)
Erbstain analog
  (methyl 2,5-dihydroxy
  cinnamate)
Genistein
HA1004
H-7 (actually Iso-H-7)
H-8
H-9
Herbimycin A
K252a
K252b
KN-04
KN-62
KT5720
KT5823
KT5926
Lavendustin A

Lavendustin B
ML-7
ML-9
PKI(6–22)amide
PKI(6–24)
Purified PKI
Sphingosine
Staurosporine
Tyrphostin
Tyrphostin 1
Tyrphostin 23 (RG 50810)
Tyrphostin A25
Tyrphostin A46
Tyrphostin A47 (RG 50864)
Tyrphostin A51
Tyrphostin A63
Tyrosine kinase inhibitor
  peptide
Wortmannin

Tocris Cookson, St. Louis, MO (800) 421–3701

A3	Tyrphostin A23
Dihydrosphingosine	Tyrphostin A46/B40
GF 109203X	Tyrphostin A47
H-7	Tyrphostin B42
H-9	Tyrphostin B44(–)
HA-1077	Tyrphostin B44(+)/B50
Erbstatin analog	Tyrphostin B46
(methyl 2,5-dihydroxy	Tyrphostin B48
cinnamate)	Tyrphostin B52
	Tyrphostin B56

Wako Bioproducts, Richmond, VA (800) 992–9256

Calphostin C	Genistein
GF 109203X	Herbimycin A
Staurosporine	

Appendix II

Sources of Protein Phosphatase Inhibitors

Alamone Labs, Jerusalem, Israel (800) 618–1644

Calyculin A	Okadaic acid
Deltamethrin	1-Norokadaone
Permethrin	

Bachem California, Torrance, CA (310) 539–4171
CaNA(467–492) Autoinhibitory Fragment

Boehringer Mannheim, Indianapolis, IN
(800) 428–5433

Calyculin A	Okadaic acid
Staurosporine	

BIOMOL, Plymouth Meeting, PA (800) 942–0430

Calyculin A	Nodularin
Microcystin-LR	Okadaic acid

Calbiochem-Novabiochem International, San Diego, CA
(800) 854–3417

Calyculin A	Nodularin
Cantharidic acid	Okadaic acid

Cantharidin
Microcystin-LR
Microcystin-RR
Microcystin-YR
Cypermethrin
Deltamethrin
Fenvalerate

Okadaic acid, sodium salt
Okadaic acid, potassium salt
Okadaic acid, ammonium salt
Methyl okadaate
1-Norokadaone
Tautomycin
Inhibitor-2,
	recombinant rabbit muscle

Gibco BRL, Gaithersburg, MD (301) 840-8000
	Okadaic acid Microcystin-LR
	Calyculin A

LC Laboratories, Woburn, MA (617) 938–1700
	Cantharidic acid 1-Norokadaone
	Cantharidin Okadaic acid, ammonium salt
	Methyl okadaate Okadaic acid, free acid
	Microcystin LR Okadaic acid, potassium salt
	Nodularin Okadaic acid, sodium salt

Promega, Madison, WI (800) 356-9526
	Inhibitor-1 Inhibitor-2

Research Biochemicals International, Natick, MA
	(800) 736–3690
	Calyculin A Nodularin
	Microcystin-LR Okadaic acid

Sigma Chemical, St. Louis, MO (800) 325–3010
	Calyculin A 1-Norokadaone
	Cantharidin Okadaic acid
	Diamide Okadaic acid, ammonium salt
	Microcystin-LR Okadaic acid, free acid
	Microcystin-RR Phenylarsine oxide

Wako Bioproducts, Richmond, VA (800) 992–9256
	Calyculin A Okadaic acid
	Nodularin Tautomycin

References

Abdul-Ghani, M., Kravitz, E. A., Meiri, H., and Rahamimoff, R. (1991) Protein phosphatase inhibitor okadaic acid enhances transmitter release at neuromuscular junctions. *Proc. Natl. Acad. Sci. USA* **88**, 1803–1807.

Abe, K. and Saito, H. (1993) Tyrosine kinase inhibitors, herbimycin A and lavendustin A, block formation of long-term potentiation in the dentate gyrus in vivo. *Brain Res.* **621**, 167–170.

Abrahams, T. P. and Schofield, G. G. (1992) Norepinephrine-induced Ca^{2+} current inhibition in adult rat sympathetic neurons does not require protein kinase C activation. *Eur. J. Pharmacol.* **227**, 189–197.

Adams, W. B. and Levitan, I. B. (1982) Intracellular injection of protein kinase inhibitor blocks the serotonin-induced increase in K^+ conductance in *Aplysia* neuron R15. *Proc. Natl. Acad. Sci. USA* **79**, 3877–3880.

Aflalo, E., Iftach, S., Segal, S., Gazit, A., and Priel E. (1994) Inhibition of topoisomerase I activity by tyrphostin derivatives, protein tyrosine kinase blockers: mechanism of action. *Cancer Res.* **54**, 5138–5142.

Agbotounou, W. K., Umezawa, K., Jacquemin-Sablon, A., and Pierre, J. (1994) Inhibition by two lavendustins of the tyrosine kinase activity of pp60F527 in vitro and in intact cells. *Eur. J. Pharmacol.* **269**, 1–8.

Aitken, A. and Cohen, P. (1982) Isolation and characterisation of active fragments of protein phosphatase inhibitor-1 from rabbit skeletal muscle. *FEBS Lett.* **147**, 54–58.

Akasu, T., Ito, M., Nakano, T., Schneider, C. R., Simmons, M. A., Tanaka, T., Tokimasa, T., and Yoshida, M. (1993) Myosin light chain kinase occurs in bullfrog sympathetic neurons and may modulate voltage-dependent potassium currents. *Neuron* **11**, 1133–1145.

Akiyama, T., and Ogawara, H. (1991) Use and specificity of genistein as inhibitor of protein-tyrosine kinases. *Methods Enzymol.* **201**, 362–370.

Akiyama, T., Ishida, J., Nakagawa, S., Ogawara, H., Watanabe, S.-I., Itoh, N., Shibuya, M., and Fukami, Y. (1987) Genistein, a specific inhibitor of tyrosine-specific protein kinases. *J. Biol. Chem.* **262**, 5592–5595.

Alessi, D. R., Cuenda, A., Cohen, P., Dudley, D. T., and Saltiel, A. R. (1995) PD 098059 is a specific inhibitor of the activation of mitogen-activated protein kinase *in vitro* and *in vivo*. *J. Biol. Chem.* **270**, 27,489–27,494.

Aloyo, V. J. (1995) Modulation of mu opioid binding by protein kinase inhibitors. *Biochem. Pharmacol.* **49**, 17–21.

Anderson, M. E., Braun, A. P., Schulman, H., and Premack, B. A. (1994) Multifunctional Ca^{2+}/calmodulin-dependent protein kinase mediates Ca^{2+}-induced enhancement of the L-type Ca^{2+} current in rabbit ventricular myocytes. *Circ. Res.* **75**, 854–861.

Antoni, F. A., Shipston, M. J., and Smith, S. M. (1993) Inhibitory role for calcineurin in stimulus-secretion coupling revealed by FK506 and cyclosporin A in pituitary corticotrope tumor cells. *Biochem. Biophys. Res. Commun.* **194**, 226–233.

Aoyagi, A., Nishikawa, K., Saito, H., and Abe, K. (1994) Characterization of basic fibroblast growth factor-mediated acceleration of axonal branching in cultured rat hippocampal neurons. *Brain Res.* **661**, 117–126.

Aperia, A., Fryckstedt, J., Svensson, L., Hemmings, H. C., Jr., Nairn, A. C., and Greengard, P. (1991) Phosphorylated M$_r$ 32,000 dopamine- and cAMP-regulated phosphoprotein inhibits Na$^+$,K$^+$-ATPase activity in renal tubule cells. *Proc. Natl. Acad. Sci. USA* **88**, 2798–2801.

Aperia, A., Ibarra, F., Svensson, L.-B., Klee, C., and Greengard, P. (1992) Calcineurin mediates α-adrenergic stimulation of Na$^+$,K$^+$-ATPase activity in renal tubule cells. *Proc. Natl. Acad. Sci. USA* **89**, 7394–7397.

Aprelikova, O., Xiong, Y., and Liu, E. T. (1995) Both p16 and p21 families of cyclin-dependent kinase (CDK) inhibitors block the phosphorylation of cyclin-dependent kinases by the CDK-activating kinase. *J. Biol. Chem.* **270**, 18,195–18,197.

Armstrong, D. L. (1995) Environmental toxins reveal ion channel regulation by protein phosphatases. *Neuroprotocols* **6**, 62–71.

Balbi, D. and Allen, J. M. (1994) Role of protein kinase C in mediating NGF effect on neuropeptide Y expression in PC12 cells. *Mol. Brain Res.* **23**, 310–316.

Baorto, D. M., Mellado, W., and Shelanski, M. L. (1992) Astrocyte process growth induction by actin breakdown. *J. Cell Biol.* **117**, 357–367.

Bare, D. J., Ghetti, B., and Richter, J. A. (1995) The tyrosine kinase inhibitor genistein increases endogenous dopamine release from normal and weaver mutant mouse striatal slices. *J. Neurochem.* **65**, 2096–2104.

Barret, J. M., Ernould, A. P., Ferry, G., Genton, A., and Boutin, J. A. (1993) Integrated system for the screening of the specificity of protein kinase inhibitors. *Biochem. Pharmacol.* **46**, 439–448.

Bartschat, D. K. and Rhodes, T. E. (1995) Protein kinase C modulates calcium channels in isolated presynaptic nerve terminals of rat hippocampus. *J. Neurochem.* **64**, 2064–2072.

Basudev, H., Jones, P. M., and Howell, S. L. (1995) Protein phosphorylation in the regulation of insulin secretion: the use of site-directed inhibitory peptides in electrically permeabilised islets of Langerhans. *Acta Diabetol.* **32**, 32–37.

Berg, M. M., Sternberg, D. W., Parada, L. F., and Chao, M. V. (1992) K-252a inhibits nerve growth factor-induced *trk* proto-oncogene tyrosine phosphorylation and kinase activity. *J. Biol. Chem.* **267**, 13–16.

Bishop, W. R., Petrin, J., Wang, L., Ramesh, U., and Doll, R. J. (1990) Inhibition of protein kinase C by the tyrosine kinase inhibitor erbstatin. *Biochem. Pharmacol.* **40**, 2129–2135.

Bixby, J. L. and Jhabvala, P. (1992) Inhibition of tyrosine phosphorylation potentiates substrate-induced neurite growth. *J. Neurobiol.* **23,** 468–480.

Botelho, L. H. P., Rothermel, J. D., Coombs, R. V., and Jastorff, B. (1988) cAMP analog antagonists of cAMP action. *Methods Enzymol.* **159,** 159–172.

Bowlby, M. R. and Levitan I. B. (1995) Block of cloned voltage-gated potassium channels by the second messenger diacylglycerol independent of protein kinase C. *J. Neurophysiol.* **73,** 2221–2229.

Braun, A. P. and Schulman, H. (1995) The multifunctional calcium/calmodulin-dependent protein kinase: from form to function. *Ann. Rev. Physiol.* **57,** 417–445.

Brown, N. A., and Seabrook, G. R. (1995) Phosphorylation- and voltage-dependent inhibition of neuronal calcium currents by activation of human D2(short) dopamine receptors. *Br. J. Pharmacol.* **115,** 459–466.

Bruns, R. F., Miller, F. D., Merriman, R. L., Howbert, J. J., Heath, W. F., Kobayashi, E., Takahashi, I., Tamaoki, T., and Nakano, H. (1991) Inhibition of protein kinase C by calphostin C is light-dependent. *Biochem. Biophys. Res. Commun.* **176,** 288–293.

Buchdunger, E., Zimmerman, J., Mett, H., Meyer, T., Müller, M., Regenass, U., and Lydon, N. B. (1995) Selective inhibition of the platelet-derived growth factor signal transduction pathway by a protein–tyrosine kinase inhibitor of the 2-phenylaminopyrimidine class. *Proc. Natl. Acad. Sci. USA* **92,** 2558–2562.

Buchler, W., Walter, U., Jastoroff, B., and Lohmann, S. M. (1988) Catalytic subunit of cAMP-dependent protein kinase is essential for cAMP-mediated mammalian gene expression. *FEBS Lett.* **228,** 27–32.

Budworth, J. and Gescher, A. (1995) Differential inhibition of cytosolic and membrane-derived protein kinase C activity by staurosporine and other kinase inhibitors. *FEBS Lett.* **362,** 139–142.

Butt, E., Bemmelen, M., Fischer, L., Walter, U., and Jastorff, B. (1990) Inhibition of cGMP-dependent protein kinase by (Rp)-guanosine 3',5'-monophosphorothioates. *FEBS Lett.* **263,** 47–50.

Butt, E., Eigenthaler, M., and Genieser, H.-G. (1994) (Rp)-8-pCPT-cGMPS, a novel cGMP-dependent protein kinase inhibitor. *Eur. J. Pharmacol.* **269,** 265–268.

Butt, E., Pöhler, D., Genieser, H.-G., Huggins, J. P., and Bucher, B. (1995) Inhibition of cyclic GMP-dependent protein kinase-mediated effects by (Rp)-8-bromo-PET-cyclic GMPS. *Br. J. Pharmacol.* **116,** 3110–3116.

Cabell, L. and Audesirk, G. (1993) Effects of selective inhibition of protein kinase C, cyclic AMP-dependent protein kinase, and Ca^{2+}-calmodulin-dependent protein kinase on neurite development in cultured rat hippocampal neurons. *Int. J. Dev. Neurosci.* **11,** 357–368.

Calabresi, P., Pisani, A., Mercuri, N. B., and Bernardi, G. (1994) Post-receptor mechanism underlying striatal long-term depression. *J. Neurosci.* **14,** 4871–4881.

Campenot, R. B., Draker, D. D., and Senger, D. L. (1994) Evidence that protein kinase C activates in regulating neurite growth are localized to distal neurites. *J. Neurochem.* **63**, 868–878.

Castellucci, V. F., Nairn, A., Greengard, P., Schwartz, J. H., and Kandel, E. R. (1982) Inhibitor of adenosine 3':5'-monophosphate-dependent protein kinase blocks presynaptic facilitation in *Aplysia. J. Neurosci.* **2**, 1673–1681.

Cataldi, M., Taglialatela, M., Guerriero, S., Amoroso, S., Lombardi, G., Di Renzo, G., and Annunziato, L. (1996) Protein-tyrosine kinases activate while protein-tyrosine phosphatases inhibit L-type calcium channel activity in pituitary GH3 cells. *J. Biol. Chem.* **271**, 9441–9446.

Catarsi, S., Ching, S., Merz, D. C., and Drapeau, P. (1995) Tyrosine 1 phosphorylation during synapse formation between identified leech neurons. *J. Physiol.* **485**, 775–786.

Chang, H. Y., Takei, K., Sydor, A. M., Born, T., Rusnak, F., and Jay, D. G. (1995) Asymmetric retraction of growth cone filopodia following focal inactivation of calcineurin. *Nature* **376**, 686–690.

Chen, G.-G., Chalazonitis, A., Shen, K.-F., and Crain, S. M. (1988) Inhibitor of cyclic AMP-dependent protein kinase blocks opioid-induced prolongation of the action potential of mouse sensory ganglion neurons in dissociated cell cultures. *Brain Res.* **462**, 372–377.

Chijiwa, T., Mishima, A., Hagiwara, M., Sano, M., and Hayashi, T. (1990) Inhibition of forskolin-induced neurite outgrowth and protein phosphorylation by a newly synthesized inhibitor of cyclic AMP-dependent protein kinase, N-(2-[p- bromocinnamyl-amino)ethyl]-5-isoquinolinesulfonamide (H-89), of PC12D pheochromocytoma cells. *J. Biol. Chem.* **265**, 5267–5272.

Chiou, J.-Y. and Westhead, E. W. (1992) Okadaic acid, a protein phosphatase inhibitor, inhibits nerve growth factor-directed neurite outgrowth in PC12 cells. *J. Neurochem.* **59**, 1963–1966.

Cobb, M. H. and Goldsmith, E. J. (1995) How MAP kinases are regulated. *J. Biol. Chem.* **270**, 14,843–14,846.

Coffey, E. T., Sihra, T. S., and Nicholls, D. G. (1993) Protein kinase C and the regulation of glutamate exocytosis from cerebrocortical synaptosomes. *J. Biol. Chem.* **268**, 21,060–21,065.

Cohen, P. (1991) Classification of protein-serine/threonine phosphatases: identification and quantitation in cell extracts. *Methods Enzymol.* **201**, 389–398.

Cohen, P., Klumpp, S., and Schelling, D. L. (1989) An improved procedure for identifying and quantitating protein phosphatases in mammalian tissues. *FEBS Lett.* **250**, 596–600.

Cohen, P., Foulkes, J. G., Holmes, C. F., Nimmo, G. A., and Tonks, N. K. (1990) Protein phosphatase inhibitor-1 and inhibitor-2 from rabbit skeletal muscle. *Methods Enzymol.* **159**, 427–437.

Colbran, R. J., Smith, M. K., Fong, Y. L., Schworer, C. M., and Soderling, T. R. (1989) Regulatory domain of calcium/calmodulin-dependent protein kinase II. Mechanism of inhibition and regulation by phosphorylation. *J. Biol. Chem.* **264,** 4800–4804.

Conrad, R., Keranen, L. M., Ellington, A. D., and Newton, A. C. (1994) Isozyme-specific inhibition of protein kinase C by RNA aptamers. *J. Biol. Chem.* **269,** 32,051–32,054.

Conroy, L. A., Merritt, J. E., and Hallam, T. J. (1994) Regulation of T-cell-receptor-stimulated bivalent-cation entry in Jurkat E6 cells: role of protein kinase C. *Biochem. J.* **303,** 671–677.

Courage, C., Budworth, J., and Gescher, A. (1995) Comparison of ability of protein kinase C inhibitors to arrest cell growth and to alter cellular protein kinase C localisation. *Br. J. Cancer* **71,** 697–704.

Crepel, F., Audinat, E., Daniel, H., Hemart, N., Jaillard, D., Rossier, J., and Lambolez, B. (1994) Cellular locus of the nitric oxide-synthase involved in cerebellar long-term depression induced by high external potassium concentration. *Neuropharmacology* **33,** 1399–1405.

Daub, H., Weiss, F. U., Wallasch, C., and Ullrich, A. (1996) Role of transactivation of the EGF receptor in signalling by G-protein-coupled receptors. *Nature* **379,** 557–560.

Davis, P. D., Hill, C. H., Keech, E., Lawton, G., Nixon, J. S., Sedgwick, A. D., Wadsworth, J., Westmacott, D., and Wilkinson, S. E. (1989) Potent selective inhibitors of protein kinase C. *FEBS Lett.* **259,** 61–63.

Davis, P. D., Elliott, L. H., Harris, W., Hill, C. H., Hurst, S. A., Keech, E., Kumar, M. K. H., Lawton, G., Nixon, J. S., and Wilkinson, S. E. (1992) Inhibitors of protein kinase C. 2. Substituted bisindolylmaleimides with improved potency and selectivity. *J. Med. Chem.* **35,** 994–1001.

Day, R. N., Walder, J. A., and Maurer, R. A. (1989) A protein kinase inhibitor gene reduces both basal and multihormone-stimulated prolactin gene transcription. *J. Biol. Chem.* **264,** 431–436.

Desdouits, F., Cheetham, J. S., and Huang, H.-B., Kwon, Y.-G., da Cruz e Silva, E. F., Denefle, P., Ehrlich, M. E., Nairn, A. C., Greengard, P., and Girault, J. A. (1995) Mechanism of inhibition of protein phosphatase-1 by DARPP-32: studies with recombinant DARPP-32 and synthetic peptides. *Biochem. Biophys. Res. Commun.* **206,** 652–658.

Dildy-Mayfield, J. E., and Harris, R. A. (1995) Ethanol inhibits kainate responses of glutamate receptors expressed in *Xenopus* oocytes: role of calcium and protein kinase C. *J. Neurosci.* **15,** 3162–3171.

Doherty, P., Furness, J., Williams, E. J., and Walsh, F. S. (1994) Neurite outgrowth stimulated by the tyrosine kinase inhibitor herbimycin A requires activation of tyrosine kinases and protein kinase C. *J. Neurochem.* **62,** 2124–2131.

Dolphin, A. C. (1992) The effect of phosphatase inhibitors and agents increasing cyclic-AMP-dependent phosphorylation on calcium channel currents in cultured rat dorsal root ganglion neurones:

interaction with the effect of G protein activation. *Pflügers Arch.* **421,** 138–145.

Dorée, M., and Galas, S. (1994) The cyclin-dependent protein kinases and the control of cell division. *FASEB J.* **8,** 1114–1121.

Dostmann, W. R. G., Taylor, S. S., Genieser, H.-G., Jastorff, B., Doskeland, S. O., and Øgreid, D. (1990) Probing the cyclic nucleotide binding sites of cAMP-dependent protein kinase I and II with analogs of adenosine 3',5'-cyclic phosphorothioates. *J. Biol. Chem.* **265,** 10,484–10,491.

Dudley, D. T., Pang, L., Decker, S. J., Bridges, A. J., and Saltiel, A. R. (1995) A synthetic inhibitor of the mitogen-activated protein kinase cascade. *Proc. Natl. Acad. Sci. USA* **92,** 7686–7689.

Eboli, M. L., Mercanti, D., Ciotti, M. T., Aquino, A., and Castellani, L. (1994) Glutamate-induced protein phosphorylation in cerebellar granule cells: role of protein kinase C. *Neurochem. Res.* **19,** 1257–1264.

Edelman, A. M., Higgins, D. M., Bowman, C. L., Haber, S. N., Rabin, R. A., and Cho-Lee, J. (1992) Myosin light chain kinase is expressed in neurons and glia: immunoblotting and immunocytochemical studies. *Mol. Brain Res.* **14,** 27–34.

Eichholtz, T., de Bont, D. B. A., de Widt, J., Liskamp, R. M. J., and Ploegh, H. L. (1993) A myristoylated pseudosubstrate peptide, a novel protein kinase C inhibitor. *J. Biol. Chem.* **268,** 1982–1986.

Enan, E. and Matsumura, F. (1992) Specific inhibition of calcineurin by type II synaptic pyrethroid insecticides. *Biochem. Pharmacol.* **43,** 1777–1784.

Endo, S., Critz, S. D., Byrne, J. H., and Shenolikar, S. (1995) Protein phosphatases-1 regulates outward K^+ currents in sensory neurons of *Aplysia* californica. *J. Neurochem.* **64,** 1833–1840.

Enslen, H. and Soderling, T. R. (1994) Roles of calmodulin-dependent protein kinases and phosphatase in calcium-depenent transcription of immediate early genes. *J. Biol. Chem.* **33,** 20,872–20,877.

Enslen, H., Sun, P., Brickey, D., Soderling, T. H., Klamo, E., and Soderling, T. R. (1994) Characterization of Ca^{2+}/calmodulin-dependent protein kinase IV. Role in transcriptional regulation. *J. Biol. Chem.* **269,** 15,520–15,527.

Erneux, C., Van Sande, J., Jastorff, B., and Dumont, J. E. (1986) Modulation of cyclic AMP action in the dog thyroid by its agonist and antagonist *Sp*- and *Rp*-adenosine 3',5'-monophosphorothioate. *Biochem. J.* **234,** 193–197.

Evans, G. A., Garcia, G. G., Erwin, R., Howard, O. M., and Farrar, W. L. (1994) Pervanadate stimulates the effects of interleukin-2 (IL-2) in human T cells and provides evidence for the activation of two distinct tyrosine kinase pathways by IL-2. *J. Biol. Chem.* **269,** 23,407–23,412.

Fackler, M. J., Civin, C. I., and May, W. S. (1992) Up-regulation of surface CD34 is associated with protein kinase C-mediated hyperphosphorylation of CD34. *J. Biol. Chem.* **267,** 17,540–17,546.

Fantl, W. J., Johnson, D. E., and Williams, L. T. (1993) Signalling by receptor tyrosine kinases. *Ann. Rev. Biochem.* **62,** 453–481.

Feinstein, D. L., Galea, E., Cermak, J., Chugh, P., Lyandvert, L., and Reis, D. J. (1994) Nitric oxide synthase expression in glial cells: suppression by tyrosine kinase inhibitors. *J. Neurochem.* **62,** 811–814.

Felipo, V., Minana, M. D., and Grisolia, S. (1993) Inhibitors of protein kinase C prevent the toxicity of glutamate in primary neuronal cultures. *Brain Res.* **604,** 192–196.

Fernandez, A., Mery, J., Vandromme, M., Basset, M., and Cavadore, J.-C. (1991) Effective intracellular inhibition of the cAMP-dependent protein kinase by microinjection of a modified form of the specific inhibitor peptide PKI in living fibroblasts. *Exp. Cell Res.* **195,** 468–477.

ffrench-Mullen, J. M., Danks, P., and Spence, K. T. (1994) Neurosteroids modulate calcium currents in hippocampal CA1 neurons via a pertussis toxin-sensitive G-protein-coupled mechanism. *J. Neurosci.* **14,** 1963–1977.

Figurov, A., Boddeke, H., and Muller, D. (1993) Enhancement of AMPA-mediated synaptic transmission by the protein phosphatase inhibitor calyculin A in rat hippocampal slices. *Eur. J. Neurosci.* **5,** 1035–1041.

Foulkes, J. G., Strada, S. J., Henderson, P. J., and Cohen, P. (1983) A kinetic analysis of the effects of inhibitor-1 and inhibitor-2 on the activity of protein phosphatase-1. *Eur. J. Biochem.* **132,** 309–313.

Fry, D. W., Kraker, A. J., McMichael, A., Ambroso, L. A., Nelson, J. M., Leopold, W. R., Connors, R. W., and Bridges, A. J. (1994) A specific inhibitor of the epidermal growth factor receptor tyrosine kinase. *Science* **265,** 1093–1095.

Funauchi, M., Haruta, H., and Tsumoto, T. (1994) Effects of an inhibitor for calcium/calmodulin-dependent protein phosphatase, calcineurin, on induction of long-term potentiation in rat visual cortex. *Neurosci. Res.* **19,** 269–278.

Gadbois, D. M., Hamaguchi, J. R., Swank, R. A., and Bradbury, E. M. (1992) Staurosporine is a potent inhibitor of p34cdc2 and p34cdc2-like kinases. *Biochem. Biophys. Res. Commun.* **184,** 80–85.

Garcia-Morales, P., Minami, Y., Luong, E., Klausner, R. D., and Smaelsin, L. E. (1990) Tyrosine phosphorylation in T cells is regulated by phosphatase activity: studies with phenylarsine oxide. *Proc. Natl. Acad. Sci. USA* **87,** 9255–9259.

Garver, T. D., Oyler, G. A., Harris, K. A., Polavarapu, R., Damuni, Z., Lehman, R. A. W., and Billingsley, M. L. (1995) Tau phosphorylation in brain slices: pharmacological evidence for convergent effects of protein phosphatases on tau and mitogen-activated protein kinase. *Mol. Pharmacol.* **47,** 745–756.

Gazit, A., Yaish, P., Gilon, C., and Levitazki, A. (1989) Tyrphostins I: synthesis and biological activity of protein tyrosine kinase inhibitors. *J. Med. Chem.* **32,** 2344–2352.

Gazit, A., Osherov, N., Posner, I., Yaish, P., Poradosu, E., Gilon, C., and Levitzki, A. (1991) Tyrphostins. 2. Heterocyclic and α-substituted benzylidenemalononitrile tyrphostins as potent inhibitors of EGF receptor and ErbB2/neu tyrosine kinases. *J. Med. Chem.* **34,** 1896–1907.

Gjertsen, B. T., Mellgren, G., Otten, A., Maronde, E., Genieser, H.-G., Jastorff, B., Vintermyr, O. K., McKnight, G. S., and Doskeland, S. O. (1995) Novel (*R*p)-cAMPS analogs as tools for inhibition of cAMP-kinase in cell culture. *J. Biol. Chem.* **270,** 20,599–20,607.

Glass, D. B., Cheng, H.-C., Kemp, B. E., and Walsh, D. A. (1986) Differential and common recognition of the catalytic sites of cGMP-dependent and cAMP-dependent protein kinases by inhibitory peptides derived from the heat-stable inhibitor protein. *J. Biol. Chem.* **261,** 12,166–12,171.

Glass, D. B., Cheng, H.-C., Mende-Mueller, L., Reed, J., and Walsh, D. A. (1989) Primary structural determinants essential for potent inhibition of cAMP-dependent protein kinase by inhibitory peptides corresponding to the active portion of the heat-stable inhibitor protein. *J. Biol. Chem.* **264,** 8802–8810.

Golard, A., Role, L., and Siegelbaum, S. A. (1994) Substance P potentiates calcium channel modulation by somatostatin in chick sympathetic ganglia. *J. Neurophysiol.* **72,** 2683–2690.

Goldberg, D. J. and Wu, D. Y. (1995) Inhibition of formation of filopodia after axotomy by inhibitors of protein tyrosine kinases. *J. Neurobiol.* **27,** 553–560.

Golowash, J., Paupardin-Tritsch, D., and Gerschenfeld, H. M. (1995) Enhancement by muscarinic agonists of a high voltage-activated Ca^{2+} current via phosphorylation in a snail neuron. *J. Physiol.* **485,** 21–28.

Gong, M. C., Cohen, P., Kitazawa, T., Ikebe, M., Masuo, M., Somlyo, A. P., and Somlyo, A. V. (1992) Myosin light chain phosphatase activities and the effects of phosphatase inhibitors in tonic and phasic smooth muscle. *J. Biol. Chem.* **267,** 14,662–14,668.

Gordon, J. A. (1991) Use of vanadate as protein-tyrosine phosphatase inhibitor. *Methods Enzymol.* **201,** 477–482.

Gordon, S. E., Brautigan, D. L., and Zimmerman, A. L. (1992) Protein phosphatases modulate the apparent agonist affinity of the light-regulated ion channel in retinal rods. *Neuron* **9,** 739–748.

Grider, J. R. (1993) Interplay of VIP and nitric oxide in regulation of the descending relaxation phase of peristalsis. *Am. J. Physiol.* **264,** G334–G340.

Griffith, L. C., Verselis, L. M., Aitken, K. M., Kyriacou, C. P., Danho, W., and Greenspan, R. J. (1993) Inhibition of calcium/calmodulin-dependent protein kinase in Drosophila disrupts behavioral plasticity. *Neuron* **10,** 501–509.

Groschner, K., Schuhmann, K., Baumgartner, W., Pastushenko, V., Schindler, H., and Romanin, C. (1995) Basal dephosphorylation controls slow gating of L-type Ca²⁺ channels in human vascular smooth muscle. *FEBS Lett.* **373**, 30–34.

Grove, J. R., Deutsch, P. J., Price, D. J., Habener, J. F., and Avruch, J. (1989) Plasmids encoding PKI(1-31), a specific inhibitor of cAMP-stimulated gene expression, inhibit the basal transcriptional activity of some but not all cAMP-regulated DNA response elements in JEG-3 cells. *J. Biol. Chem.* **264**, 19,506–19,513.

Gschwendt, M., Leibersperger, H., Kittstein, W., and Marks, F. (1992) Protein kinase Cζ and η in murine epidermis. TPA induces down-regulation of PKCη but not PKCζ. *FEBS Lett.* **307**, 151–155.

Gschwendt, M., Fürstenberger, G., Leibersperger, H., Kittstein, W., Lindner, D., Rudolph, C., Barth, H., Kleinschroth, J., Marmé, D., Schächtele, C., and Marks, F. (1995) Lack of an effect of novel inhibitors with high specificity for protein kinase C on the action of the phorbol ester 12-O-tetradecanoylphorbol-13-acetate on mouse skin in vivo. *Carcinogenesis* **16**, 107–111.

Gutierrez, L. M., Viniegra, S., Quintanar, J. L., Reig, J. A., and Sala, F. (1994) Calyculin A blocks bovine chromaffin cell calcium channels independently of phosphatase inhibition. *Neurosci. Lett.* **178**, 55–58.

Haavik, J., Schelling, D. L., Campbell, D. G., Andersson, K. K., Flatmark, T., and Cohen, P. (1989) Identification of protein phosphatase 2A as the major tyrosine hydroxylase phosphatase in adrenal medulla and corpus striatum: evidence from the effects of okadaic acid. *FEBS Lett.* **251**, 36–42.

Hack, N., Hidaka, H., Wakefield, M. J., and Balázs, R. (1993) Promotion of granule cell survival by high K⁺ or excitatory amino acid treatment and Ca²⁺/calmodulin-dependent protein kinase activity. *Neuroscience* **57**, 9–20.

Hajimohammadreza, I., Probert, A. W., Coughenour, L. L., Borosky, S. A., Marcoux, F. W., Boxer, P. A., and Wang, K. K. W. (1995) A specific inhibitor of calcium/calmodulin-dependent protein kinase-II provides neuroprotection against NMDA- and hypoxia/hypoglycemia-induced cell death. *J. Neurosci.* **15**, 4093–4104.

Han, Y. F. and Dokas, L. A. (1991) Okadaic acid-induced inhibition of B-50 dephosphorylation by presynaptic membrane-associated protein phosphatases. *J. Neurochem.* **57**, 1325–1331.

Hanks, S. K., Quinn, A. M., and Hunter, T. (1988) The protein kinase family: conserved features and deduced phylogeny of the catalytic domains. *Science* **241**, 42–52.

Hannun, Y. A., Merrill, A. H., Jr., and Bell, R. M. (1991) Use of sphingosine as inhibitor of protein kinase C. *Methods Enzymol.* **201**, 316–328.

Hardie, D. G., Haystead, T. A. J., and Sim, A. T. R. (1991) Use of okadaic acid to inhibit protein phosphatases in intact cells. *Methods Enzymol.* **201**, 469–477.

Hashimoto, Y., Perrino, B. A., and Soderling, T. R. (1990) Identification of an autoinhibitory domain in calcineurin. *J. Biol. Chem.* **265,** 1924–1927.

Helmeste, D. M., and Tang, S. W. (1994) Kinase inhibitors compete with imipramine for binding and inhibition of serotonin transport. *Eur. J. Pharmacol.* **267,** 239–242.

Hemmings, H. C., Jr. and Adamo, A. I. B. (1996) Activation of endogenous protein kinase C by halothane in synaptosomes. *Anesthesiology* **84,** 652–662.

Hemmings, H. C., Jr., Greengard, P., Tung, H. Y. L., and Cohen, P. (1984) DARPP-32, a dopamine-regulated neuronal phosphoprotein, is a potent inhibitor of protein phosphatase-1. *Nature* **310,** 503–505.

Hemmings, H. C., Jr., Nairn, A. C., Elliott, J. I., and Greengard, P. (1990) Synthetic peptide analogs of DARPP-32 (M_r 32,000 dopamine- and cAMP-regulated phosphoprotein), an inhibitor of protein phosphatase-1: phosphorylation, dephosphorylation and inhibitory activity. *J. Biol. Chem.* **265,** 20,369–20,376.

Hemmings, H. C., Jr., Desdouits, F., and Girault, J.-A. (1995) DARPP-32 (dopamine- and cyclic AMP-regulated phosphoprotein, M_r 32,000), a neuronal protein phosphatase-1 inhibitor: preparation and biochemical analysis. *Neuroprotocols* **6,** 35–45.

Herbert, J. M., Augereau, J. M., Gleye, J., and Maffrand, J. P. (1990) Chelerythrine is a potent and specific inhibitor of protein kinase C. *Biochem. Biophys. Res. Commun.* **172,** 993–999.

Herzig, S., Meier, A., Pfeiffer, M., and Neumann, J. (1995) Stimulation of protein phosphatases as a mechanism of the muscarinic-receptor-mediated inhibition of cardiac L-type Ca^{2+} channels. *Pflügers Arch.* **429,** 531–538.

Hidaka, H. and Kobayashi, R. (1992) Pharmacology of protein kinase inhibitors. *Ann. Rev. Pharmacol. Toxicol.* **32,** 377–397.

Hingtgen, C. M. and Vasko, M. R. (1994) The phosphatase inhibitor, okadaic acid, increases peptide release from rat sensory neurons in culture. *Neurosci. Lett.* **178,** 135–138.

Hoffmann, F., Bechtel, P. J., and Krebs, E. G. (1977) Concentrations of cyclic AMP-dependent protein kinase subunits in various tissues. *J. Biol. Chem.* **252,** 1441–1447.

Honkanen, R. E. (1993) Cantharidin, another natural toxin that inhibits the activity of serine/threonine protein phosphatases types 1 and 2A. *FEBS Lett.* **330,** 283–286.

Honkanen, R. E., Zwillers, J., Moore, R. E., Daily, S. L., Khatra, B. S., Dukelow, M., and Boynton, A. L. (1990) Characterization of microcystin-LR, a potent inhibitor of type 1 and type 2A protein phosphatases. *J. Biol. Chem.* **265,** 19,401–19,404.

Hsu, C.-Y. J., Persons, P. E., Spada, A. P., Bednar, R. A., Levitzki, A., and Zilberstein, A. (1991) Kinetic analysis of the inhibition of the epidermal growth factor receptor tyrosine kinase by Lavendustin-A and its analogue. *J. Biol. Chem.* **266**, 21,105–21,112.

Hu, B. R., Yang, Y. B., and Wieloch, T. (1993) Depression of neuronal protein synthesis initiation by protein tyrosine kinase inhibitors. *J. Neurochem.* **61**, 1789–1794.

Hug, H. and Sarre, T. F. (1993) Protein kinase C isoforms: divergence in signal transduction? *Biochem. J.* **291**, 329–343.

Huggenvik, J. I., Collard, M. W., Stofko, R. S., Seasholtz, A. F., and Uhler, M. D. (1991) Regulation of the human enkephalin promotor by two isoforms of the catalytic subunit of cyclic adenosine 3',5'-monophosphate-dependent protein kinase. *Mol. Endocrinol.* **5**, 921–930.

Hvalby, Ø., Hemmings, H. C., Jr., Paulsen, O., Czernik, A. J., Nairn, A. C., Godfraind, J.-M., Jensen, V., Raastad, M., Strom, J. F., Andersen, P., and Greengard, P. (1994) Specificity of protein kinase inhibitor peptides and induction of long-term potentiation. *Proc. Natl. Acad. Sci. USA* **91**, 4761–4765.

Ichinose, M., Endo, S., Critz, S. D., Shenolikar, S., and Byrne, J. H. (1990) Microcystin-LR, a potent phosphatase inhibitor, prolongs the serotonin- and cAMP-induced currents in sensory neurons of *Aplysia californica*. *Brain Res.* **533**, 137–140.

Ishida, A., Kameshita, I., Okuno, S., Kitani, T., and Fujisawa, H. (1995) A novel highly specific and potent inhibitor of calmodulin-dependent protein kinase II. *Biochem. Biophys. Res. Commun.* **212**, 806–812.

Ishihara, H., Martin, B. L., Brautigan, D. L., Karaki, H., Ozaki, H., Kato, Y., Fusetani, N., Watabe, S., Hashimoto, K., Uemura, D., and Hartshorne, D. J. (1989) Calyculin A and okadaic acid: inhibitors of protein phosphatase activity. *Biochem. Biophys. Res. Commun.* **159**, 871–877.

Ishii, H., Jirousek, M. R., Koya, D., Takagi, C., Xia, P., Clermont, A., Bursell, S.-E., Kern, T. S., Ballas, L. M., Heath, W. F., Stramm, L. E., Feener, E. P., and King, G. L. (1996) Amelioration of vascular dysfunctions in diabetic rats by an oral PKCβ inhibitor. *Science* **272**, 728–731.

Ishii, A., Kiuchi, K., Kobayashi, R., Sumi, M., Hidaka, H., and Nagatsu, T. (1991) A selective Ca^{2+}/calmodulin-dependent protein kinase II inhibitor, KN-62, inhibits the enhanced phosphorylation and the activation of tyrosine hydroxylase by 56 mM K^+ in rat pheochromocytoma PC12h cells. *Biochem. Biophys. Res. Commun.* **156**, 1051–1056.

Ishikawa, T., Chijiwa, T., Hagiwara, M., Mamiya, S., Saitoh, M., and Hidaka, H. (1988) ML-9 inhibits the vascular contraction via the inhibition of myosin light chain phosphorylation. *Mol. Pharmacol.* **33**, 598–603.

Ito, I., Hidaka, H., and Sugiyama, H. (1991) Effects of KN-62, a specific inhibitor of calcium/calmodulin-dependent protein kinase II, on long-term potentiation in the rat hippocampus. *Neurosci. Lett.* **121**, 119–121.

Itoh, T., Ikebe, M., Kargacin, G. J., Hartshorne, D. J., Kemp, B. E., and Fay, F. S. (1989) Effect of modulators of myosin light-chain kinase activity in single smooth muscle cells. *Nature* **338,** 164–168.

Izquierdo, M., Leevers, S. J., Williams, D. H., Marshall, C. J., Weiss, A., and Cantrell, D. (1994) The role of protein kinase C in the regulation of extracellular signal-regulated kinase by the T cell antigen receptor. *Eur. J. Immunol.* **24,** 2462–2468.

Jian, X., Hidaka, H., and Schmidt, J. T. (1994) Kinase requirement for retinal growth cone motility. *J. Neurobiol.* **25,** 1310–1328.

Kase, H., Iwahashi, K., Nakanishi, S., Matsuda, Y., Yamada, K., Takahashi, M., Murakata, C., Sato, A., and Kaneko, M. (1987) K-252 compounds, novel and potent inhibitors of protein kinase C and cyclic nucleotide-dependent protein kinases. *Biochem. Biophys. Res. Commun.* **142,** 436–440.

Kedzierski, W., Aguila-Mansilla, N., Kozlowski, G. P., and Porter, J. C. (1994) Expression of tyrosine hydroxylase gene in cultured hypothalamic cells: roles of protein kinase A and C. *J. Neurochem.* **62,** 431–437.

Keller, H. U. and Niggli, V. (1993) The PKC-inhibitor Ro 31–8220 selectively suppresses PMA- and diacylglycerol-induced fluid pinocytosis and actin polymerization in PMNS. *Biochem. Biophys. Res. Commun.* **194,** 1111–1116.

Kemp, B. E. and Pearson, R. B. (1991) Design and use of peptide substrates for protein kinases. *Methods Enzymol.* **200,** 121–134.

Kemp, B. E., Pearson, R. B., and House, C. M. (1991) Pseudosubstrate-based peptide inhibitors. *Methods Enzymol.* **201,** 287–304.

Keyser, D. O. and Alger, B. E. (1990) Arachidonic acid modulates hippocampal calcium current via protein kinase C and oxygen radicals. *Neuron* **5,** 545–553.

Kim, K.-S., Tinti, C., Song, B., Cubells, J. F., and Joh, T. H. (1994) Cyclic AMP-dependent protein kinase regulates basal and cyclic AMP-stimulated but not phorbol ester-stimulated transcription of the tyrosine hydroxylase gene. *J. Neurochem.* **63,** 834–842.

Kindy, M. S. (1993) Inhibition of tyrosine phosphorylation prevents delayed neuronal death following cerebral ischemia. *J. Cereb. Blood Flow Metab.* **13,** 372–377.

Kitagawa, M., Okabe, T., Ogino, H., Matsumoto, H., Suzuki-Takahashi, I., Kokubo, T., Higashi, H., Saitoh, S., Taya, Y., Yasuda, H., Ohba, Y., Nishimura, S., Tanaka, N., and Okuyama, A. (1993) Butyrolactone I, a selective inhibitor of cdk2 and cdc2 kinase. *Oncogene* **8,** 2425–2432.

Klann, E., Chen, S. J., and Sweatt, J. D. (1991) Persistent protein kinase activation in the maintenance phase of long-term potentiation. *J. Biol. Chem.* **266,** 24,253–24,256.

Klann, E., Chen, S. J., and Sweatt, J. D. (1992) Increased phosphorylation of a 17-kDa protein kinase C substrate (P-17) in long-term potentiation. *J. Neurochem.* **58,** 1576–1579.

Klee, C. B., Draetta, G. F., and Hubbard, M. J. (1988) *Calcineurin. Adv. Enzymol.* **61**, 149–200.

Knüsel, B. and Hefti, F. (1992) K-252 compounds: modulators of neurotrophin signal transduction. *J. Neurochem.* **59**, 1987–1996.

Kobayashi, E., Nakano, H., Morimoto, M., and Tamaoki, T. (1989) Calphostin C (UCN-1028C), a novel microbial compound, is a highly potent and specific inhibitor of protein kinase C. *Biochem. Biophys. Res. Commun.* **159**, 548–553.

Kolaj, M., Cerne, R., Cheng, G., Brickley, D. A., and Randic, M. (1994) Alpha subunit of calcium/calmodulin-dependent protein kinase enhances excitatory amino acid and synaptic responses of rat spinal dorsal horn neurons. *J. Neurophysiol.* **72**, 2525–2531.

Kovalenko, M., Gazit, A., Böhmer, A., Rorsman, C., Rönnstrand, L., Heldin, C.-H., Waltenberger, J., Böhmer, F.-D., and Levitzki, A. (1995) Selective platelet-derived growth factor receptor kinase blockers reverse *sis*-transformation. *Cancer Res.* **54**, 6106–6114.

Kumakura, K., Sasaki, K., Sakurai, T., Ohara-Imaizumi, M., Misonou, H., Nakamura, S., Matsuda, Y., and Nonomura, Y. (1994) Essential role of myosin light chain kinase in the mechanism for MgATP-dependent priming of exocytosis in adrenal chromaffin cells. *J. Neurosci.* **14**, 7695–7703.

Kurino, M., Fukunage, K., Ushio, Y., and Miyamoto, E. (1995) Activation of mitogen-activated protein kinase in cultured rat hippocampal neurons by stimulation of glutamte receptors. *J. Neurochem.* **65**, 1282–1289.

Le Cabec, V. and Maridonneau-Parini, I. (1995) Complete and reversible inhibition of NADPH oxidase in human neutrophils by phenylarsine oxide at a step distal to membrane translocation of the enzyme subunits. *J. Biol. Chem.* **270**, 2967–2073.

Levitzki, A. and Gazit, A. (1995) Tyrosine kinase inhibition: an approach to drug development. *Science* **267**, 1782–1788.

Lew, J. and Wang, J. H. (1995) Neuronal cdc2-like kinase. *Trends Biochem. Sci.* **20**, 33–37.

Li, G., Hidaka, H., and Wollheim, C. B. (1992) Inhibition of voltage-gated Ca^{2+} channels and insulin secretion in HIT cells by the Ca^{2+}/calmodulin-dependent protein kinase II inhibitor KN-62: comparison with antagonists of calmodulin and L-type Ca^{2+} channels. *Mol. Pharmacol.* **42**, 489–498.

Li, Y.-M., MacKintosh, C., and Casida, J. E. (1993) Protein phosphatase 2A and its [³H]cantharidin/[³H]endothall thioanhydride binding site: inhibitor specificity of cantharidin and ATP analogues. *Biochem. Pharmacol.* **46**, 1435–1443.

Liao, K., Hoffman, R. D., and Lane, M. D. (1991) Phosphotyrosyl turnover in insulin signaling. Characterization of two membane-bound pp15 protein tyrosine phosphatases from 3T3–L1 adipocytes. *J. Biol. Chem.* **266**, 6544–6553.

Lieberman, D. N. and Mody, I. (1994) Regulation of NMDA channel function by endogenous Ca^{2+}-dependent phosphatase. *Nature* **369,** 235–239.

Linden, D. J. and Conner, J. A. (1991) Participation of postsynaptic PKC in cerebellar long-term depression in culture. *Science* **254,** 1656–1659.

Liu, J., Farmer, J. D., Jr., Lane, W. S., Friedman, J., Weissman, I., and Schreiber, S. L. (1991) Calcineurin is a common target of cyclophilin-cyclosporin A and FKBP-FK506 complexes. *Cell* **66,** 807–815.

Liu, J.-P., Sim, A. T. R., and Robinson, P. J. (1994) Calcineurin inhibition of dynamin I GTPase activity coupled to nerve terminal depolarization. *Science* **265,** 970–973.

Lopez-Molina, L., Boddeke, H., and Muller, D. (1993) Blockade of long-term potentiation and of NMDA receptors by the protein kinase C antagonist calphostin C. *Naunyn Schmiedebergs Arch. Pharmacol.* **348,** 1–6.

Losiewicz, M. D., Carlson, B. A., Kaur, G., Sausville, E. A., and Worland, P. J. (1994) Potent inhibition of cdc2 kinase activity by the flavonoid L86–8275. *Biochem. Biophys. Res. Commun.* **201,** 589–595.

MacKintosh, C. and Klumpp, S. (1990) Tautomycin from the bacterium *Streptomyces verticillatus,* another potent and specific inhibitor or protein phosphatases 1 and 2A. *FEBS Lett.* **277,** 137–140.

MacKintosh, C. and MacKintosh, R. W. (1994) Inhibitors of protein kinases and phosphatases. *Trends Biochem. Sci.* **19,** 444–448.

MacKintosh, C., Beattie, K. A., Klumpp, S., Cohen, P., and Codd, G. A. (1990) Cyanobacterial microcystin-LR is a potent and specific inhibitor of protein phosphatases 1 and 2A from both manmmals and higher plants. *FEBS Lett.* **264,** 187–192.

Mahon, T. M. and O'Neill, L. A. J. (1995) Studies into the effect of the tyrosine kinase inhibitor herbimycin A on NF-κB activation in T lymphocytes. *J. Biol. Chem.* **270,** 28,557–28,564.

Majumdar, S., Kane, L. H., Rossi, M. W., Volpp, B. D., Nauseef, W. M., and Korchak, H. M. (1993) Protein kinase C isotypes and signal-transduction in human neutrophils: selective substrate specificity of calcium-dependent beta-PKC and novel calcium-independent nPKC. *Biochim. Biophys. Acta* **1176,** 276–286.

Malenka, R. C., Kauer, J. A., Perkel, D. J., Mauk, M. D., and Kelly, P. T. (1989) An essential role for postsynaptic calmodulin and protein kinase activity in long-term potentiation. *Nature* **340,** 554–557.

Malinow, R., Schulman, H., and Tsien, R. W. (1989) Inhibition of postsynaptic PKC or CaMKII blocks induction but not expression of LTP. *Science* **245,** 862–866.

Mamiya, N., Goldenring, J. R., Tsunoda, Y., Modlin, I. M., Yasui, K., Usuda, N., Ishikawa, T., Natsume, A., and Haidaka, H. (1993) Inhibition of acid secretion in gastric parietal cells by the Ca^{2+}/calmodulin-dependent protein kinase II inhibitor KN-93. *Biochem. Biophys. Res. Commun.* **195,** 608–615.

Margolis, B., Rhee, S. G., Felder, S., Mervic, M., Lyall, R., Levitzki, A., Ullrich, A., Zilberstein, A., and Schlessinger, J. (1989) EGF induces tyrosine phosphorylation of phospholipase C-II: a potential mechanism for EGF receptor signaling. *Cell* **57**, 1101–1107.

Markovits, J., Larsen, A. K., Segal-Bendirdjian, E., Fosse, P., Saucier, J. M., Gazit, A., Levitzki, A., Umezawa, K., and Jacquemin-Sablon, A. (1994) Inhibition of DNA topoisomerases I and II and induction of apoptosis by erbstatin and tyrphostin derivatives. *Biochem. Pharmacol.* **48**, 549–560.

Martiny-Baron, G., Kazanietz, M. G., Mischak, H., Blumberg, P. M., Kochs, G., Hug, H., Marmé, D., and Schächtele, C. (1993) Selective inhibition of protein kinase C isozymes by the indolocarbazole Gö 6976. *J. Biol. Chem.* **268**, 9194–9197.

Mathis, C., Lehmann, J., and Ungerer, A. (1992) The selective protein kinase C inhibitor, NPC 15437, induces specific deficits in memory retention in mice. *Eur. J. Pharmacol.* **220**, 1101–1107.

Matsushima, R., Yoshizawa, S., Watanabe, M. F., Harada, K., Furusawa, M., Carmichael, W. W., and Fujiki, H. (1990) *In vitro* and *in vivo* effects of protein phosphatase inhibitors, microcystins and nodularin, on mouse skin and fibroblasts. *Biochem. Biophys. Res. Commun.* **171**, 867–874.

Mattiazzi, A., Hove-Madsen, L., and Bers, D. M. (1994) Protein kinase inhibitors reduce SR Ca^{2+} transport in permeabilized myocytes. *Am. J. Physiol.* **267**, H812-H820.

McCarron, J. G., McGeown, J. G., Reardon, S., Ikebe, M., Fay, F. S., and Walsh, J. V., Jr. (1992) Calcium-dependent enhancement of calcium current in smooth muscle by calmodulin-dependent protein kinase II. *Nature* **357**, 74–77.

Meydan, N., Grunberger, T., Dadi, H., Shahar, M., Arpaia, E., Lapidot, Z., Leeder, J. S., Freedman, M., Cohen, A., Gazit, A., Levitzki, A., and Roifman, C. M. (1996) Inhibition of acute lymphoblastic leukaemia by a Jak-2 inhibitor. *Nature* **379**, 645–648.

Miller, D. R., Lee, G. M., and Maness, P. F. (1993) Increased neurite outgrowth induced by inhibition of protein tyrosine kinase activity in PC12 pheochromocytoma cells. *J. Neurochem.* **60**, 2134–2144.

Mizuno, K., Noda, K., Udea, Y., Hanaki, H., Saido, T. C., Ikuta, T., Kuroki, T., Tamaoki, T., Hirai, S.-I., Osada, S., and Ohno, S. (1995) UCN-01, an anti-tumor drug, is a selective inhibitor of the conventional PKC family. *FEBS Lett.* **359**, 259–261.

Mochida, S., Nonomura, Y., and Kobayashi, H. (1994) Analysis of the mechanism for acetylcholine release at the synapse formed between rat sympathetic neurons in culture. *Microsc. Res. Tech.* **29**, 94–102.

Monteiro, H. P., Ivaschenko, Y., Fischer, R., and Stern, A. (1991) Inhibition of protein tyrosine phosphatase activity by diamide is reversed by epidermal growth factor in fibroblasts. *FEBS Lett.* **295**, 146–148.

Morgan, D. O. (1995) Principles of CDK regulation. *Nature* **374**, 131–134.

Muid, R. E., Dale, M. M., Davis, P. D., Elliott, L. H., Hill, C. H., Kumar, H., Lawton, G., Twomey, B. M., Wadsworth, J., Wilkinson, S. E., and Nixon, J. S. (1991) A novel conformationally restricted protein kinase C inhibitor, Ro 31-8425, inhibits human neutrophil superoxide generation by soluble, particulate and post-receptor stimuli. *FEBS Lett.* **293**, 169–172.

Mulkey, R. M., Herron, C. E., and Malenka, R. C. (1993) An essential role for protein phosphatases in hippocampal long-term depression. *Science* **261**, 1051–1055.

Mulkey, R. M., Endo, S., Shenolikar, S., and Malenka, R. C. (1994) Involvement of a calcineurin/inhibitor-1 phosphatase cascade in hippocampal long-term depression. *Nature* **369**, 486–488.

Müller, W., Petrozzino, J. J., Griffith, L. C., Danho, W., and Conner, J. A. (1992) Specific involvement of Ca^{2+}-calmodulin kinase II in cholinergic modulation of neuronal responsiveness. *J. Neurophysiol.* **68**, 2264–2269.

Murata, K., Sakon, M., Kambayashi, J., Yukawa, M., Yano, Y., Fujitani, K., Kawasaki, T., Shiba, E., and Mori, T. (1993) The possible involvement of protein phosphatase 1 in thrombin-induced Ca^{2+} influx of human platelets. *J. Cell Biochem.* **51**, 442–445.

Nagatsu, T., Suzuki, H., Kiuchi, K., Saitoh, M., and Hidaka, H. (1987) Effects of myosin light-chain kinase inhibitor on catecholamine secretion from rat pheochromocytoma PC12h cells. *Biochem. Biophys. Res. Commun.* **143**, 1045–1048.

Nairn, A. C. and Picciotto, M. R. (1994) Calcium/calmodulin-dependent protein kinases. *Cancer Biol.* **5**, 295–303.

Nairn, A. C. and Shenolikar, S. (1992) The role of protein phosphatases in synaptic transmission, plasticity and neuronal development. *Curr. Opinion Neurobiol.* **2**, 296–301.

Nakanishi, S., Kakita, S., Takahashi, I., Kawahara, K., Tsukuda, E., Sano, T., Yamada, K., Yoshida, M., Kase, H., and Matsuda, Y. (1992) Wortmannin, a microbial product inhibitor of myosin light chain kinase. *J. Biol. Chem.* **267**, 2157–2163.

Nakazawa, K., Mikawa, S., Hashikawa, T., and Ito, M. (1995) Transient and persistent phosphorylation of AMPA-type glutamate receptor subunits in cerebellar Purkinje cells. *Neuron* **15**, 697–709.

Neumann, J., Herzig, S., Boknik, P., Apel, M., Kaspareit, G., Schmitz, W., Scholz, H., Tepel, M., and Zimmermann, N. (1995) On the cardiac contractile, biochemical and electrophysiological effects of cantharidin, a phosphatase inhibitor. *J. Pharmacol. Exp. Ther.* **274**, 530–539.

Nichols, R. A., Sihra, T. S., Czernik, A. J., Nairn, A. C., and Greengard, P. (1990) Calcium/calmodulin-dependent protein kinase II increases glutamate and noradrenaline release from synaptosomes. *Nature* **343**, 647–651.

Nichols, R. A., Suplick, G. R., and Brown, J. M. (1994) Calcineurin-mediated protein dephosphorylation in brain nerve terminals regulates the release of glutamate. *J. Biol. Chem.* **269,** 23,817–23,823.

Nichols, R. A., Suplick, G. R., and Golemme, S. S. (1996) Inhibition of phosphatases 1 and 2A reduces stimulation-induced glutamate release from synaptosomes by inhibiting Ca^{2+} influx. *J. Neurochem.,* in press.

Niki, I., Okazaki, K., Saitoh, M., Niki, A., Niki, H., Tamagawa, T., Iguchi, A., and Hidaka, H. (1993) Presence and possible involvement of Ca^{2+} / calmodulin dependent protein kinases in insulin release from the rat pancreatic β cell. *Biochem. Biophys. Res. Commun.* **191,** 255–261.

Nixon, J. S., Bishop, J., Bradshaw, D., Davis, P. D., Hill, C. H., Elliott, L. H., Kumar, H., Lawton, G., Lewis, E. J., Mulqueen, M., Westmacott, D., Wadsworth, J., and Wilkinson, S. E. (1992) The design and biological properties of potent and selective inhibitors of protein kinase C. *Biochem. Soc. Trans.* **20,** 419–425.

Noronha, K. F. and Mercier, A. J. (1995) A role for calcium/calmodulin-dependent protein kinase in mediating synaptic modulation by a neuropeptide. *Brain Res.* **673,** 70–74.

Nye, S. H., Squinto, S. P., Glass, D. J., Stitt, T. N., Hantzopoulos, P., Macchi, M. J., Lindsay, N. S., Ip, N. Y., and Yancopoulos, G. D. (1992) K-252a and staurosporine selectively block autophosphorylation of neurotrophin receptors and neurotrophin-mediated responses. *Mol. Biol. Cell* **3,** 677–686.

O'Dell, T. J., Kandel, E. R., and Grant, S. G. N. (1991) Long-term potentiation in the hippocampus is blocked by tyrosine kinase inhibitors. *Nature* **353,** 558–560.

Ogiwara, T., Murdoch, G., Chik, C. L., and Ho, A. K. (1995) Tyrosine kinase inhibitors enhance cGMP production in rat pinealocytes. *Biochem. Biophys. Res. Commun.* **207,** 994–1002.

Øgreid, D., Dostmann, W., Genieser, H.-G., Niemann, P., Doskeland, S. O., and Jastorff, B. (1994) (Rp)- and (Sp)-8-piperidino-adenosine 3',5'-(cyclic)thiophosphatases discriminate completely between site A and B of the regulatory subunits of cAMP-dependent protein kinase type I and II. *Eur. J. Biochem.* **221,** 1089–1094.

Ohara-Imaizumi, M., Sakurai, T., Nakamura, S., Nakanishi, S., Matsuda, Y., Muramatsu, S., Nonmura, Y., and Kumakura, K. (1992) Inhibition of Ca^{2+}-dependent catecholamine release by myosin light chain kinase inhibitor, wortmannin, in adrenal chromaffin cells. *Biochem. Biophys. Res. Commun.* **185,** 1016–1021.

Ohmichi, M., Pang, L., Ribon, V., Gazit, A., Levitzki, A., and Saltiel, A. R. (1993) The tyrosine kinase inhibitor tyrphostin blocks the cellular actions of nerve growth factor. *Biochemistry* **32,** 4650–4658.

Okada, D. (1995) Protein kinase C modulates calcium sensitivity of nitric oxide synthase in cerebellar slices. *J. Neurochem.* **64,** 1298–1304.

Okazaki, K., Ishikawa, T., Inui, M., Tada, M., Goshima, K., Okamoto, T., and Hidaka, H. (1994) KN-62, a specific Ca^{2+}/calmodulin-dependent protein kinase inhibitor, reversibly depresses the rate of beating of cultured fetal mouse cardiac myocytes. *J. Pharmacol. Exp. Ther.* **270**, 1319–1324.

Olsen, S. R. and Uhler, M. D. (1991) Inhibition of protein kinase-A by overexpression of the cloned human protein kinase inhibitor. *Mol. Endocrinol.* **5**, 1246–1247.

Onoda, T., Linuma, H., Sasaki, Y., Hamada, M., Isshiki, K., Naganawa, H., Takeuchi, T., Tatsuta, K., and Umezawa, K. (1989) Isolation of a novel tyrosine kinase inhibitor, lavendustin A, from Streptomyces griseolavendus. *J. Nat. Prod.* **52**, 1252–1257.

Onozuka, M., Watanabe, K., Nagata, K.-Y., and Imai, S. (1994) Involvement of a Ca^{++}/calmodulin-dependent protein kinase II-associated mechanism in the induction of an outward potassium current by quisqualate. *Brain Res.* **650**, 336–340.

Osborne, P. B. and Williams, J. T. (1995) Characterization of acute homologous desensitization of μ-opioid receptor-induced currents in locus coeruleus neurons. *Br. J. Pharmacol.* **115**, 925–932.

Osherov, N. and Levitzki, A. (1994) Epidermal-growth-factor-dependent activation of the Src-family kinases. *Eur. J. Biochem.* **225**, 1047–1053.

Otero, A. D. S. and Sweitzer, N. M. (1993) Benzoquinoid tyrosine kinase inhibitors are potent blockers of cardiac muscarinic receptor function. *Mol. Pharmacol.* **44**, 595–604.

Pang, L., Sawada, T., Decker, S. J., and Saltiel, A. R. (1995) Inhibition of MAP kinase kinase blocks the differentiation of PC-12 cells induced by nerve growth factor. *J. Biol. Chem.* **270**, 13,585–13,588.

Park, D. S., Farinelli, S. E., and Greene, L. A. (1996) Inhibitors of cyclin-dependent kinases promote survival of post-mitotic neuronally differentiated PC12 cells and sympathetic neurons. *J. Biol. Chem.* **271**, 8161–8169.

Parsons, J. N., Wiederrecht, G. J., Salowe, S., Burbaum, J. J., Rokosz, L. L., Kincaid, R. L., and O'Keefe, S. J. (1994) Regulation of calcineurin phosphatase activity and interaction with the FK-506•FK-506 binding protein complex. *J. Biol. Chem.* **269**, 19,610–19,616.

Pavlakovic, G., Eyer, C. L., and Isom, G. E. (1995) Neuroprotective effects of PKC inhibition against chemical hypoxia. *Brain Res.* **676**, 205–211.

Pawson, T. (1995) Protein modules and signalling networks. *Nature* **373**, 573–579.

Payne, M. E., Fong, Y. L., Ono, R. J., Colbran, R. J., and Kemp, B. E. (1988) Calcium/calmodulin-dependent protein kinase II. Characterization of distinct calmodulin binding and inhibitory domains. *J. Biol. Chem.* **263**, 7190–7195.

Peng, H. B., Baker, L. P., and Dai, Z. (1993) A role of tyrosine phosphorylation in the formation of acetylcholine receptor clusters induced by electric fields in cultured *Xenopus* muscle cells. *J. Cell Biol.* **120**, 197–204.

Posner, I., Engel, M., Gazit, A., and Levitzki, A. (1993) Kinetics of inhibition by tyrphostins of the tyrosine kinase activity of the epidermal growth factor receptor and analysis by a new computer program. *Mol. Pharmacol.* **45,** 673–683.

Powis, G. (1991) Signalling targets for anticancer drug development. *TIPS* **12,** 188–194.

Powis, G., Bonjouklian, R., Berggren, M. M., Gallego, A., Abraham, R., Ashendel, C., Zalkow, L., Matter, W. F., Dodge, J., Grindey, G., and Vlahos, C. J. (1994) Wortmannin, a potent and selective inhibitor of phosphatidylinositol-3-kinase. *Cancer Res.* **54,** 2419–2423.

Quick, J., Ware, J. A., and Driedger, P. E. (1992) The structure and biological activities of the widely used protein kinase inhibitor, H7, differ depending on the commercial source. *Biochem. Biophys. Res. Commun.* **187,** 657-663.

Reber, B. F. and Bouron, A. (1995) Calyculin-A-induced fast neurite retraction in nerve growth factor-diferentiated rat pheochromocytoma (PC12) cells. *Neurosci. Lett.* **183,** 198–201.

Reeve, H. L., Vaughan, P. F., and Peers, C. (1995) Enhancement of Ca^{2+} channel currents in human neuroblastoma (SH-SY5Y) cells by phorbol esters with and without activation of protein kinase C. *Pflügers Arch.* **429,** 729–737.

Reinhart, P. H. and Levitan, I. B. (1995) Kinase and phosphatase activities intimately associated with a reconstituted calcium-dependent potassium channel. *J. Neurosci.* **15,** 4572–4579.

Reisine, T., Rougon, G., and Barbet, J. (1986) Liposome delivery of cyclic AMP-dependent protein kinase inhibitor into intact cells; specific blockade of cyclic AMP-mediated adrenocorticotropin release from mouse anterior pituitary tumor cells. *J. Cell Biol.* **102,** 1630–1637.

Reisine, T., Rougon, G., Barbet, J., and Affolter, H. U. (1985) Corticotropin-releasing factor-induced adrenocorticotropin hormone release and synthesis is blocked by incorporation of the inhibitor of cyclic AMP-dependent protein kinase into anterior pituitary tumor cells by liposomes. *Proc. Natl. Acad. Sci. USA* **82,** 8261–8265.

Robinson, P. J. (1992) Potencies of protein kinase C inhibitors are dependent on the activators used to stimulate the enzyme. *Biochem. Pharmacol.* **44,** 1325–1334.

Rodriguez, J., Quignard, J.-F., Fagni, L., Lafon-Cazal, M., and Bockaert, J. (1994) Blockade of nitric oxide synthesis by tyrosine kinase inhibitors in neurones. *Neuropharmacology* **33,** 1267–1274.

Rüegg, U. T. and Burgess, G. M. (1989) Staurosporine, K-252 and UCN-01: potent but nonspecific inhibitors of protein kinases. *TIPS* **10,** 218–220.

Saitoh, M., Naka, M., and Hidaka, H. (1986) The modulatory role of myosin light chain phosphorylation in human platelet activation. *Biochem. Biophys. Res. Commun.* **140,** 280–287.

Saitoh, M., Ishikawa, T., Matsushima, S., Naka, M., and Hidaka, H. (1987) Selective inhibition of catalytic activity of smooth muscle myosin light chain kinase. *J. Biol. Chem.* **262,** 7796–7801.

Sargeant, P., Farndale, R. W., and Sage, S. O. (1993) ADP- and thapsigargin-evoked Ca^{2+} entry and protein-tyrosine phosphorylation are inhibited by the tyrosine kinase inhibitors genistein and methyl-2,5-dihydroxycinnamate in fura-2–loaded human platelets. *J. Biol. Chem.* **268,** 18,151–18,156.

Sato, K.-I., Miki, S., Tachibana, H., Hayashi, F., Akiyama, T., and Fukami, Y. (1990) A synthetic peptide corresponding to residues 137 to 157 of pp60$^{v\text{-}src}$ inhibits tyrosine-specific protein kinases. *Biochem. Biophys. Res. Commun.* **171,** 1152–1159.

Shigeri, Y. and Fujimoto, M. (1994) Y2 receptors for neuropeptide Y are coupled to three intracellular signal transduction pathways in a human neuroblastoma cell line. *J. Biol. Chem.* **269,** 8842–8848.

Shimokado, K., Yokota, T., Umezawa, K., Sasaguri, T., and Ogata, J. (1994) Protein tyrosine kinase inhibitors inhibit chemotaxis of vascular smooth muscle cells. *Arteriosclerosis Thromb.* **14,** 973–981.

Sihra, T. S. and Pearson, H. A. (1995) Ca/Calmodulin-dependent kinase II inhibitor KN62 attenuates glutamate release by inhibiting voltage-dependent Ca^{2+}-channels. *Neuropharmacology* **34,** 731–741.

Sihra, T. S., Bogonez, E., and Nicholls, D. G. (1992) Localized Ca^{2+} entry preferentially effects protein dephosphorylation, phosphorylation, and glutamate release. *J. Biol. Chem.* **267,** 1983–1989.

Sim, A. T. R., Lloyd, H. G. E., Jarvie, P. E., Morrison, M., Rostas, J. A. P., and Dunkley, P. R. (1993) Synaptosomal amino acid release: effect of inhibiting protein phosphatases with okadaic acid. *Neuroscience Lett.* **160,** 181–184.

Sim, A. T. R., Ratcliffe, E., Mumby, M. C., Villa-Moruzzi, E., and Rostas, J. A. P. (1994) Differential activities of protein phosphatase types 1 and 2A in cytosolic and particulate fractions from rat forebrain. *J. Neurochem.* **62,** 1552–1559.

Simpson, C. S. and Morris, B. J. (1995) Induction of c-fos and zif/268 gene expression in rat striatal neurons, following stimulation of D_1–like dopamine receptors, involves protein kinase A and protein kinase C. *Neuroscience* **68,** 97–106.

Sitges, M., Dunkley, P. R., and Chiu, L. M. (1995) A role for calcium/calmodulin kinase(s) in the regulation of GABA exocytosis. *Neurochem. Res.* **20,** 245–252.

Smith, M. K., Colbran, R. J., and Soderling, T. R. (1990) Specificities of autoinhibitory domain peptides for four protein kinases: implications for intact cell studies of protein kinase function. *J. Biol. Chem.* **265,** 1837–1840.

Smith, M. K., Colbran, R. J., Brickey, D. A., and Soderling, T. R. (1992) Functional determinants in the autoinhibitory domain of calcium/

calmodulin-dependent protein kinase II: role of His[282] and multiple basic residues. *J. Biol. Chem.* **267,** 1761–1768.

Snell, L. D., Iorio, K. R., Tabakoff, B., and Hoffman, P. L. (1994) Protein kinase C activation attenuates N-methyl-D-aspartate-induced increases in intracellular calcium in cerebellar granule cells. *J. Neurochem.* **62,** 1783–1789.

Soderling, T. R. (1990) Protein kinases: regulation by autoinhibitory domains. *J. Biol. Chem.* **265,** 1823–1826.

Stone, R. L. and Dixon, J. E. (1994) Protein-tyrosine phosphatases. *J. Biol. Chem.* **269,** 31,323–31,326.

Sullivan, J. A., Merritt, J. E., Budd, J. M., Booth, R. F., and Hallam, T. J. (1994) Effect of a selective protein kinase C inhibitor, Ro 31-8425, on Mac-1 expression and adhesion of human neutrophils. *Eur. J. Immunol.* **24,** 621–626.

Sullivan, J. P., Connor, J. R., Tiffany, C., Shearer, B. G., and Burch, R. M. (1991) NPC 15437 interacts with the C1 domain of protein kinase C. An analysis using mutant PKC constructs. *FEBS Lett.* **285,** 120–123.

Sullivan, J. P., Connor, J. R., Shearer, B. G., and Burch, R. M. (1992) 2,6-Diamino-N-([1-(1-oxotridecyl)-2-piperidinyl] methyl)hexanamide (NPC 15437): a novel inhibitor of protein kinase C interacting at the regulatory domain. *Mol. Pharmacol.* **41,** 38–44.

Sumi, M., Kiuchi, K., Ishikawa, T., Ishii, A., Hagiwara, M., Nagatsu, T., and Hidaka, H. (1991) The newly synthesized selective Ca^{2+}/calmodulin-dependent protein kinase II inhibitor KN-93 reduces dopamine contents in PC12h cells. *Biochem. Biophys. Res. Commun.* **181,** 968–975.

Surmeier, D. J., Bargas, J., Hemmings, H. C., Jr., Nairn, A. C., and Greengard, P. (1995) Modulation of calcium currents by a D_1 dopaminergic protein kinase/phosphatase cascade in rat neostriatal neurons. *Neuron* **14,** 385–397.

Takano, K., Stanfield, P. R., Nakajima, S., and Nakajima, Y. (1995) Protein kinase C-mediated inhibition of an inward rectifier potassium channel by substance P in nucleus basalis neurons. *Neuron* **14,** 999–1008.

Tamaoki, T. (1991) Use and specificity of staurosporine, UCN-01, and calphostin C as protein kinase inhibitors. *Methods Enzymol.* **201,** 340–347.

Tan, S.-E., Wenthold, R. J., and Soderling, T. R. (1994) Phsophorylation of AMPA-type glutamate receptors by calcium/calmodulin-dependent protein kinase II and protein kinase C in cultured hippocampal neurons. *J. Neurosci.* **14,** 1123–1129.

Tanaka, C. and Nishizuka, Y. (1994) The protein kinase C family for neuronal signaling. *Ann. Rev. Neurosci.* **17,** 551–567.

Tanaka, H., Katagiri, M., Arima, S., Matsuzaki, K., Inokoshi, J., and Omura, S. (1995) Neuronal differentiation of Neuro 2a cells by lactacystin and its partial inhibition by the protein phosphatase inhibitors calyculin A and okadaic acid. *Biochem. Biophys. Res. Commun.* **216,** 291–297.

Tansey, M. G., Luby-Phelps, K., Kamm, K. E., and Stull, J. T. (1994) Ca²⁺-dependent phosphorylation of myosin light chain kinase decreases the Ca²⁺ sensitivity of light chain phosphorylation within smooth muscle cells. *J. Biol. Chem.* **269,** 9912–9920.

Teng, K. K. and Greene, L. A. (1994) KT5926 selectively inhibits nerve growth factor-dependent neurite elongation. *J. Neurosci.* **14,** 2624–2635.

Terbush, D. R. and Holz, R. W. (1990) Activation of protein kinase C is not required for exocytosis from bovine adrenal chromaffin cells. The effects of protein kinase C(19–31), Ca/CaM kinase II(291–317), and staurosporine. *J. Biol. Chem.* **265,** 21,179–21,184.

Tokimasa, T., Ito, M., Simmons, M. A., Schneider, C. R., Tanaka, T., Nakano, T., and Akasu, T. (1995) Inhibition by wortmannin of M-current in bullfrog sympathetic neurones. *Br. J. Pharmacol.* **114,** 489–495.

Tokumitsu, H., Chijiwa, T., Hagiwara, M., Mizutani, A., and Terasawa, M. (1990) KN-62, 1-(N,O-Bis[5-isoquinolinesulfonyl]-N-methyl-L-tyrosyl]-4-phenylpiperazine, a specific inhibitor of Ca²⁺/calmodulin-dependent protein kinase II. *J. Biol. Chem.* **265,** 4315–4320.

Tong, G. and Jahr, C. E. (1994) Regulation of glycine-insensitive desensitization of the NMDA receptor in outside-out patches. *J. Neurophysiol.* **72,** 754–761.

Torii, N., Kamishita, T., Otsu, Y., and Tsumoto, T. (1995) An inhibitor for calcineurin, FK506, blocks induction of long-term depression in rat visual cortex. *Neurosci. Lett.* **185,** 1–4.

Toullec, D., Pianetti, P., Coste, H., Bellevergue, P., Grand-Perret, T., Ajakane, M., Baudet, V., Boissin, P., Boursier, E., Loriolle, F., Duhamel, L., Charon, D., and Kirilovsky, J. (1991) The bis-indolylmaleimide GF 109203X is a potent and selective inhibitor of protein kinase C. *J. Biol. Chem.* **266,** 15,771–15,781.

Ueda, H., Miyamae, T., Hayashi, C., Watanabe, S., Fukushima, N., Sasaki, Y., Iwamura, T., and Misu, Y. (1995) Protein kinase C involvement in homologous desensitization of δ-opioid receptor coupled to Gᵢₗ-phospholipase C activation in *Xenopus* oocytes. *J. Neurosci.* **15,** 7485–7499.

Uehara, Y. and Fukazawa, H. (1991) Use and selectivity of herbimycin A as inhibitor of protein-tyrosine kinases. *Methods Enzymol.* **201,** 370–379.

Uehara, Y., Fukazawa, H., Murakami, Y., and Mizuno, S. (1989) Irreversible inhibition of v-*src* tyrosine kinase activity by herbimycin A and its abrogation by sulfhydryl compounds. *Biochem. Biophys. Res. Commun.* **163,** 803–809.

Umezawa, K. and Imoto, M. (1991) Use of erbstatin as protein-tyrosine kinase inhibitor. *Methods Enzymol.* **201,** 379–385.

Umezawa, K., Hori, T., Tajima, H., Imoto, M., Isshiki, K., and Takeuchi, T. (1990) Inhibition of epidermal growth factor-induced DNA synthesis by tyrosine kinase inhibitors. *FEBS Lett.* **260,** 198–200.

Umezawa, K., Sugata, D., Yamashita, K., Johtoh, N., and Shibuya, M. (1992) Inhibition of epidermal growth factor receptor functions by tyrosine kinase inhibitors in NIH3T3 cells. *FEBS Lett.* **314,** 289–292.

Valenzuela, C. F., Machu, T. K., McKernan, R. M., Whiting, P., Van Renterghem, B. B., McManaman, J. L., Brozowski, S. J., Smith, G. B., Olsen, R. W., and Harris, R. A. (1995) Tyrosine kinase phosphorylation of GABA$_A$ receptors. *Mol. Brain Res.* **31,** 165–172.

Vera, J. C., Reyes, A. M., Cárcamo, J. G., Velásquez, F. V., Rivas, C. I., Zhang, R. H., Strobel, P., Iribarren, R., Scher, H. I., Slebe, J. C., and Golde, D. W. (1996) Genistein is a natural inhibitor of hexose and dehydroascorbic acid transport through the glucose transporter, GLUT1. *J. Biol. Chem.* **271,** 8719–8724.

Veselý, J., Havlicek, L., Strnad, M., Blow, J. J., Donella-Deana, A., Pinna, L., Letham, D. S., Kato, J.-Y., Detivaud, L., Leclerc, S., and Meijer, L. (1994) Inhibition of cyclin-dependent kinases by purine analogues. *Eur. J. Biochem.* **224,** 771–786.

Vickroy, T. W., Malphurs, W. L., and Carriger, M. L. (1995) Regulation of stimulus-dependent hippocampal acetylcholine release by okadaic acid-sensitive phosphoprotein phosphatases. *Neurosci. Lett.* **191,** 200–204.

Victor, R. G., Thomas, G. D., Marban, E., and O'Rourke, B. (1995) Presynaptic modulation of cortical synaptic activity by calcineurin. *Proc. Natl. Acad. Sci. USA* **92,** 6269–6273.

Walaas, S. I. and Greengard, P. (1991) Protein phosphorylation and neuronal function. *Pharmacol. Rev.* **43,** 299–349.

Walker, D. L. and Gold, P. E. (1994) Intra-amygdala kinase inhibitors disrupt retention of a learned avoidance response in rats. *Neurosci. Lett.* **176,** 255–258.

Walsh, D. A. and Glass, D. B. (1991) Utilization of the inhibitor protein of adenosine cyclic monophosphate-dependent protein kinase, and peptides derived from it, as tools to study adenosine cyclic monophosphate-mediated cellular processes. *Methods Enzymol.* **201,** 304–316.

Walterova, D., Ulrichova, Preininger, V., Simanek, V., Lenfeld, J., and Lasovsky, J. (1981) Inhibition of liver alanine aminotransferase activity by some benzophenanthridine alkaloids. *J. Med. Chem.* **24,** 1100–1103.

Wang, J.-H. and Feng, D. P. (1992) Postsynaptic protein kinase C essential to induction and maintenance of long-term potentiation in the hippocampal CA1 region. *Proc. Natl. Acad. Sci. USA* **89,** 2576–2580.

Wang, J.-H. and Kelly, P. T. (1995) Postsynaptic injection of Ca^{2+}/CaM induces synaptic potentiation requiring CaMKII and PKC activity. *Neuron* **15,** 443–452.

Wang, L.-Y., Dudek, E. M., Browning, M. D., and MacDonald, J. F. (1994a) Modulation of AMPA/kainate receptors in cultured murine hippocampal neurones by protein kinase C. *J. Physiol.* **475,** 431–437.

Wang, L.-Y., Orser, B. A., Brautigan, D. L., and MacDonald, J. F. (1994b) Regulation of NMDA receptors in cultured hippocampal neurons by protein phosphatases 1 and 2A. *Nature* **369**, 230–232.

Wang, L.-Y., Salter, M. W., and MacDonald, J. F. (1991) Regulation of kainate receptors by cAMP-dependent protein kinase and phosphatases. *Science* **253**, 1132–1135.

Wang, S. S., Mathes, C., and Thompson, S. H. (1993) Membrane toxicity of the protein kinase C inhibitor calpohostin A by a free-radical mechanism. *Neurosci. Lett.* **157**, 25–28.

Wang, Y.-T. and Salter, M. W. (1994) Regulation of NMDA receptors by tyrosine kinases and phosphatases. *Nature* **369**, 233–235.

Wenham, R. M., Landt, M., Walters, S. M., Hidaka, H., and Easom, R. A. (1992) Inhibition of insulin secretion by KN-62, a specific inhibitor of the multifunctional Ca^{2+}/calmodulin-dependent protein kinase II. *Biochem. Biophys. Res. Commun.* **189**, 128–133.

Wenzel-Seifert, K., Schächtele, C., and Seifert, R. (1994) N-Protein kinase C isoenzymes may be involved in the regulation of various neutrophil functions. *Biochem. Biophys. Res. Commun.* **200**, 1536–1543.

Wera, S. and Hemmings, B. A. (1995) Serine/threonine protein phosphatases. *Biochem. J.* **311**, 17–29.

Wetsel, W. C., Eraly, S. A., Whyte, D. B., and Mellon, P. L. (1993) Regulation of gonadotropin-releasing hormone by protein kinase-A and-C in immortalized hypothalamic neurons. *Endocrinology* **132**, 2360–2370.

Wiechen, K., Yue, D. T., and Herzig, S. (1995) Two distinct functional effects of protein phosphatase inhibitors on guinea-pig cardiac L-type Ca^{2+} channels. *J. Physiol.* **484**, 583–592.

Wijetunge, S., Aalkjaer, C., Schachter, M., and Hughes, A. D. (1992) Tyrosine kinase inhibitors block calcium channel currents in vascular smooth muscle cells. *Biochem. Biophys. Res. Commun.* **189**, 1620–1623.

Wilkinson, S. E., Parker, P. J., and Nixon, J. S. (1993) Isoenzyme specificity of bisindolylmaleimides, selective inhibitors of protein kinase C. *Biochem. J.* **294**, 335–337.

Williams, E. J., Walsh, F. S., and Doherty, P. (1994) Tyrosine kinase inhibitors can differentially inhibit integrin-dependent and CAM-stimulated neurite outgrowth. *J. Cell Biol.* **124**, 1029–1037.

Wolbring, G., Hollenberg, M. D., and Schnetkamp, P. P. M. (1994) Inhibition of GTP-utilizing enzymes by tyrphostins. *J. Biol. Chem.* **269**, 22,470–22,472.

Worland, P. J., Stetler-Stevenson, M., Sebers, S., Sartor, O., and Sausville, E. A. (1993) Alteration of the phosphorylation state of p34edc2 kinase by the flavone L86-8275 in breast carcinoma cells. Correlation with decreased H1 kinase activity. *Biochem. Pharmacol.* **46**, 1831–1840.

Wu, Y. N. and Wagner, P. D. (1991) Effects of phosphatase inhibitors and a protein phosphatase on norepinephrine secretion by permeabilized bovine chromaffin cells. *Biochim. Biophys. Acta* **1092**, 384–390.

Wyllie, D. J. A. and Nicoll, R. A. (1994) A role for protein kinases and phosphatases in the Ca^{2+}-induced enhancement of hippocampal AMPA receptor-mediated synaptic responses. *Neuron* **13,** 635–643.

Yan, X., Corbin, J. D., Francis, S. H., and Lawrence, D. S. (1996) Precision trageting of protein kinases: an affinity label that inactivates the cGMP- but not the cAMP-dependent protein kinase. *J. Biol. Chem.* **271,** 1845–1848.

Yang, J. and Tsien, R. W. (1993) Enhancement of N- and L-type calcium channel currents by protein kinase C in frog sympathetic neurons. *Neuron* **10,** 127–136.

Yashpal, K., Pitcher, G. M., Parent, A., Quirion, R., and Coderre, T. J. (1995) Noxious thermal and chemical stimulation induce increases in ^{3}H-phorbol 12,13-dibutyrate binding in spinal cord dorsal horn as well as persistent pain and hyperalgesia, which is reduced by inhibition of protein kinase C. *J. Neurosci.* **15,** 3263–3272.

Zirrgiebel, U., Ohga, Y., Carter, B., Berninger, B., Inagaki, N., Thoenen, H., and Lindholm, D. (1995) Characterization of TrkB receptor-mediated signaling pathways in rat cerebellar granule neurons: involvement of protein kinase C in neuronal survival. *J. Neurochem.* **65,** 2241–2250.

Phosphorylation State-Specific Antibodies

Andrew J. Czernik, Jeffrey Mathers, and Sheenah M. Mische

1. Introduction

Over the past four decades, studies directed toward the elucidation of mechanisms involved in the hormonal regulation of metabolism have burgeoned into the field we now know as cellular signal transduction. Both then and now, the role of protein phosphorylation has been central to these investigations, and most physiological processes appear to be subject to phosphorylation-dependent modulation. Detection and quantitation of changes in the state of phosphorylation of specific proteins is of great utility in the quest to establish the function of a given protein and the consequences of its reversible phosphorylation. Two methods commonly used to measure protein phosphorylation and dephosphorylation in cell preparations employ prelabeling with $^{32}P_i$ or back phosphorylation (*see* Chapters 1–3). These methods continue to be very effective and have advantages for many test systems, but they do have several practical and theoretical limitations (Nestler and Greengard, 1984; Chapters 1–3, this volume). Based in large part on the successful use of short synthetic peptides to produce epitope-targeted antibodies (Lerner, 1982; Sutcliffe et al., 1983), an immunochemical approach became an attractive alternative for detecting changes in the state of phosphorylation of specific proteins at a specific site. The use of phosphorylation state-specific antibodies takes

From: *Neuromethods, Vol. 30: Regulatory Protein Modification: Techniques and Protocols*
Ed: H. C. Hemmings, Jr. Humana Press Inc.

advantage of the sensitivity and selectivity afforded by immunochemical methodology, combined with relatively simple preparation and potentially broad applications.

The first report of phosphorylation-dependent antibodies appeared in 1981, when polyclonal antibodies that could detect phosphotyrosine-containing proteins were produced by immunization with benzyl phosphonate conjugated to keyhole limpet hemocyanin (KLH) (Ross et al., 1981). Shortly thereafter, Nairn and colleagues reported the production of serum antibodies that distinguished between the phospho and dephospho forms of G-substrate, a protein localized to cerebellar Purkinje cells and phosphorylated by cGMP-dependent protein kinase (Nairn et al., 1982). A synthetic heptapeptide, Arg-Lys-Asp-Thr-Pro-Ala-Leu, corresponding to a repeated sequence surrounding two phosphorylated threonyl residues in the intact protein, served as antigen. Rabbit antisera against a peptide-KLH conjugate were specific for the dephospho form of G-substrate. Phospho-specific antibodies were prepared by immunization of rabbits with the purified phosphoprotein, phosphorylated in vitro to a stoichiometry of 2 mol/mol with cGMP-dependent protein kinase. Despite this initial success, other attempts in our laboratory to produce phospho-specific polyclonal antisera by immunization with the phospho form of intact proteins were not very successful, probably because of two significant factors. First, many phosphorylated proteins are believed to undergo rapid dephosphorylation during immunization, regardless of the route of injection, leading to the loss of the desired phosphoepitope. Second, holoproteins generally contain multiple immunogenic epitopes; this decreases the probability that clonal dominance for a phospho-specific epitope will be obtained. Despite these potential problems, reports began to appear describing the production of phosphorylation state-dependent monoclonal antibodies (MAbs), demonstrating the power of the monoclonal technique (Davis et al., 1983; Sternberger and Sternberger, 1983; Kosik et al., 1984; Grundke-Iqbal et al., 1986; Luca et al., 1986; Lee et al., 1987). Each of these MAbs was isolated from larger pools derived from holoprotein immunization protocols. Although the phosphorylation-dependent epitopes were not known at the time and were not specifically targeted, the results clearly confirmed the feasibility and the potential of the phosphorylation state-specific antibody approach to study protein phosphorylation.

Taking a more directed approach utilizing phosphorylated and unphosphorylated forms of synthetic phosphopeptides, we developed a general protocol for the production of phosphorylation state-specific antibodies for substrates with established site(s) of phosphorylation (Czernik et al., 1991; Snyder et al., 1992). The popularity of this technique continues to grow rapidly, and many other laboratories have utilized this methodology (*see* Section 3.). Perhaps this is best exemplified by the fact that many phosphorylation state-specific antibodies can now be obtained through commercial sources, and New England Biolabs (Beverly, MA) and Promega (Madison, WI) have produced their own collections of phospho-specific antibodies.

2. Phosphorylation State-Specific Antibody Production

In early stages of our development of this methodology, phosphopeptides were routinely prepared by enzymatic phosphorylation (Czernik et al., 1991; Snyder et al., 1992). Although this approach remains perfectly valid today, the preparation of synthetic phosphopeptides by chemical phosphorylation has become the state-of-the-art (Czernik et al., 1995). Likewise, we have examined the use of both polyclonal and monoclonal techniques for antibody production. Given the high success rate that we and others have obtained with the polyclonal technique, it has become the method of choice, because it is an easier and less costly method for the average laboratory.

2.1. Design of Phosphopeptides

Specific targeting of antibody production to a phosphorylated epitope is accomplished by the use of a short synthetic phosphopeptide corresponding to the sequence surrounding a given phosphorylation site in the protein of interest. Peptides of 8–10 residues in length appear to be optimal. This length is short enough to maximize the probability of including the phosphorylatable residue within the epitope while minimizing the chances of providing other, phosphorylation-independent epitopes. Inclusion of the acidic amino acids aspartic acid and glutamic acid, and the basic amino acids arginine and lysine, in the synthetic peptide will improve its antigenicity. Fortunately, these amino acids comprise part of the primary sequence determinants for substrate recognition by a number of protein kinases (Kemp and Pearson, 1990; Kennedy and Krebs, 1991), and are therefore often found in close proximity to the phosphorylated residue to be targeted. When designing a peptide to be phosphorylated chemically, there are two factors that dramatically affect the yield. The position of the phosphorylation site relative to the N-terminus is critical. Factors related to steric hinderance, caused by protecting groups, amino acid side-chains, and peptide folding, begin to reduce the yield when the location of the phosphorylated residue exceeds eight residues from the N-terminal. The nature of the N-terminal residue is also of prime importance in determining successful chemical phosphorylation. Bulky residue and residues of relatively high basicity can significantly reduce yield. The inclusion of a smaller residue, such as alanine, at the N-terminal can be the difference between success and failure for some peptides. This is clearly illustrated with the peptide NH_2-R-R-N-S(P)-V-F-Q-Q-C-COOH. Attempts to chemically phosphorylate this peptide were unsuccessful. After resynthesis with the addition of an alanyl residue at the N-terminus, the peptide was phosphorylated with an overall yield of 21%.

Depending on the type of coupling chemistry one plans to use, additional amino acid residues containing functional groups not present in the native sequence can also be included at the N- or C-terminus of the synthetic peptide. This will direct the coupling to occur at one end of the molecule, leaving the residues surrounding the phosphorylation site unmodified. This directionality also facilitates the preparation of a peptide affinity column for subsequent purification of antisera (*see* Section 2.5.2.). We routinely place a cysteinyl residue at the C-terminus of the peptide to enable conjugation to the carrier protein via the peptide sulfhydryl group using *m*-maleimidodibenzoyl sulfosuccinimide ester. Other options include the addition of a tyrosyl residue to facilitate subsequent coupling with bis-diazotized benzidine or addition of a lysyl residue to facilitate coupling to the peptide ε-amino group with glutaraldehyde (Walter, 1986).

2.2. Preparation of Phosphopeptides

Two approaches have been used for the chemical synthesis of phosphorylated peptides by solution or solid-phase peptide synthesis (SPPS). One method involves the use of either tert-butyloxycarbonyl (*t*-Boc) or 9-fluorenylmethyloxycarbonyl (Fmoc) chemistry and the incorporation of phosphorylated/protected amino acid derivatives during synthesis of the peptide (Frank, 1984; Schlesinger et al., 1987; Arendt et al., 1989). An alternative method, global phosphorylation, has significant advantages over methods incorporating protected derivatives. Perich and Johns (1988) first described this method, which has since been modified (Mathers et al., 1993; Czernik et al., 1995) for the incorporation of phosphate into unprotected side-chains of seryl- or threonyl-containing peptides. It employs a two-step procedure comprised of manual phosphitylation of the unprotected hydroxy group of serine or threonine, followed by oxidation to derivatize the free side-chain.

2.2.1. Fmoc Solid Phase Synthesis of Peptides

The following reagents are used: chromatography grade N-methylpyrrolidone (NMP) (Burdick and Jackson, Muskegon, MI); dichloromethane (DCM) (Mallinckrodt, St. Louis, MO); piperidine and diisopropylethylamine (DIEA) (Applied Biosystems, Foster City, CA); dimethylformamide (DMF) (EM Science, Gibbstown, NJ); 1-hydroxybenzotriazole (HOBT) and 2-(1H-benzotriazol-1-yl)1,1,3,3-tetramethyluroniumhexafluorophosphate (HBTU) (AnaSpec, San Jose, CA; Advanced ChemTech, Louisville, KY); preloaded hydroxy-methylphenoxy (HMP) resins (Novabiochem, San Jose, CA; BioServe, Laurel, MD; Bachem, Torrance, CA; Amino Tech, Ogdensburg, NY); and protected amino acids (BioServe, Novabiochem/Calbiochem, Bachem, Amino Tech, and AnaSpec). Fmoc amino acid protecting groups are as follows: *t*-butyl (Asp, Glu, Ser, Thr, Tyr); Pmc (Arg); Boc (Lys, Trp); and trityl (Asn, Cys, Gin, His). Peptides are synthesized using an Applied Biosystems Model 430A peptide synthesizer using 0.25-mmol FASTMOC chemistry (Fields et al., 1990; Applied Biosystems User Bulletin No. 32, 1990), and a modified Fmoc/NMP chemistry using HBTU/HOBT/DIEA (to activate amino acids [Fields et al., 1990]) and HMP preloaded resins. An initial substitution value of >0.4 mmol/g is required for all resins. Residues to be phosphorylated are incorporated with free hydroxyl side-chain functional groups. Single coupling is employed exclusively for all peptides. The recommended NH_2-terminal protecting group is *t*-Boc, which is removed in conjunction with the cleavage procedure.

2.2.2. Global Phosphorylation, Cleavage, and Phosphopeptide Purification

The following reagents are used: DNA synthesis grade 1-H-tetrazole, *N,N*-dimethylacetamide (DMA), 70% (v/v) aqueous *t*-butylhydroperoxide, and molecular-biology-grade phenol (Sigma, St. Louis, MO); Di-*t*-butyl-*N,N*-diisopropyl-phosphoramidite (AnaSpec and Novabiochem); HPLC-grade methanol and acetonitrile, reagent-grade *t*-amyl alcohol, and reagent-grade acetic acid (Fisher Scientific, Pittsburgh, PA); chromatography-grade DMF and trifluoroacetic acid (TFA) (EM Science); 80% *t*-butylhydroperoxide/di-*t*-butylperoxide (Fluka, Ronkonkama, NY); thioanisole, 1,2-ethanedithiol, and *m*-cresol (Aldrich, Milwaukee, WI); and DCM (Mallinckrodt). Milli-Q water is used in all steps. Global phosphorylation of peptides prepared in Section 2.2.1. is performed using a modification of the procedure previously described (Mathers et al., 1993; Czernik et al., 1995). A stepwise description of the procedure is given in Table 1. An unphosphorylated sample of the peptide should be reserved for subsequent cleavage and purification. The dephospho peptide is used as a control for some of the screening protocols to determine the phospho-selectivity of the antibodies. The dephospho peptide may also be useful as an affinity ligand during subsequent antibody purification.

Deprotection of side-chain functional groups, removal of the N-terminal protecting group, and cleavage of the peptide from the resin are simultaneously accomplished as described (King et al., 1990). Briefly, the peptide is cleaved from the resin by gently stirring the resin sample (200–300 mg) in 5% (w/v) phenol/TFA (3.5 mL), ethanedithiol (100 µL), thioanisole (200 µL), and water (200 µL) for 2.5 h. The amounts of scavengers and length of cleavage are varied depending on the composition of the peptide. The resin is filtered and rinsed three times with a total of 30 mL DMF and then the peptide is precipitated with diethyl ether and air-dried overnight. The peptide is dissolved in acetonitrile/water (10–50% [v/v], as required for peptide solubility) and analyzed for purity by mass spectroscopy and HPLC. Peptides are then purified by preparative HPLC, typically using Vydac C_{18}-reverse-phase columns and gradient elution using buffers containing 0.1% (v/v) TFA and increasing concentrations of acetonitrile.

By the use of the phosphorylation method outlined in Table 1, we have successfully prepared more than 50 peptides containing phospho-seryl, phospho-threonyl, and phospho-tyrosyl residues, including several that contain multiple phosphorylated residues. Standard Fmoc amino acids are compatible with this phosphorylation method, including methionyl and trityl-protected cysteinyl residues, which do not undergo oxidation during this phosphorylation method.

There are a number of critical points that should be mentioned regarding the specific reagents and quantities used in the above procedure:

1. The amount of peptide resin used for the procedure has been reduced from 450 to 200 mg. We have found that

Table 1
Global Phosphorylation Method

---

1. Phosphitylation:
   a. In a 150-mL round-bottom flask covered with aluminum foil add 200 mg peptide resin (approx 0.125 mmol peptide); 400 mg 1-H tetrazole (sublimed); 0.7 mL di-*t*-butyl-N,N-diisopropyl-phosphoramidite; and 2.5 mL N,N-dimethylacetamide.
   b. Cover with parafilm and stir at room temperature for 1 h.
2. Wash 1:
   a. Transfer suspension to a medium pore fritted glass filter and apply vacuum.
   b. Completely transfer all of the peptide resin using repetitive washes with N,N-dimethylacetamide until crystals of 1-H-tetrazole are no longer visible in the round bottom flask. Total volume for step 1 and 2 should be approx 20–30 mL.
   c. Dry on medium pore fritted glass filter under vacuum for 20 min.
3. Oxidation:
   a. Transfer peptide resin to a clean 150-mL round-bottom flask.
   b. Add 0.325 mL 80% *t*-butylhydroperoxide/*t*-butylperoxide; and 0.8 mL N,N-dimethylformamide.
   c. Cover with parafilm and stir at room temperature for 1 h.
4. Wash 2:
   a. Transfer peptide resin to a medium pore fritted glass filter and apply vacuum.
   b. For each of the following reagents, disconnect the vacuum, add the reagent, swirl the suspension for 10–15 s, and reapply the vacuum: 10–15 mL *t*-amyl alcohol; 10–15 mL N,N-dimethylacetamide; 2–3 mL glacial acetic acid; 10–15 mL amyl alcohol; and 20–25 mL methanol.
   c. Dry on medium pore fritted glass filter under vacuum for 20 min and continue with cleavage procedure.

---

by scaling down the phosphorylation and cleavage, the overall peptide yield has increased two- to fourfold. This is caused by a more efficient cleavage, with the target peptide recovered in higher yield.

2. In order to drive the reaction to completion, proper solvation of the resin during the phosphitylation and oxidation steps is critical. During the phosphitylation and oxidation steps, the resin swells on addition of *N-N*-dimethylacetamide or dimethylformamide, and may

require additional solvent for proper agitation during the reaction period. In this case, an additional aliquot of 250–500 μL of solvent should be added following the addition of the other reagents in the respective step. Appropriate adjustments should be made to maintain the same relative concentrations of all reagents.

3. The use of 80% *t*-butylhydroperoxide/di-*t*-butylperoxide, instead of the same peroxide in aqueous solution, dramatically increases the yield of the desired phosphopeptide.

4. The amount of tetrazole has been reduced from 800 to 400 mg and should be sublimed.

5. Because the vast majority of phosphorylated peptides we now synthesize include a C-terminal cysteine residue, we use Cys-chlorotrityl resin. This modification has improved our overall yield, in some instances by more than twofold. More importantly, the quality of the initial crude phosphorylated peptide is improved, usually appearing as a single peak in HPLC analysis. Chlorotrityl resins are more expensive, but the yield and quality of the peptide synthesized is usually much higher and can justify the higher initial cost of the resin. If HMP resins are used, employ those with lower substitution values (0.35–0.4 mmol/g).

6. The importance of using fresh or properly stored reagents must be stressed. This can contribute to significantly higher yields of phosphopeptide and concomitant reduction in the level of impurities in the crude peptide products.

Although the global phosphorylation method has worked for the majority of peptides we have synthesized, we have experienced specific instances in which the phosphorylated form has been recovered in very poor yield, or not at all. These problems have been attributed in some cases to synthesis problems, which were corrected on resynthesis, but more often remain unexplained. It appears that phosphopeptides that are difficult to produce may result from inaccessibility of the hydroxyl side-chain of the seryl or threonyl residue to reaction with the phosphoramidate. One example of this phenomenon is illustrated by the peptide NH$_2$-L-S-M-S(P)-P-R-Q-R-K-C-COOH, which had an 11% total yield of phosphorylated species. When the t-butyl-protected seryl residue at position 2 was changed to an alanyl residue, which has no protecting group for Fmoc synthesis, the total yield of phosphopeptide was increased to 47%. However, employing a similar strategy with other recalcitrant peptides has resulted in no increase in total yield. Some peptides continue to be problematic, with total yields always <10%, regardless of improvements that have aided in the synthesis of other phosphopeptides. Reagent problems have always been ruled out, because for each failure there has been a corresponding successful synthesis that has utilized the same lots of reagents. In such cases, the unphosphorylated version of the peptide can usually be phosphorylated enzymatically to produce the phosphopeptide, or more recently, has been synthesized using the new benzyl-protected Fmoc derivatives described in the next section.

2.2.3. Phosphopeptide Synthesis
Using New Benzyl-Protected Fmoc Derivatives

We have recently found that two new protected amino acid derivatives, Fmoc-Ser(PO(OBzl)OH)-OH and Fmoc-Thr(PO(OBzl)OH)-OH (AnaSpec, cat. # 20419 and 20835, respectively), show very promising results for phosphopeptide synthesis. Both derivatives have been used to produce longer and more complex phosphopeptides without many of the sequence-related problems associated with the global phosphorylation procedure. In addition, the derivatives are compatible with typical Fmoc chemistry and cleavage. Although our use of Fmoc-Ser(PO(OBzl)OH)-OH and Fmoc-Thr(PO(OBzl)OH)-OH has been limited to the synthesis of phosphopeptides that were refractory to global phosphorylation, their use has proven successful for these syntheses as well as for synthesis of phosphopeptides containing multiple phosphorylation sites. In all cases, the phosphopeptide was obtained in high yield and represented the major species in the reaction mixture. The unphosphorylated peptide was a minor species and there was no intermediate phosphorylation species present, vastly simplifying the purification of the phosphopeptide. If our initial success with these new derivatives continues, and commercial supplies can be expanded without any decrease in quality of these reagents, we expect that the use of these derivatives will quickly replace the global phosphorylation procedure and become the new state-of-the-art procedure for synthetic phosphopeptide production.

2.3. Phosphopeptide Conjugation and Immunization

Cysteine-containing phosphopeptides are conjugated to carrier proteins using the heterobifunctional crosslinker *m*-maleimidobenzoyl-*N*-hydroxysulfosuccinimide ester (sulfo-MBS, Pierce, Rockford, IL). The NHS moiety of the crosslinker is first coupled to free amino groups (N-terminal and/or ε-amino group of Lys) in the carrier protein. After excess MBS is removed by a desalting column, the Cys-containing peptide is covalently coupled via reaction with the maleimide group. We routinely use *Limulus* hemocyanin (LHC, Sigma) as the carrier protein for immunization, but other proteins, including KLH, bovine thyroglobulin (Sigma), and bovine serum albumin (BSA, Sigma), can be used. This protocol provides sufficient material for immunization and four rounds of boosting for each of two rabbits.

2.3.1. Phosphopeptide Conjugation Protocol

A buffer containing 75 mM sodium phosphate, pH 7.2, 250 mM NaCl is used throughout the procedure. Use ~2 mg carrier protein and 2 mg phosphopeptide (assume ~75% peptide by dry wt).

1. Dissolve 4 mg LHC in 2 mL buffer. If protein is not completely dissolved, filter through 0.45 µ durapore syringe filter (Millipore, Bedford, MA). Some batches of LHC will not completely dissolve. After filtration, use 1.25 mL.
2. To the 1.25-mL sample of LHC solution, add 125 µL of sulfo-MBS (from a stock solution of 2.5 mg/mL in buffer,

dissolved immediately prior to use). Incubate 60 min at room temperature.

3. During LHC/MBS incubation, equilibrate a 10-mL Speedy desalting column (Pierce) with 70 mL buffer. Apply LHC/MBS solution onto column. Elute with 0.5-mL aliquots of buffer, collect 0.5-mL fractions.

4. Assay fractions with Bradford reagent (Bio-Rad, Hercules, CA), or measure absorbance of each fraction at 280 nm. For Bradford assay, use 10 μL sample/1 mL Bradford reagent. Protein peak should elute in tubes #6–8 (tube 1 includes the volume displaced by load volume plus the first 0.5 mL aliquot of elution buffer.) Pool peak fractions as required, but no more than three tubes total (1.5 mL).

5. Dissolve 2 mg peptide in 200 μL buffer (as a check of peptide solubility) and check for the presence of free sulfhydryl groups using Ellman's reagent (5,5'-dithio-bis-(2-nitrobenzoic acid; DTNB; Pierce). Dissolve 4 mg DTNB in 1 mL $0.1M$ sodium phosphate, pH 8.0 (PB). Prepare cysteine standards (Sigma, 0.1–2 mM) in the same buffer. To assay, add 50 μL DTNB stock solution to 2.5 mL PB, and then add 250 μL cysteine standard, test sample, or assay blank. Mix and incubate at room temperature for 15 min, then measure absorbance at 412 nm. Although this step appears to be trivial, we encourage that it always be performed as a precaution. On rare occasion, sequence-related instability of -SH groups in certain peptides has been encountered. Should the Ellman's assay yield an unexpectedly low value, one can reduce the peptide with 2-mercaptoethylamine (or sodium borohydride) just prior to using it for conjugation.

6. Once the presence of the expected free sulfhydryl groups has been confirmed, the peptide solution is added to the pooled fractions from Speedy column containing the activated LHC. Check the pH of the solution; some HPLC-purified peptides can be very acidic. Adjust the pH if necessary. Incubate for 2 h at room temperature. Dilute to final volume of 1.8 mL with H_2O. Aliquot into 5 tubes: 1×0.6 mL (for immunization); 4×0.3 mL (for boosts).

Store tubes at –20°C until use. Each tube will be used for two rabbits; thus each rabbit receives 300 µg phosphopeptide at the immunization step and 150 µg for each boost.

2.3.2. Rabbit Immunization Protocol

Antibodies are raised in female New Zealand white rabbits (2.5–3 kg). Prior to injection, preimmune serum (5 mL) is obtained from each rabbit. The phosphopeptide-LHC conjugate (representing 300 µg of phosphopeptide/rabbit) is emulsified in complete Freund's adjuvant (Difco, Detroit, MI) and then injected intradermally at multiple sites on the back over the shoulder blades. Booster injections (150 µg phosphopeptide) in incomplete Freund's adjuvant (Difco) are given at 2, 4, and 6 wk. Blood is drawn at wk 5 and wk 7, and serum is prepared for testing. Further boosting is dependent on response obtained. We have had rabbits respond after one boost, whereas others require six or more boosts before phospho-specific sera can be detected. If a rabbit is still negative after four boosts, it is sometimes helpful to continue boosting with a new batch of phosphopeptide that has been conjugated to a different carrier protein. The specificity of the antiserum is verified using several screening criteria (*see* Section 2.4.). Once a rabbit is identified as having positive antiserum, blood is drawn (7.7 mL/kg body weight) on alternate weeks and the antiserum is stored at –20°C until further use. Antibody titer should be monitored on a monthly basis for the duration of the collection period. If the titer drops, additional boosting can be performed.

2.4. Primary Screening of Antiserum for Phosphorylation State-Specificity

We employ dot-immunoblotting assays to screen for phosphorylation state-specific antibody production. Based on the availability of sufficient quantities of the dephospho and phospho forms of the holoprotein, either the holoprotein or the synthetic peptides are chosen for use in the primary screen. When peptides are used for the primary screening, secondary tests using the holoprotein must be performed to confirm the immunoreactivity. ELISA assays can also be used; they are simple, sensitive, fast, and can handle large numbers of samples.

2.4.1. Dot-Immunoblotting Assay

If sufficient quantities of the dephospho and phospho forms of the holoprotein are available, antisera are screened with these samples directly. Dot-blot assays were used to identify phospho- and dephospho-specific polyclonal antisera for phosphorylation site 1 in synapsin I (Fig. 1). Serially diluted aliquots of purified phospho-synapsin I (1–50 ng/ 1 μL), which had been phosphorylated in vitro at site 1 to a stoichiometry of 0.9 mol phosphate/mol protein with the catalytic subunit of cAMP-dependent protein kinase, were spotted onto premoistened strips of pure nitrocellulose membrane (0.2–μ pore size, Schleicher and Schuell, Keene, NH), together with identical amounts of dephospho-synapsin I. As an additional control, aliquots of phospho-synapsin I that had been phosphorylated in vitro at sites 2 and 3 to a stoichiometry of 1.9 mol/mol by Ca^{2+}/calmodulin-dependent protein kinase II were also spotted. Immunoblot assays were performed in 50 mM Tris, pH 7.5, 200 mM NaCl, 0.05% (v/v) Tween-20 (Sigma) containing 2.5% (w/v) nonfat dried milk (Carnation, Nestlé, Glendale, CA). Nitrocellulose strips were preincubated in buffer, incubated with aliquots of antisera (1:50, 1:100, 1:500, 1:1000 dilution) for 90 min, washed four times and incubated for 90 min with ^{125}I-labeled protein A (Amersham, Arlington Heights, IL) diluted 1:1000 in the same buffer. Strips were then washed thoroughly, dried, and subjected to autoradiography. Initial screening with this method identified positive antiserum, which was then affinity-purified as described below. Assays employing each antiserum at 1:100 dilution are shown in Fig. 1. Note that the phospho-specific antibody, G-257, was specific for the presence of a phosphoryl group at site 1 and did not recognize the phosphorylated form of synapsin I at sites 2 and 3. Conversely, the dephospho-specific antibody, G-143, did not react with synapsin I that had been phosphorylated at site 1.

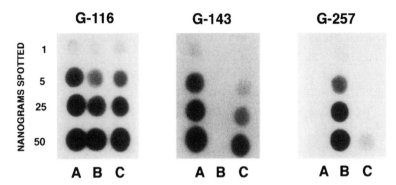

Fig. 1. Phosphorylation-state specific antisera for site 1 in synapsin I. Antisera were tested for phosphorylation state-specificity with nitrocellulose blots containing spots of increasing amounts (1–50 ng) of three forms of purified bovine synapsin I. **(A)** dephospho-synapsin I; **(B)** site 1 phospho-synapsin I; **(C)** sites 2,3 phospho-synapsin I. Site 1 of synapsin I was phosphorylated in vitro with the catalytic subunit of cAMP-dependent protein kinase; sites 2 and 3 were phosphorylated in vitro with Ca^{2+}/calmodulin-dependent protein kinase II. Antibody binding was detected with ^{125}I-protein A. Left panel, G-116 (1:100 dilution). This antiserum recognizes phosphorylation-independent epitopes in the C-terminal portion of the molecule. The autoradiograph demonstrates the equal reactivity of G-116 with three forms of synapsin I. Middle panel, G-143 (1:100 dilution). This antiserum was prepared against the dephosphopeptide corresponding to residues 3–13 in synapsin I. Right panel, G-257, (1:100 dilution). This antiserum was prepared against the phosphopeptide corresponding to residues 3–13 in synapsin I, with phosphoserine at position 9 (site 1). G-257 is specific for the site 1 phospho-form of synapsin I, and does not react with synapsin I that is phosphorylated at sites 2 and 3.

It is important to prepare holoprotein samples that have been stoichiometrically phosphorylated. If antiserum is completely phospho-specific, then holoprotein samples with lower phosphorylation stoichiometry can be utilized for screening. However, it has been our common experience that antiserum generally will contain a mixture of IgGs, directed toward phospho-specific, dephospho-specific, and phosphorylation-independent epitopes, and the relative proportions of these components can change quite dramatically during the course of months of boosting. Under these circumstances, the presence of a significant level of the dephospho form of the holoprotein in the phosphorylated sample could complicate the analysis.

If limited quantities of the holoprotein are available, then the phosphorylated and unphosphorylated forms of the synthetic peptide can be used for primary screening. Peptides are dissolved in 50 mM HEPES, pH 7.6, at a concentration of 2 mg/mL (approx 2 mM). This stock solution is serially diluted to yield additional samples of 1.0, 0.5, 0.1, and 0.05 mg/mL. Aliquots (1 µL) of each sample are directly spotted on polyvinylidene difluoride (PVDF) membrane (Immobilon P, Millipore) prewetted in methanol and then equilibrated in H$_2$O. After spotting of the peptide samples, the membrane is then dried completely before processing, as described above. Although the binding of short peptides to PVDF is not quantitative, a sufficient quantity of most peptides does remain bound, and this type of "yes-or-no" assay can be carried out reliably. Every phosphorylation site peptide we have assayed has bound to some degree to PVDF. Invariably, there will be some peptides that do not bind to PVDF. If there is doubt, the blot can be probed with protein stains like Coomassie brilliant blue, amido black, Ponceau S, or ninhydrin. If the peptide does not bind directly to PVDF, it can be conjugated to a carrier protein distinct from that used for conjugation of the immunization samples. The peptide-carrier complex can then be spotted onto nitrocellulose, and used for dot-blots as described above. The switch in carrier protein will avoid false positives caused by carrier protein-directed antibodies.

2.4.2. ELISA Assay

ELISA (enzyme-linked immunosorbant assay) is most useful when the primary screening requirements are large, as in MAb production. Because we now use rabbits exclusively, ELISA is no longer used routinely. Detailed descriptions of the methods used for ELISA can be found in our earlier work (Czernik et al., 1991; Snyder et al., 1992). If this method is chosen, we note that most peptides (dephospho and phospho forms) will bind to some extent to plastic 96-

well assay trays and can be utilized for the screening procedure without conjugation to a carrier protein.

2.5. Affinity Purification of Antisera

Depending on the quality and titer of the whole antiserum, further purification may be essential or beneficial. With a few manipulations, done in the right order, it has been possible to purify a subpopulation of a given IgG pool with the desired phosphorylation state-specificity. Once positive antiserum has been produced, immunoblots should be performed using appropriate cell and tissue extracts to determine if the antisera detects the protein of interest and if other, nonspecific bands are also detected (Czernik et al., 1991). The appearance of these contaminating bands can usually be eliminated by further affinity purification of the antiserum. For example, each of the synapsin I phospho-specific antibodies were checked using total rat brain homogenates and homogenates of primary cerebrocortical cultures. The tissue/cells are extracted in boiling 1% (w/v) sodium dodecyl sulfate (SDS), and 100 μg protein/lane is subjected to SDS/polyacrylamide gel electrophoresis and processed for immunoblotting (*see* Chapter 3). Most antisera will contain undesirable, "nonspecific" IgGs, such that an additional step of affinity purification is beneficial.

2.5.1. Purification of IgG Pool
by Protein A-Sepharose Chromatography

Although it is possible to directly employ a phosphopeptide affinity column, we recommend prior protein A-sepharose chromatography. One major concern is that repeated loading of whole antiserum over a phosphopeptide column will slowly result in the loss of phosphate and/or the degradation of the peptide ligand resulting from the exposure of the column to serum phosphatases and proteases. To circumvent this potential problem, we first prepare an IgG fraction and then proceed to the peptide affinity columns as a second step. Several procedures are available for the preparation of an IgG fraction from whole antiserum (Harlow and Lane, 1988). We routinely use fractionation with protein A-sepharose beads (Pharmacia Biotech, Piscataway, NJ). A sample (2 g dry weight) of the resin beads is reconstituted in water, poured into a 1.5 × 10 cm Flex column (Kontes, Vineland, NJ), and the column is washed with 100 mL of 50 mM HEPES, pH 7.6/200 mM NaCl. Serum samples (25–40 mL) are thawed, pooled, filtered through a 0.22 µ Millex-GV syringe filter (Millipore), and kept on ice. To reduce potential proteolysis of the antiserum, a cocktail of protease inhibitors is added to the filtered antiserum pool prior to loading (final concentrations: benzamidine, 25 mM; EDTA and EGTA, 1 mM; leupeptin and antipain, 20 µg/mL; pepstatin and chymostatin, 2 µg/mL; phenylmethylsulfonyl fluoride, 100 µM; aprotinin, 100 U/mL). Small aliquots of cold antiserum are continuously applied to the protein A-sepharose column, which is kept at room temperature, until all the antiserum has been loaded. A 0.5-mL aliquot of the antiserum load is saved for testing along with the additional fractions. The antiserum that passes through the column is collected in 5-mL aliquots. Analysis of these flow-through fractions will be used to determine when the capacity of the column has been exceeded. Flow-through fractions are stored at −20°C until it has been determined that they do not contain antibodies of interest. After the antiserum has been applied, the column is washed with 40 mL each of the following buffers:

1. 50 mM Tris, pH 7.5/1M NaCl/0.05% (v/v) Tween-20;
2. 50 mM sodium borate, pH 8.5/500 mM NaCl/0.05% (v/v) Tween-20;
3. 50 mM sodium acetate, pH 5.5/500 mM NaCl/0.05% (v/v) Tween-20; and
4. 50 mM HEPES, pH 7.6/500 mM NaCl.

Bound antibodies are then eluted with 100 mM glycine, pH 2.5. One milliliter fractions are collected and immediately neutralized with 1M Tris, pH 9.0. Protein content is monitored using Bradford reagent or absorbance at 280 nm and peak fractions are pooled. Aliquots of the IgG pool, antiserum load, and each flow-through fraction are then tested to obtain an estimate of the recovery, which at this stage should be high. After use, the protein A-sepharose column is washed thoroughly, equilibrated in 50 mM Tris, pH 7.5/1M NaCl/ 0.05% (v/v) Tween-20, 0.02% (w/v) sodium azide, and stored at 4°C. The top and bottom caps of the Flex column should be fastened securely to prevent the column resin from drying out, and a sufficient volume of buffer should be maintained above the resin for long-term storage.

2.5.2. Affinity Chromatography with Phospho- and Dephospho-Peptide Columns

At this stage, the IgG pool is ready for further purification utilizing affinity columns of the phosphopeptide and/ or the dephosphopeptide. SulfoLink coupling gel (Pierce) is used routinely. This is a support matrix comprised of 6% crosslinked agarose, to which an iodoacetyl group is immobilized on a 12-atom spacer arm. Cys-containing dephospho- and phosphopeptides are coupled to the matrix following the protocol recommended by the manufacturer. Briefly, 3–5 mL of resin are brought to room temperature in a 15-mL plastic tube and washed with 5 × 15 mL of 50 mM Tris, pH 8.5, 5 mM EDTA. During the washing, the peptide to be coupled to the resin should be checked for the presence of free sulfhydryl groups with Ellman's reagent (*see* Section 2.3.1.). The resin is allowed to settle and the wash buffer is removed. Peptide (2–5 mg, dissolved in 2 mL of the same buffer), is added to the resin. A small aliquot of the peptide solution is saved for the determination of coupling efficiency. The suspension is incubated for 1 h with gentle mixing on a tube rotator. The suspension is briefly centrifuged (1000g, 30 s) to gently compact the resin, and the supernatant solution containing unreacted peptide is removed and saved. Coupling efficiency is determined by the measurement of the content of free sulfhydryl groups in the original peptide solution and in the solution after incubation with the resin, using Ellman's reagent (*see* Section 2.3.1.). A comparison of the values obtained for the original peptide solution with that for the supernatant after the resin incubation gives the relative extent of coupling. If the coupling efficiency is poor, the resin may be too old and partially inactivated. If a large amount of free sulflhydryl reactivity remains in the supernatant, it should be saved for reuse with a fresh lot of resin. Although not explicitly stated on the packaging, in our hands, SulfoLink resin appears to have a limited shelf life of perhaps 6 mo, when stored as directed by the manufacturer.

If peptide coupling was satisfactory, the resin is washed with 5 × 15 mL 50 mM Tris, pH 8.5, 5 mM EDTA. The remaining reactive sites on the resin are then blocked with excess cysteine. The resin is incubated with a solution of 50 mM cysteine in 50 mM Tris, pH 8.5, 5 mM EDTA, for 60 min at room temperature on a tube rotator. Use 1 mL cysteine/mL of resin. The suspension is then poured into a Flex column (1.5 × 10 cm), washed with 200 mL 1M NaCl, and stored at 4°C in 50 mM Tris, pH 7.5/1M NaCl/0.05% (v/v) Tween-20, 0.05% (w/v) sodium azide.

The strategy for further purification of IgG pools is variable. Typically, there are nonspecific IgGs (unrelated to the protein of interest but reactive with other proteins in the tissue or cell type under study), as well as dephospho-specific and/or phosphorylation-independent IgGs that react with the protein of interest. All these "contaminating" IgGs require removal. This is most easily done by first passing the IgG pool over a dephosphopeptide column to deplete the pool of dephospho-specific and phospho-independent IgGs, and then passing the depleted flow-through fraction over a phosphopeptide affinity column. After loading the depleted IgG pool over the phosphopeptide column, the steps of washing and elution are identical to those described above in Section 2.5.1. for the protein A-sepharose column. An accurate record of the recoveries of reactive IgGs should be kept. All flow-through fractions should be tested. It may be necessary to adjust the elution conditions to gain full recovery of the antibodies of interest. For example, it may be necessary to elute the peptide column with glycine buffer at pH 2.3, or at pH 2.0, or even lower, before all the antibodies are recovered. In some instances, elution at basic pH in 100 mM triethylamine, pH 10.0 or 11.0, might be more effective than the elution at low pH. The IgG pool eluted from the phosphopeptide affinity column should be concentrated by centrifugation to a protein concentration of approx 1 mg/mL with the use of a Centriprep concentrator (30,000 mol-wt cutoff, Amicon). The IgG pool is then dialyzed in 10 mM MOPS, pH 7.5/154 mM NaCl. A final absorbance reading at 280 nm is taken and sodium azide is added to yield a final concentration of 0.02% (w/v). The samples are divided in appropriate aliquots, snap-frozen in liquid nitrogen, and stored at –80°C. One vial is kept at 4°C and used as the working stock.

The objective of this series of purification steps is to recover, in an active state and in a sufficient quantity, a subpopulation of IgGs that exhibit the desired specificity for use in the experiments one has planned. The protocols detailed above provide a place to start. However, each antibody pool is different, and there are many variations one can try. For example, it may prove better to do the preadsorption to a column of the dephospho-holoprotein, rather than of the dephosphopeptide. Or perhaps final purification of the depleted, phospho-selected IgG pool with a phospho-holoprotein column will be the step that yields the pool of antibodies that one requires. Some trial-and-error experiments may be required if you do not obtain the desired pool on the first attempts at purification.

3. Advantages, Limitations, and Applications of Phosphorylation State-Specific Antibodies

Several practical advantages are afforded by chemical phosphorylation of peptides over the enzymatic procedure. First, the need to purify, purchase, or otherwise obtain a sufficient supply of a protein kinase for enzymatic phosphorylation is eliminated. Although many protein kinases can now be obtained commercially, they can be very expensive. Preparing the required enzyme yourself can be a daunting task, especially for those uninitiated with techniques of protein purification. Second, the enzymatic phosphorylation procedure requires two purification steps: one to isolate the pure synthetic peptide after cleavage and deprotection and a second step to separate phosphopeptide from unreacted peptide after the enzymatic phosphorylation reaction. Since the global phosphorylation reaction precedes cleavage and deprotection, a single purification step is required for chemical phosphorylation of peptides.

Based on the promising results with the new protected amino acid derivatives, Fmoc-Ser(PO(OBzl)OH)-OH and Fmoc-Thr(PO(OBzl)OH)-OH, it now appears possible to prepare phospho-specific antibodies for any phosphorylation site in any phosphoprotein, including those with multiple sites in close proximity. The use of small synthetic phosphopeptides, 8–10 residues in length, improves the probability of targeting antibody production to phosphorylation-dependent epitopes of the native protein. Larger amounts of the phosphopeptide antigens can be prepared for immunization, and the shorter phosphopeptides are more resistant to dephosphorylation than the corresponding holoproteins (Hemmings et al., 1990).

One limitation often mentioned is the potential for crossreactivity with other unrelated substrates, because of the probability that the sequence of a particular phosphorylation site will be homologous with other substrates having a similar consensus sequence for substrate recognition by a specific protein kinase. In practice, this potential problem has not proven to be a serious one. The antibodies that have been produced are generally quite specific for the protein of interest. Also, this is not usually a problem for immunoblot assays, because crossreacting species can usually be resolved by SDS-polyacrylamide gel electrophoresis prior to assay. However, crossreactivity of the antibodies with additional bands may preclude their use for immunocytochemical studies.

The value and utility of phospho-specific antibodies is highlighted using synapsin I as an example. Synapsin I is a prominent neuron-specific phosphoprotein, localized to presynaptic terminals in adult neurons, where it plays an important role in vesicle clustering and in the regulation of neurotransmitter release (Greengard et al., 1993; Li et al., 1995; Pieribone et al., 1995). Six distinct phosphorylation sites in synapsin I have now been identified, and synapsin I serves as a substrate for at least nine distinct protein kinases (Greengard et al., 1993; Jovanovic et al., 1996). On depolarization, all six phosphorylation sites are subject to rapid and transient regulation: At 15–30 s poststimulus, phosphorylation at sites 1, 2, and 3 is increased and phosphorylation at sites 4, 5, and 6 is decreased (Greengard et al., 1993; Jovanovic et al., 1996). Given the number of signal transduction pathways that can converge at the level of synapsin I, and the complexity of the regulation of the state of phosphorylation at each of six sites, it would be very difficult to characterize and quantitate these changes by conventional techniques.

Phosphorylation state-specific antibodies can provide a direct means to quantitate the regulation of the state of phosphorylation of a protein in intact cells in response to various physiological and pharmacological manipulations. When phospho-specific antibodies are used in combination with other phosphorylation-independent antibodies (to measure the total amount of protein), and known standards of the phospho (or dephospho) forms of the protein, it is possible to quantitate the in vivo stoichiometry of phosphorylation (Czernik et al., 1995; Iwata et al., 1996). Immunocytochemical studies using these antibodies can also provide useful information on the regulation and subcellular location of the phosphorylation events, and whether phosphorylation-dependent changes in the subcellular distribution of protein might occur. These approaches can be useful in the elucidation of the functional role of a particular phosphoprotein. Table 2 gives a summary of published reports describing the production and utilization of phospho-specific antibodies for a large number of proteins containing Ser/Thr phosphorylation sites. Many classes of proteins have been studied, including membrane and soluble proteins, cell surface receptors, ion channels, cytoskeletal proteins, enzymes, protein kinases, and transcription factors. This list should continue to expand at a rapid pace.

Table 2
Phospho-Specific Antibodies for Proteins
Containing Ser/Thr Phosphorylation Sites

Phosphoprotein	Ab name	Ab type	P-site residue	Protein kinase specificity	Uses[a]	Reference
Amyloid precursor protein (APP)	PAbT668	Polyclonal	P-Thr-668	cdc2 kinase	Blot	Suzuki et al., 1994
	PAbT654	Polyclonal	P-Thr-654	CaM kinase II, PKC	Blot	Oishi et al., 1996
	PAbT655	Polyclonal	P-Ser-655	PKC	Blot	Oishi et al., 1996
Ca^{2+} channel cardiac L-type, α1 subunit	Anti-CH (1923–1932)-P	Polyclonal	P-Ser-1928	PKA	Blot	De Jongh et al., 1996
CaM kinase II	Anti-PY66	Polyclonal	P-Thr-286 (α) P-Thr-287 (β)	Autophosphorylation site	Blot, ICC	Suzuki et al., 1992
	22B1	MAb	P-Thr-286 (α) P-Thr-287 (β)	Autophosphorylation site	Blot, ICC	Patton et al., 1993
CREB	Anti-PCREB	Polyclonal	P-Ser-133	PKA	Blot, ICC	Ginty et al., 1993
	#9191	Polyclonal	P-Ser-133	PKA	Blot, ICC	New England Biolabs catalog
DARPP-32	MAb23	MAb	P-Thr-34	PKA, PKG	Blot, IP	Snyder et al., 1992
eEF-2	CC81	Polyclonal	P-Thr-56	eEF-2 kinase	Blot	Marin et al., 1996
Elk1	#9181	Polyclonal	P-Ser-383	MAP kinase	Blot	New England Biolabs catalog
GAP-43	2G12/C7	MAb	P-Ser-41	PKC	Blot, ICC	Meiri et al., 1991
GFAP	YC10	MAb	P-Ser-8	PKA, PKC	Blot	Yano et al., 1991
G-substrate	Phospho-Ab	Polyclonal	RKDT(P)PAL 2 repeats	PKG	IP, RIA	Nairn et al., 1982
Insulin receptor	Anti-PS1327	Polyclonal	P-Ser-1327	PKC	Blot	Coghlan et al., 1994

Antigen	Antibody	Type	Epitope/site	Kinase specificity	Application	Reference
c-*Jun*	Anti-*P*T1348	Polyclonal	P-Thr-1348	PKC	Blot	Coghlan et al., 1994
	#9161	Polyclonal	P-Ser-63	SAP kinase/JNK	Blot, IP, ICC	New England Biolabs catalog
MAP 1A	MAP 1B-3	MAb	?	?	Blot	Luca et al., 1986
MAP 1B	Ab 150	MAb	?	MAP kinase, cdc kinases	Blot, ICC	Gordon-Weeks et al., 1993
	Clone 125	MAb	?	Casein kinase II	Blot, ICC	Riederer, 1995
	MAP 1B-3	MAb	?	?	Blot	Luca et al., 1986
MAP kinases (ERK1/ERK2)	#9102	Polyclonal	P-Thr-202 and P-Tyr-204	MEK kinases	Blot, IP, ICC	New England Biolabs catalog
	# V6671	Polyclonal	P-Thr-202 and P-Tyr-204	MEK kinases	Blot, IP, ICC	Promega catalog
MPP1, MPP2 and M-phase phosphoproteins	MPM2	MAb	LT(*P*)PLK as consensus sequence	cdc2 kinase, mitotic kinases	Blot, expression library screening	Westendorf et al., 1994
Na$^+$,K$^+$-ATPase α1 subunit	Phospho-Ab "Group (ii)"	Polyclonal	P-Ser-943	PKA	Blot	Fisone et al., 1994
Neurofilaments	NF-H (200 kDa)	MAbs 06-17, 03-44, 06-68, 04-7, 07-5	KS(*P*)PV sites	Proline-directed kinases	Blot, ICC	Sternberger and Sternberger, 1983
	Ta51, RMO24 RMO217		C-terminal epitopes	Proline-directed kinases	Blot, ICC	Lee et al., 1987
	MAP 1B-3	MAbs	Crossreact with MAP 1A/B MAb	?	Blot	Luca et al., 1986

243

(continued)

Table 2 (*continued*)

Phosphoprotein	Ab name	Ab type	P-site residue	Protein kinase specificity	Uses[a]	Reference
NF-M (150 kDa)	"Group (ii)" see above	MAbs	?	?	Blot, ICC	Sternberger and Sternberger, 1983
	RMO45, 54, 55, 93, 108	MAbs	C-terminal epitopes	?	Blot, ICC	Lee et al., 1987
	MAP 1B-3	MAb	?	?	Blot	Luca et al., 1986
Phospholamban	PS-16	Polyclonal	P-Ser-16	PKA	Biot, ICC	Drago and Colyer, 1994
	PT-17	Polyclonal	P-Thr-17	CaM kinase II	Blot, ICC (+/-)	
Phenylalanine hydroxylase	PH7	MAb	P-Ser-16	PKA	Blot	Smith et al., 1987
Protein phosphatase-1Cα	G-97	Polyclonal	P-Thr-320	Cdc2 kinase	Blot, ICC	Kwon et al., submitted for publication
SEK1/MKK4	#9151	Polyclonal	P-Thr-223	MEKK1	Blot, IP	New England Biolabs catalog
Synapsins I/II	G-257	Polyclonal	P-Ser-9(syn I) P-Ser-10(syn II) "P-site 1"	PKA, CaM kinases I, IV	Blot	Czernik et al., 1991
Synapsin I	RU 19	Polyclonal	P-Ser-603 (rat) "P-site 3"	CaM kinases II, IV	Blot, ICC	Yamagata et al., 1995
	G-526	Polyclonal	P-Ser-62/67 "P-site 4/5"	MAP kinase	Blot	Jovanovic et al., 1996

	Antibody	Type	Site	Kinases	Applications	Reference
	G-555	Polyclonal	P-Ser-549 (rat) "P-site 6"	MAP kinase, cdk1, cdk5	Blot	Jovanovic et al., 1996
Tau (PHF) paired helical filaments	AT8	MAb	P-Ser-202	MAP kinases, proline-directed protein kinases	Blot, ICC	Goedert et al., 1993; Biernat et al., 1992
	PHF-1	MAb	P-Ser-396 and P-Ser-404	MAP kinase, proline-directed protein kinases	Blot, ICC	Otvos et al., 1994
	Anti-T3P	Polyclonal	P-Ser-396	MAP kinase, proline-directed protein kinases	Blot, ICC	Lee et al., 1991
	SMI31	MAb	P-Ser-396 and P-Ser-404	MAP kinase, proline-directed protein kinases	Blot, ICC	Lichtenberg-Kraag et al., 1992
	SMI34	MAb	"Kinked" conformation	MAP kinase, proline-directed protein kinases	Blot, ICC	Lichtenberg-Kraag et al., 1992
Rat tyrosine hydroxylase	Anti-pTH	Polyclonal	P-Ser-40	PKA, PKC, PKG, CaM kinase II	Blot, IP, ICC	Goldstein et al., 1995

[a]Applications: CaM = calmodulin; blot = immunoblot; ICC = immunocytochemistry; IP = immunoprecipitation; RIA = radio-immunoassay.

Acknowledgments

We thank James Bibb for his help in preparing Fig. 1, and thank all who responded to our request for information via the bionet.user groups on the Internet.

References

Arendt, A., Palczewski, K., Moore, W. T., Caprioli, R. M., McDowell, J. H., and Hargrave, P. A. (1989) Synthesis of phosphopeptides containing O-phosphoserine and O-phosphothreonine. *Int. J. Pep. Protein Res.* **33**, 468–476.

Biernat, J., Mandelkow, E. M., Schröter, C., Lichtenberg-Kraag, B., Steiner, B., Berling, B., Meyer, H., Mercken, M., Vandermeeren, A., Goedert, M., and Mandelkow, E. (1992) The switch of tau protein to an Alzheimer-like state includes the phosphorylation of two serine-proline motifs upstream of the microtubule-binding region. *EMBO J.* **11**, 1593–1597.

Coghlan, M. P., Pillay, T. S., Tavaré, J. M., and Siddle, K. (1994) Site-specific anti-phosphopeptide antibodies: use in assessing insulin receptor serine/threonine phosphorylation state and identification of serine-1327 as a novel site of phorbol ester-induced phosphorylation. *Biochem. J.* **303**, 893–899.

Czernik, A. J., Girault, J.-A., Nairn, A. C., Chen, J., Snyder, G., Kebabian, J., and Greengard, P. (1991) Production of phosphorylation state-specific antibodies, in *Methods in Enzymology, Protein Phosphorylation, Part B.* vol. 201 (Hunter, T. and Sefton, B. M., eds.), Academic, San Diego, CA, pp. 264–283.

Czernik, A. J., Mathers, J., Tsou, K., Greengard, P., and Mische, S. M. (1995) Phosphorylation state-specific antibodies: preparation and applications. *Neuroprotocols* **6**, 56–61.

Davis, F. M., Tsao, T. Y., Fowler, S. K., and Rao, P. N. (1983) Monoclonal antibodies to mitotic cells. *Proc. Natl. Acad. Sci. USA* **80**, 2926–2930.

De Jongh, K. S., Murphy, B. J., Colvin, A. A., Hell, J. W., Takahashi, M., and Catterall, W. A. (1996) Specific phosphorylation of a site in the full-length form of the α1 subunit of the cardiac L-type calcium channel by adenosine 3', 5'-cyclic monophosphate-dependent protein kinase. *Biochemistry* **35**, 10,392–10,402.

Drago, G. A. and Colyer, J. (1994) Discrimination between two sites of phosphorylation on adjacent amino acids by phosphorylation site-specific antibodies to phospholamban. *J. Biol. Chem.* **269**, 25,073–25,077.

Fields, C. G., Lloyd, D. H., Macdonald, R. L., Ottesen, K. M., and Noble, R. L. (1990) HBTU activation for automated Fmoc solid phase peptide synthesis. *Peptide Res.* **4**, 95–101.

Fisone, G., Cheng, S. X.-J., Nairn, A. C., Czernik, A. J., Hemmings, H. C., Jr., Höög, J.-O., Bertorello, A. M., Kaiser, R., Bergman, T., Jörnvall, H., Aperia, A., and Greengard, P. (1994) Identification of the phosphorylation site for cAMP-dependent protein kinase on the Na$^+$, K$^+$-ATPase and effects of site-directed mutagenesis. *J. Biol. Chem.* **269**, 9368–9373.

Frank, A. W. (1984) Synthesis and properties of N-, O-, and S-phospho-derivatives of amino acids, peptides and proteins. *CRC Crit. Rev. Biochem.* **16**, 51–101.

King, D. S., Fields, C. G., and Fields, G. B. (1990) A cleavage method which minimizes side reactions following Fmoc solid phase peptide synthesis. *Int. J. Pep. Protein Res.* **36**, 255–266.

Ginty, D. D., Kornhauser, J. M., Thompson, M. A., Bading, H., Mayo, K. E., Takahashi, J. S., and Greenberg, M. E. (1993) Regulation of CREB phosphorylation in the suprachiasmatic nucleus by light and a circadian clock. *Science* **260**, 238–241.

Goedert, M., Jakes, R., Crowther, R. A., Six, J., Lübke, U., Vandermeeren, M., Cras, P., Trojanowski, J. Q., and Lee, V. M.-Y. (1993) The abnormal phosphorylation of tau protein at Ser 202 in Alzheimer disease recapitulates phosphorylation during development. *Proc. Natl. Acad Sci. USA* **90**, 5066–5070.

Goldstein, M., Lee, K. Y., Lew, J. Y., Harada, K., Wu, J., Haycock, J. W., Hokfelt, T., and Deutch, A. Y. (1995) Antibodies to a segment of tyrosine hydroxylase phosphorylated at serine-40. *J. Neurochem.* **64**, 2281–2287.

Gordon-Weeks, P. R., Mansfield, S. G., Alberto, C., Johnstone, M., and Moya, F. (1993) A phosphorylation epitope on MAP 1B that is transiently expressed in growing axons in the developing rat nervous system. *Eur. J. Neurosci.* **5**, 1302–1311.

Greengard, P., Valtorta, F., Czernik, A. J., and Benfenati, F. (1993) Synaptic vesicle phosphoproteins and regulation of synaptic function. *Science* **259**, 780–785.

Grundke-Iqbal, I., Iqbal, K., Tung, Y.-C., Quinlan, M., Wisniewski, H. M., and Binder, L. I. (1986) Abnormal phosphorylation of the microtubule-associated protein τ (tau) in Alzheimer cytoskeletal pathology. *Proc. Natl. Acad. Sci. USA* **83**, 4913–4917.

Harlow, E. and Lane, D. (1988) *Antibodies: A Laboratory Manual.* Cold Spring Harbor Laboratory Press, Cold Spring Harbor, NY, pp. 288–318.

Hemmings, H. C., Jr., Nairn, A. C., Elliott, J. I., and Greengard, P. (1990) Synthetic peptide analogs of DARPP-32 (Mr 32,000 dopamine- and cAMP-regulated phosphoprotein), an inhibitor of protein phosphatase-1. Phosphorylation, dephosphorylation, and inhibitory activity. *J. Biol. Chem* **265**, 20,369–20,376

Iwata, S.-I., Hewlett, G. H. K., Ferrell, S. T., Czernik, A. J., Meiri, K. F., and Gnegy, M. E. (1996) Increased in vivo phosphorylation state of neuromodulin and synapsin I in striatum from rats treated with repeated amphetamine. *J. Pharmacol. Exp. Therap.* **278,** 1428–1434.

Jovanovic, J. N., Benfenati, F., Siow, Y. L., Sihra, T. S., Sanghera, J. S., Pelech, S. L., Greengard, P., and Czernik, A. J. (1996) Neurotrophins stimulate phosphorylation of synapsin I by MAP kinase and regulate synapsin I-actin interactions. *Proc. Natl. Acad. Sci. USA* **93,** 3679–3683.

Kemp, B. E. and Pearson, R. B. (1990) Protein kinase recognition sequence motifs. *Trends Biochem. Sci.* **15,** 342–346.

Kennelly, P. J. and Krebs, E. G. (1991) Consensus sequences as substrate specificity determinants for protein kinases and protein phosphatases. *J. Biol. Chem.* **266,** 15,555–15,558.

Kosik, K. S., Duffy, L. K., Dowling, M. M., Abraham, C., McCluskey, A., and Selkoe, D. J. (1984) Microtubule-associated protein 2: monoclonal antibodies demonstrate the selective incorporation of certain epitopes into Alzheimer neurofibrillary tangles. *Proc. Natl. Acad. Sci. USA* **81,** 7941,7942.

Kwon, Y.-G., Lee, S.-Y., Choi, Y., Nairn, A. C., and Greengard, P. (1997) Cell cycle-dependent regulation of mammalian protein phosphatase-1 by threonine-230 phosphorylation. (Submitted for publication).

Lee, V. M., Carden, M. J., Schlaepfer, W. W., and Trojanowski, J. Q. (1987) Monoclonal antibodies distinguish several differentially phosphorylated states of the two largest rat neurofilament subunits (NF-H and NF-M) and demonstrate their existence in the normal nervous system of adult rats. *J. Neurosci.* **7,** 3474–3488.

Lee, V. M.-Y., Balin, B. J., Otvos, L., and Trojanowski, J. Q. (1991) A68: a major subunit of paired helical filaments and derivatized forms of normal tau. *Science* **251,** 675–678.

Lerner, R. A. (1982) Tapping the immunological repertoire to produce antibodies of predetermined specificity. *Nature (London)* **299,** 593–596.

Li, L., Chin, L.-S., Shupliakov, O., Brodin, L., Sihra, T. S., Hvalby, Ø., Jensen, V., Zheng, D., McNamara, J. O., Greengard, P., and Andersen, P. (1995) Impairment of synaptic vesicle clustering and of synaptic transmission, and increased seizure propensity, in synapsin I-deficient mice. *Proc. Natl. Acad. Sci. USA* **92,** 9235–9239.

Lichtenberg-Kraag, B., Mandelkow, E.-M., Biernat, J., Steiner, B., Schröter, C., Gustke, N., Meyer, H. E., and Mandelkow, E. (1992) Phosphorylation-dependent epitopes of neurofilament antibodies on tau protein and relationship with Alzheimer tau. *Proc. Natl. Acad. Sci. USA* **89,** 5384–5388.

Luca, F. C., Bloom, G. S., and Vallee, R. B. (1986) A monoclonal antibody that cross-reacts with phosphorylated epitopes on two microtubule-associated proteins and two neurofilament polypeptides. *Proc. Natl. Acad. Sci. USA* **83,** 1006–1010.

Marin, P., Nastiuk, K. L., Daniel, N., Girauit, J.-A., Czernik, A. J., Glowinski, J., Nairn, A. C., and Prémont, J. (1996) Glutamate-dependent phosphorylation of elongation factor-2 and inhibition of protein synthesis in neurons, submitted for publication.

Mathers, J. C., Gharahdaghi, F., and Mische, S. M. (1994) FMOC solid phase synthesis of phosphopeptides, in *Techniques in Protein Chemistry V* (Crabb, J., ed.), Academic, San Diego, CA, pp. 477–484.

Meiri, K. F., Bickerstaff, L. E., and Schwob, J. E. (1991) Monoclonal antibodies show that kinase C phosphorylation of GAP-43 during axonogenesis is both spatially and temporally restricted in vivo. *J. Cell Biol.* **112,** 991–1005.

Nairn, A. C., Detre, J. A., Casnellie, J. E., and Greengard, P. (1982) Serum antibodies that distinguish between the phospho- and dephospho-forms of a phosphoprotein. *Nature (London)* **299,** 734–736.

Nestler, E. J. and Greengard, P (1984) *Protein Phosphorylation in the Nervous System,* Wiley, NY, pp. 96–98; 284–285.

Oishi, M., Nairn, A. C., Czernik, A. J., Lim, G. S., Isohara, T., Gandy, S. E., Greengard, P., and Suzuki, T. (1996) The cytoplasmic domain of the Alzheimer β-amyloid precursor protein is phosphorylated at Thr-654, Ser-655 and Thr-668 in adult rat brain and cultured cells. *Mol. Med.* in press.

Otvos, L., Jr., Feiner, L., Lang, E., Szendrei, G. I., Goedert, M., and Lee, V. M.-Y. (1994) Monoclonal antibody PHF-1 recognizes tau protein phosphorylated at serine residues 396 and 404. *J. Neurosci. Res.* **39,** 669–673.

Patton, B. L., Molloy, S. S., and Kennedy, M. B. (1993) Autophosphorylation of type II CaM kinase in hippocampal neurons: localization of phospho- and dephosphokinase with complementary phosphorylation site-specific antibodies. *Mol. Biol. Cell* **4,** 159–172.

Perich, J. W. and Johns, R. B. (1988) Di-tert-butyl N,N-diethylphosphoramidite. A new phosphitylating agent for the efficient phosphorylation of alcohols. *Tetrahedron Lett.* **29,** 2369–2372.

Pieribone, V., Shupliakov, O., Brodin, L., Hilfiker-Rothenfluh, S., Czernik, A. J., and Greengard, P. (1995) Distinct pools of vesicles in neurotransmitter release. *Nature (London)* **375,** 493–497.

Riederer, B. M. (1995) Differential phosphorylation of MAP1b during postnatal development of the cat brain. *J. Neurocytol.* **24,** 45–54.

Ross, A. H., Baltimore, D., and Eisen, H. N. (1981) Phosphotyrosine-containing proteins isolated by affinity chromatography with antibodies to a synthetic hapten. *Nature (London)* **294,** 654–656.

Schlesinger, D. H., Buku, A., Wyssbrod, H. R., and Hay, D. I. (1987) Chemical synthesis of phosphoseryl-phosphoserine, a partial analogue of human salivary statherin, a protein inhibitor of calcium phosphate precipitation in human saliva. *Int. J. Pep. Protein Res.* **30,** 257–262.

Smith, S. C., McAdam, W. J., Kemp, B. E., Morgan, F. J., and Cotton, R. G. H. (1987) A monoclonal antibody to the phosphorylated form of phenylalanine hydroxylase. *Biochem. J.* **244,** 625–631.

Snyder, G. L., Girault, J.-A., Chen, J. Y. C., Czernik, A. J., Kebabian, J. W., Nathanson, J. A., and Greengard, P. (1992) Phosphorylation of DARPP-32 and protein phosphatase inhibitor-1: regulation by factors other than dopamine. *J. Neurosci.* **12,** 3071–3083.

Sternberger, L. A. and Sternberger, N. H. (1983) Monoclonal antibodies distinguish phosphorylated and nonphosphorylated forms of neurofilaments *in situ. Proc. Natl. Acad. Sci. USA* **80,** 6126–6130.

Sutcliffe, J. G., Shinnick ,T. M., Green, N., and Lerner, R. A. (1983) Antibodies that react with predetermined sites on proteins. *Science* **219,** 660–666.

Suzuki, T., Oishi, M., Marshak, D. R., Czernik, A. J., Nairn, A. C., and Greengard, P. (1994) Cell cycle-dependent regulation of the phosphorylation and metabolism of the Alzheimer amyloid precursor protein. *EMBO J.* **13,** 1114–1122.

Suzuki, T., Okumura-Noji, K., Ogura, A., Kudo, Y., and Tanaka, R. (1992) Antibody specific for the Thr-286-autophosphorylated α subunit of Ca^{2+}/calmodulin-dependent protein kinase II. *Proc. Natl. Acad. Sci.-USA* **89,** 109–113.

Walter, G. (1986) Production and use of antibodies against synthetic peptides. *J. Immunol. Methods,* **88,** 149–161.

Westendorf, J. M, Rao, P. N., and Gerace, L. (1994) Cloning of cDNAs for M-phase phosphoproteins recognized by the MPM2 monoclonal antibody and determination of the phosphorylated epitope. *Proc. Natl. Acad. Sci. USA* **91,** 714–718.

Yamagata, Y., Obata, K., Greengard, P., and Czernik, A. J. (1995) Increase in synapsin I phosphorylation implicates a presynaptic component in septal kindling. *Neuroscience* **64,** 1–4.

Yano, T., Taura, C., Shibata, M., Hirono, Y., Ando, S., Kusubata, M., Takahashi, T., and Inagaki, M. (1991) A monoclonal antibody to the phosphorylated form of glial fibrillary acidic protein: application to a non-radioactive method for measuring protein kinase activities. *Biochem. Biophys. Res. Commun.* **175,** 1144–1151.

Protein Tyrosine Phosphorylation

Pascal Derkinderen and Jean-Antoine Girault

1. Introduction

Phosphorylated tyrosyl residues in proteins are quantitatively much less abundant in normal cells than phosphoserines or phosphothreonines, yet tyrosine phosphorylation plays a crucial role in cellular regulation, including cell transformation by viral oncogenes and the action of growth factors (Hunter and Cooper, 1985). The role of tyrosine phosphorylation in signal transduction has been neglected for some time in mature neurons, which are postmitotic cells, although the situation has changed dramatically over the past few years. For example, it has been discovered that receptors for neurotrophic factors, the chief example being nerve growth factor (NGF), are transmembrane protein-tyrosine kinases (PTKs) (Ip and Yancopoulos, 1994). In addition, molecular cloning techniques have allowed the identification of many protein-tyrosine kinases and phosphatases that are exclusively or preferentially expressed in nerve cells, in vertebrates as well as in *Drosophila*. Several PTKs and protein-tyrosine phosphatases (PTPases) are critical in the regulation of neural differentiation, neurite growth, and pathfinding (Temple and Qian, 1995; Tessier-Lavigne, 1995). However, the wealth of structural information has not been fully exploited, since the function and regulation of many of the cloned enzymes are not yet known. A number of important neuronal proteins, including ion channels and neurotransmitter receptors, are phosphorylated on tyrosine and, in several

From: *Neuromethods, Vol. 30: Regulatory Protein Modification: Techniques and Protocols*
Ed: H. C. Hemmings, Jr. Humana Press Inc.

cases, phosphorylation alters their functional properties (*see* Section 2.4.). Finally, recent work in several laboratories has revealed that classical neurotransmitters and membrane depolarization can stimulate tyrosine phosphorylation pathways in neurons (Woodrow et al., 1992; Huang et al., 1993; Siciliano et al., 1994; Lev et al., 1995). Thus, it appears that tyrosine phosphorylation is a universal signaling mechanism that is likely to play a critical role in neuronal functions, during development as well as in the adult, in response to trophic factors, extracellular matrix, cell-surface molecules, neurotransmitters, and the action potential. Perturbations of tyrosine phosphorylation signaling are likely to be involved in the physiopathology of some psychiatric or neurologic diseases. Although this area is just beginning to be explored, there are already suggestions of its role in epilepsy, ischemia, and Alzheimer's disease, although it is too early to judge whether such observations will provide interesting clues concerning the causes or treatments of these diseases.

2. Overview of Tyrosine Phosphorylation in the Nervous System

2.1. Protein-Tyrosine Kinases

PTKs possess a catalytic domain that is structurally homologous to that of protein-serine/threonine kinases (Hanks and Hunter, 1995). PTKs form a subgroup of closely related enzymes in the vast family of protein kinases, and they contain a few specific conserved residues that allow them to be distinguished from protein-serine/threonine kinases based on their deduced amino acid sequences. The conservation in protein kinase primary structure is matched by a high degree of conservation in the three-dimensional structures of eukaryotic protein kinases determined by X-ray crystallography for several protein-serine/threonine kinases (Taylor et al., 1993) and for a PTK, the catalytic domain of the insulin receptor (Hubbard et al., 1994). In addition to a catalytic domain, protein kinases contain a number of other domains that are responsible for their targeting, regulation, and specificity of action. PTKs fall into two categories: receptor tyrosine kinases, which possess a single transmembrane domain, and nonreceptor tyrosine kinases, which do not. Activation of receptor tyrosine kinases occurs following binding of a ligand, usually a peptidic growth factor, to the extracellular domain, dimerization of the receptor, and auto-phosphorylation of its intracellular kinase domain. These reactions, as well as the cascade of signaling events that they trigger, have been clarified recently (*see* Schlessinger and Ullrich, 1992; Schlessinger, 1993; Van der Geer et al., 1994). Nonreceptor tyrosine kinases, on the other hand, form a relatively heterogenous group of enzymes (Bolen, 1993). Many of these enzymes contain well-defined domains, initially described in the product of the proto-oncogene *c-src*, which are important for targeting and protein-protein interactions. SH2 (Src homology 2) domains interact specifically with peptides phosphorylated on tyrosine, whereas Src homology 3 (SH3) domains interact with proline-rich peptides (Pawson, 1995). Additional conserved domains have been identified, including the pleckstrin homology (PH) domain, which is found in the Btk subfamily of PTKs and which allows interactions with $\beta\gamma$-subunits of heterotrimeric G-proteins and with phosphatidyl-3,4,5P (Riddihough, 1994). Activation of nonreceptor tyrosine kinases can be achieved by interaction with membrane receptors that do not possess a kinase domain, as for the JAK1, JAK2, and Tyk2 PTKs, which associate with receptors for cytokines and behave, to some extent, as separate subunits of these receptors (Taniguchi, 1995). PTKs of the Src subfamily, including Fyn, Yes, and many others, are regulated by phosphorylation of a carboxy-terminal tyrosine (Cooper and Howell, 1993). When this residue is phosphorylated, it inhibits the activity of the enzyme by interacting with its SH2 domain. Activation is achieved by dephosphorylation or by interaction with other proteins. Another example of a nonreceptor tyrosine kinase is focal adhesion kinase (FAK), a PTK without SH2, SH3, or PH domains, which is enriched in focal adhesions, where it interacts with β integrins and other proteins associated with actin filaments (Schaller and Parsons, 1994). FAK transduces signals in response to cell interactions with extracellular matrix components and is highly expressed in neurons, including in growth cones (Burgaya et al., 1995). Recently, a kinase homologous to FAK, called PYK2 (Lev et al., 1995) or Cakβ (Sasaki et al., 1995), has been shown to be enriched in nervous tissue and to be activated in response to Ca^{2+}.

2.2. Protein-Tyrosine Phosphatases (PTPases)

On the basis of amino acid sequence comparison in the catalytic domain, PTPs belong to two distinct families of enzymes unrelated to other protein phosphatases (Walton and Dixon, 1993; Brady-Kalnay and Tonks, 1994; Stone and Dixon, 1994). The largest family comprises most of the known PTPs, as well as dual-specificity tyrosine phosphatases, which are able to dephosphorylate seryl and threonyl as well as tyrosyl residues. The smallest family corresponds to low-mol-wt PTPs. Interestingly, in spite of the lack of identified sequence homology between these two families, their three-dimensional structures appear similar, suggesting a possible common origin (Barford et al., 1994; Su et al., 1994). Like PTKs, many PTPs possess an extracellular domain, a single transmembrane domain, and an intracellular catalytic domain, often found in tandem. This has led to the suggestion that this class of enzymes may function as receptors, although no bona fide ligand activating these PTPs has yet been identified. Other PTPs are devoid of transmembrane domains, but contain various targeting domains, including SH2 domains (Mauro and Dixon, 1994). Purified PTPs often possess a very high specific activity, and little is known about their control in intact cells, although proper targeting is thought to play an important role in their regulation and function.

2.3. Biochemical Consequences
of Tyrosine Phosphorylation

Phosphorylation of a tyrosyl residue changes its charge and its bulk size. In several instances, phosphorylation of specific tyrosines has been found to modify the catalytic activity of enzymes (Nishibe et al., 1990) or the properties of ion channels (*see* Section 2.4.). In the absence of the knowledge of the three-dimensional structure of phosphorylated and nonphosphorylated forms of these proteins, the exact mechanisms of these effects are not known with certainty. However, as in the case of seryl phosphorylation, local changes in charge and steric hindrance and global conformational changes are likely to contribute in various proportions to the effects of tyrosyl phosphorylation (Johnson and Barford, 1993). In addition, phosphorylated tyrosyl residues have a unique property; they mediate regulated protein–protein interactions with a high degree of affinity and specificity. Many enzymes, including PTKs, PTPs, phospholipase Cγ, phosphatidyl inositol-3 kinase, adapter proteins, transcription factors, and cytoskeletal proteins, contain SH2 domains (Pawson, 1995). These domains comprise ~100 amino acids and provide a high-affinity binding site for specific peptidic sequences that contain a phosphorylated tyrosyl residue. Phosphorylation of the tyrosyl residue is required for the interaction, but specificity is provided by recognition of several amino acid residues located on the carboxy-terminal side of the phosphorylated tyrosyl residue. A second type of domain of ~200 residues, termed phosphotyrosine binding (PTB), has been identified (Van der Geer and Pawson, 1995). In addition to the phosphorylated tyrosyl residue, PTB domains recognize specific amino acids located on its amino-terminal side. The importance of tyrosine phosphorylation in promoting regulated protein–protein interactions is very apparent in the cascade of reactions that follows the activation of growth factor receptors (Van der Geer et al., 1994). Autophosphorylated tyrosyl residues provide specific docking sites for various proteins containing SH2 and/or PTB domains, including phospholipase Cγ, phosphatidylinositol-3 kinase, and adapter proteins. Some of these proteins may become directly activated by tyrosine phosphorylation; others are mainly translocated to their site of action. For instance, autophosphorylation of the EGF receptor recruits to the membrane Grb2, an adapter protein bound to SOS, a guanine nucleotide exchange factor, which transforms the membrane-attached, small G-protein Ras from an inactive GDP-bound form into an active GTP-bound form (Boguski and McCormick, 1993). Active Ras triggers a cascade of protein-serine/threonine kinases, leading to pleiotropic effects, including changes in gene transcription (Marshall, 1994; Mahadevan and Cano, 1995). It should be noted that components of this cascade, mitogen-activated protein (MAP) kinases are phosphorylated on tyrosine as a result of the action of a particular dual specificity kinase, named MAP-kinase kinase.

2.4. Neuronal Proteins Phosphorylated on Tyrosine

Many proteins regulated by tyrosine phosphorylation that have been described in nonneuronal tissues are known to be, or likely to be, present in neurons. In addition, there are several examples of tyrosine-phosphorylated proteins of specific significance for neurobiology, including neurotransmitter receptors and ion channels. The first to be identified was the nicotinic acetylcholine receptor, in *Torpedo* electric organ and at the neuromuscular junction (Huganir et al., 1984; Qu et al., 1990). Phosphorylation of the receptor is regulated by innervation and by agrin, an extracellular matrix protein (Qu and Huganir, 1994), and alters its desensitization rate (Hopfield et al., 1988). It may also be involved in the clustering of receptors (Baker and Peng, 1993). NMDA glutamate receptor subunits NR2A and NR2B are also phosphorylated on tyrosine in brain (Moon et al., 1994; Lau and Huganir, 1995; Menegoz et al., 1995). Stimulation of tyrosine phosphorylation in intact neurons increases NMDA responses (Wang and Salter, 1994), although it is not known whether this effect is owing to the direct phosphorylation of the receptor. Phosphorylation of $GABA_A$ receptors on tyrosine increases their conductance (Moss et al., 1995; Valenzuela et al., 1995). On the other hand, RAK K^+ channels are inhibited by phosphorylation on tyrosine (Huang et al., 1993). Additional neuronal proteins have been reported to be phosphorylated on tyrosine, including the synaptic vesicle protein synaptophysin (Barnekow et al., 1990), but the physiological consequences of these observations are not known.

2.5. Regulation of Protein Tyrosine Phosphorylation in the Nervous Tissue

As in other tissues, growth factors represent a major factor controlling tyrosine phosphorylation in the nervous system (Chao, 1992; Barbacid, 1993). In addition, recent work has revealed that classical neurotransmitters and neuronal activity are also able to activate tyrosine phosphorylation pathways. The K^+ channel RAK is phosphorylated on tyrosine in response to stimulation of muscarinic acetylcholine receptors and activation of protein kinase C (PKC) (Huang et al., 1993). In synaptosomes, depolarization induces a Ca^{2+}-dependent increase in protein tyrosine phosphorylation (Woodrow et al., 1992). In rat hippocampal slices and in neurons in primary culture, stimulation of several G protein-coupled receptors, including muscarinic acetylcholine, noradrenergic, and glutamate metabotropic receptors, as well as activation of ionotropic glutamate receptors or depolarization, activates tyrosine phosphorylation (Siciliano et al., 1994). This effect requires activation of PKC and increases in intracellular Ca^{2+}. Recent work has revealed that FAK is phosphorylated in neurons in response to these agents (Siciliano et al., submitted) or to anandamide (Derkinderen et al., 1996). A related kinase, PYK2/Cakβ, which is highly responsive to stimuli that increase intracellular Ca^{2+} and PKC activity in PC12 cells (Lev et al., 1995), may also be involved.

In addition to these results obtained in vitro, there is evidence that protein tyrosine phosphorylation can be altered in vivo in physiological or pathological circumstances. A marked increase in the phosphorylation of various substrates has been observed in the insular gustatory cortex of rats following learning in a paradigm of conditioned taste aversion (Rosenblum et al., 1995). Tyrosine phosphorylation of several proteins has been found to increase in the striatum of rats following lesions of dopamine neurons or chronic neuroleptic treatment (Girault et al., 1992). One of the proteins altered is the NMDA receptor NR2B subunit (Menegoz et al., 1995). Finally, changes in protein phosphorylation have been reported in rat brain following electroconvulsive shock (Lee et al., 1993), seizures (Jope et al., 1991), and ischemia (Hu and Wieloch, 1993). Although the biochemical mechanisms underlying these effects and the identity of the phosphorylated proteins remain to be established, they suggest a possible role of protein tyrosine phosphorylation in the consequences of or the recovery from various types of brain insults.

3. Methods to Study Protein Tyrosine Phosphorylation and Dephosphorylation in Nervous Tissue

3.1. Overview of the Methods Available

There are many useful and complementary approaches to the study of protein tyrosine phosphorylation in the nervous system. The state of phosphorylation of proteins in intact cells can be studied by antiphosphotyrosine immunoblotting. The levels and activities of identified PTKs and of PTPs can be measured. In addition, it may be desirable to detect interactions, usually in immunoprecipitates, between phosphorylated proteins and other proteins mediated by SH2 or PTB domains, or to measure changes in the activity of associated enzymes. Another approach is to test the role of protein tyrosine phosphorylation in specific regulatory pathways or functions. To do that, one can use PTK or phosphatase inhibitors (*see* Chapter 4). In the course of electrophysiological experiments, it is also possible to inject into recorded cells purified PTKs or PTPs (Wang and Salter, 1994; Moss et al., 1995). Finally, interesting information about the physiological function of PTKs or PTPs can be obtained using transgenic mice in which a specific gene is overexpressed or has been knocked-out by homologous recombination in embryonic stem cells (Grant et al., 1992). It is well beyond the scope of this chapter to describe all of these approaches, some of which are described in detail in other chapters of this book. Instead, we focus on specific methodological aspects of the measurement of tyrosine phosphorylation and PTK activities in the nervous system.

3.2. Model Systems for Studying the Regulation of Tyrosine Phosphorylation

3.2.1. Intact Animals

From a physiological standpoint, it would be very important to determine the state of phosphorylation of identified proteins in the brain of intact animals, in various physiological or pathological circumstances. Unfortunately, this approach is hampered by the problems related to the collection of tissues. Protein phosphorylation is a highly reversible process and is under dynamic control by protein kinases and phosphatases. This equilibrium may be severely disturbed during the sacrifice and dissection of the brain, because of the biochemical changes subsequent to postmortem ischemia, including, but not limited to, massive release of neurotransmitters, increases in cytoplasmic cAMP and Ca^{2+} concentrations, and a drop in ATP levels. These problems can be minimized by the use of a micropunching technique (Hervé et al., 1989). Rat or mouse brains are rapidly dissected. Forebrain is separated from brainstem and cerebellum, dropped in liquid nitrogen or isopentane cooled to –30°C with dry ice, and stored at –80°C until use. To prevent dephosphorylation, which may occur during brain dissection, depending on the experimental problems that are investigated, freezing the brain *in situ* in anesthetized animals (Haycock and Haycock, 1991) or using microwave irradiation (O'Callaghan et al., 1983; Jope et al., 1991) should also be considered. Frozen brains, forebrain, or hindbrain/cerebellum blocks are warmed to approx –15°C and placed on the platine of a refrigerated microtome. Coronal sections (~500 μm) are cut and placed on the refrigerated platine, without thawing. Microdisks are punched out from identified regions of these sections, using a cannula (0.5-mm id), and transferred to polypropylene tubes placed on dry ice.

3.2.2. Brain Slices

Brain slices provide a model as close as possible to the mature tissue, despite clear limitations owing to the damage occurring during slice preparation. A procedure designed to minimize tissue damage includes preincubation at 4°C and rewarming in low Ca^{2+} buffer (Halpain et al., 1990; Halpain and Girault, 1995). Rat brains are dissected and placed immediately in saline solution at 4°C. Hippocampal formations or caudate-putamens are dissected and 300-μm thick slices are obtained using a McIlwain tissue chopper. Slices are dissociated in Ca^{2+}-free artificial cerebrospinal fluid (CF-ACSF) at 4°C containing (in mM): NaCl: 125; KCl: 2.4; $MgCl_2$: 0.83; KH_2PO_4: 0.5; Na_2SO_4: 0.5; $NaHCO_3$: 27; glucose: 10; HEPES 10, pH 7.4. Individual slices are transferred to polypropylene tubes (3 slices/tube) containing 1 mL CF-ACSF at 35°C for 10 min. CF-ACSF is gently aspirated and replaced with 0.9 mL of ACSF containing 1.1 mM $CaCl_2$. Tubes are incubated at 35°C under a humidified atmosphere of 95% O_2/5% CO_2 with moderate shaking for 60 min until pharmacological treatments. After treatments, ACSF is aspirated and slices are immediately homogenized or placed in dry ice and kept at −80°C until analysis.

3.2.3. Primary Neuronal or Astrocytic Cultures

Cell cultures present a number of advantages, including better accessibility to drugs or toxins and ease of rapidly solubilizing proteins under conditions that inactivate protein kinases and phosphatases. However, several drawbacks should be kept in mind. First, neurons in culture may not express all the proteins that are present in mature neurons, or express them at very low levels. Second, levels of basal phosphorylation are often relatively high, making more difficult the detection of pharmacological effects of treatments.

3.3. Antiphosphotyrosine Antibodies

It is relatively easy to obtain antibodies that react specifically with phosphotyrosine, and such antibodies have been used extensively for immunoblotting, immunocytochemistry, and immunoprecipitation. Several excellent monoclonal antibodies (MAbs) are available commercially (4G-10, UBI, Lake Pleniol, NJ, P420; Santa Cruz Biotechnology, Santa Cruz, CA). A summarized protocol is provided below for preparing and purifying polyclonal rabbit antiserum specific for phosphotyrosine (further details can be found in Siciliano et al., 1992).

3.3.1. Preparation of Antiphosphotyrosine Rabbit Antiserum

O-Phospho-L-tyrosine (P-Tyr, Sigma, St. Louis, MO) is conjugated to thyroglobulin (Sigma) by mixing 1 mL of P-Tyr solution (50 mg/mL in 150 mM NaCl) with 1 mL of thyroglobulin (20 mg/mL in 150 mM NaCl). The coupling reaction is started by the addition of 120 mg of 1-ethyl,3-(3 dimethylaminopropyl) carbodiimide-HCl (Sigma). The final pH is adjusted to 5.5 using NaOH. The mixture is gently stirred for 18 h at room temperature and then dialyzed for 48 h against 4 L of 150 mM NaCl and 10 mM NaHPO$_4$, pH 7.4 (PBS), with one buffer change. The final volume is adjusted to 3 mL, to give a final concentration of thyroglobulin of 6 mg/mL (~10 μM). Before immunization, a preimmune serum sample is obtained from each rabbit and stored at –20°C. Rabbits, sedated with acepromazine (1.5 mg/kg, im), are immunized with 450 μL of thyroglobulin-P-Tyr conjugate (2.7 mg) mixed with 1550 μL of complete Freund's adjuvant (Sigma). The antigen emulsion is injected intradermally in the shaved skin of the paraspinal region (~100 μL/site), using a Luer-type glass syringe fitted with a 26-gage needle (0.45 × 12 mm). On d 21, animals are boosted with 200 μL of antigen (1.2 mg) mixed with 200 μL of incomplete Freund's adjuvant (Sigma, St. Louis, MO). Further boosts are administered at 2–4 wk intervals, as required. Serum is collected weekly and stored at –20°C.

3.3.2. Affinity-Purification
of Antiphosphotyrosine Antibodies

Rabbit anti-P-Tyr antibodies are purified by affinity chromatography using phosphotyrosine coupled to Sepharose 4B (Pharmacia, Piscataway, NJ), according to the instructions provided by the manufacturer. Approximately 45 μmol of phosphoamino acid are coupled with 1 g of Sepharose 4B. The column (4–5 mL bed volume) is washed with 5 vol of 50 mM Tris-HCl, 150 mM NaCl, and 0.1% (v/v) Tween-20, pH 7.5. The serum sample is filtered through a 0.22-μm Millex-GV filter, and protease inhibitors are added to give final concentrations of 25 mM benzamidine, 5 mM EDTA, 1 mM EGTA, 100 μM phenylmethylsulfonyl fluoride (dissolved in anhydrous 2-propanol), 20 μg/mL leupeptin, and 5 μg/mL pepstatin (dissolved in DMSO). A serum sample (20–40 mL) is loaded on the column and the flowthrough is recirculated once. The column is washed successively with the following buffers (20 mL):

1. 50 mM Tris-HCl, pH 7.5, 150 mM NaCl, 0.1% (v/v) Tween-20;
2. 25 mM Na borate/boric acid, pH 8.3, 1M NaCl, 0.1% Tween-20;
3. 50 mM Na acetate, pH 5.5, 1M NaCl, 0.1% Tween-20; and
4. 50 mM Tris-HCl, pH 7.5, 150 mM NaCl, pH 7.4.

Bound antibodies are then eluted with 6 mL of 4.3M MgCl$_2$, pH 6.5. The eluate is diluted 1:1 with H$_2$O to prevent MgCl$_2$ precipitation and dialyzed for 48 h against 5 L of 10 mM HEPES and 150 mM NaCl, pH 7.6, with four buffer changes. The protein concentration is adjusted to 1 mg/mL, and NaN$_3$ is added to give a final concentration of 0.02% (w/v). Aliquots are stored at –4°C. Screening of serum samples and affinity-purification fractions is done by dot immunoblotting as described elsewhere (Siciliano et al., 1992).

3.4. Immunoblotting
with Antiphosphotyrosine Antibodies

3.4.1. Sample Preparation,
Electrophoresis, and Transfer to Nitrocellulose

Samples from cell culture or intact tissue are homogenized by sonication in a hot (~100°C) solution of 1% sodium dodecyl sulfate (SDS) (w/v) and 1 mM sodium orthovanadate (Na_3VO_4), and placed in a boiling water bath for 5 min in air-tight, screw-cap tubes. Tubes are then stored at –20°C until biochemical analysis. Protein concentration in homogenates is measured with a bicinchoninic acid-based assay (Smith et al., 1985), using bovine serum albumin (BSA) as standard, and adjusted to the desired value. An aliquot of 5X sample buffer is added to give final concentrations of 3% SDS (w/v), 60 mM Tris-HCl, 5% glycerol (v/v), and 0.3M β-mercaptoethanol, and samples are boiled for 5 min. Equal amounts of protein (50–100 μg) are subjected to SDS-polyacrylamide gel electrophoresis (PAGE) and transferred to nitrocellulose sheets (nitrocellulose 0.2-μm pore size, Schleicher & Schuell [Keene, NH] or Hybond-C ECL, Amersham [Arlington Heights, IL]). Electrophoretic transfer is performed in a vertical, liquid-phase apparatus (Hoeffer-Pharmacia Biotech, San Francisco, CA) for 8 h at 500 mA. Immediately after transfer, proteins are fixed in a solution of 10% isopropanol and 10% acetic acid (v/v) in water for 10 min, and then washed extensively in deionized water. Membranes are processed immediately or stored dried at 4°C.

3.4.2. Immunoblotting
with Antiphosphotyrosine Antibodies

All incubations of nitrocellulose membranes are performed in plastic boxes at room temperature with gentle shaking. The use of milk-containing buffers as blocking solutions for anti-P-Tyr antibodies may result in an intense background. This problem can be avoided by using a nonionic detergent as blocking reagent. Membranes are incubated for 1 h in a solution of TBS-Tween (200 mM NaCl, 50 mM Tris-HCl, pH 7.4, and 0.1% Tween-20, v/v). After blocking, nitrocellulose sheets are incubated for 2 h with anti-P-Tyr affinity-purified polyclonal antibodies or MAb in TBS-Tween. Membranes are then washed three times in TBS-Tween for 10 min.

3.4.3. Detection of Antiphosphotyrosine Antibodies

Various methods can be used for revealing bound antibodies. We find that the most sensitive and convenient one is the use of horseradish peroxidase-(HRP-)coupled secondary antibodies and chemiluminescence autoradiographic detection. In our experience, however, the presence of Tween-20 is not sufficient to prevent the appearance of nonspecific bands resulting from the direct binding of the HRP moiety itself to proteins on the membrane. To overcome this problem, following incubation with anti-P-Tyr antibody and washes, membranes are incubated for 1 h in TBS-milk (a solution of 5% [w/v] of nonfat dry milk in TBS) (Johnson et al., 1984). After blocking, membranes are incubated for 2 h with secondary antibodies coupled to HRP in TBS-milk, and then washed 3 × 15 min in TBS-milk followed by 3 × 15 min in TBS. Overextension of the times of blocking or washing with TBS-milk may lead to a reduction in signal intensity. Washed membranes are incubated for 1 min in chemiluminescent reagent (ECL; Amersham or Renaissance; Dupont-NEN [Wilmington, DE]) prepared extemporaneously according to the instructions of the manufacturer. Excess liquid is drained gently; membranes are wrapped in plastic film (Saran, Dow Chemical Company) and used immediately to expose an autoradiographic film. Any film can be used; the cheaper, probably the better. We use routinely Reflection from DuPont NEN. Exposure times vary from a few seconds to dozens of minutes, depending on signal intensity. When the signal is very intense, it is better to decrease the amount of protein loaded on the gel, since too short exposure times give unreliable results. Immunoreactive bands can be quantified by measuring their optical density with a computerized image analyzer. If the amount of protein loaded on the gel, the

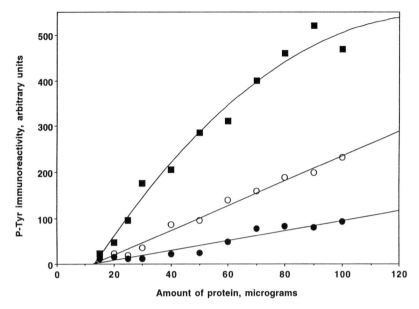

Fig. 1. Linear variation of the optical density of phosphotyrosine immunoreactive bands as a function of the amount of protein loaded. Rat hippocampal slices were homogenized by sonication in SDS/orthovanadate as described in the text and protein concentration was measured with a bicinchoninic acid-based assay, using BSA as a standard. Various amounts of protein were separated by SDS/PAGE (8% acrylamide) and analyzed by immunoblotting with an MAb against phosphotyrosine (4G10, UBI). Primary antibodies were detected with HRP-coupled antimouse immunoglobulins and chemiluminescence autoradiography (ECL, Amersham, exposure time 90 s, Reflection film). Optical density of immunoreactive bands was measured with a computer-assisted scanner (Agfa) and NIH Image software. The results, expressed as uncalibrated OD × 100, correspond to three different protein bands (closed squares: 180 kDa, open circles: 120 kDa, closed circles: 60 kDa).

immunoblotting conditions, and the exposure times are carefully selected, it is possible to obtain a signal that varies linearly in the range studied (Fig. 1). After detection, membranes are washed in TBS prior to drying and storage in plastic film. If antibodies are to be stripped, it is preferable not to dry the membrane, but to keep it in TBS at 4°C. The specificity of anti-P-Tyr antibodies can be verified in competition experiments in which the antibodies are preincubated with *O*-phospho-L-tyrosine, O-phospho-L-serine, or O-phospho-DL-threo-nine, before being placed in contact with the nitrocellulose membranes (final concentrations of phosphoamino acids: 50–100 μM).

3.4.4. Reprobing Nitrocellulose Membranes with Different Antibodies

In some instances, it can be very useful to reprobe nitro-cellulose membranes with other antibodies for identification of specific proteins or quantification of glial or neuronal markers. When sequential immunoblotting is planned, it is advisable to use at first the antibody that gives the weakest signal. To remove primary and secondary antibodies, mem-branes are incubated for 1 h in a solution of $8M$ urea, $0.1M$ β-mercaptoethanol, and 1% (v/v) Tween-20 at 60°C (modi-fied from Erickson et al., 1982). Membranes are then exten-sively washed in water and re-equilibrated in TBS. This procedure may result in the loss of proteins from the mem-brane. A milder, although less efficient procedure consists of incubating the membrane for 1 h in the presence of 100 mM glycine, pH 2.5, 200 mM NaCl, $0.1M$ β-mercaptoethanol, and 0.1% (v/v) Tween-20 at room temperature for 1 h. The effi-ciency of the elution of the antibodies should always be checked by reincubating the membrane in secondary anti-bodies and revealing them with chemiluminescence.

3.5. Immunoprecipitation

To demonstrate that an identified protein is phosphory-lated on tyrosine, it is necessary to use immunoprecipation with specific antibodies, followed by anti-P-Tyr immunoblot-ting. Various procedures for immunoprecipitation can be used, each of which has very distinct disadvantages. In most instances, it is preferable to use nondenaturing conditions of homogenization. This allows the use of any immunoprecip-itating antibodies and measurement of enzyme activities fol-lowing immunoprecipitation. However, in some instances, especially when using tissue pieces or slices, it may be neces-sary to use denaturing conditions (i.e., ionic detergent at 100°C) for complete solubilization of the protein of interest and total inactivation of phosphatases. Although this treatment precludes the study of any enzymatic activity, it is nevertheless possible to carry out immunoprecipitation from such samples.

3.5.1. Nondenaturing Homogenization for Immunoprecipitation

Brain slices or tissue pieces are stored at –80°C until analysis. All procedures should be carried out at 0–4°C, and great care should be taken to avoid warming at any step. Homogenization is carried out in the cold room, using a tissue homogenizer (Heidolph or Polytron) with a probe sufficiently small to fit in 1.5-mL tubes. Just before homogenization, each tube is partly thawed manually, and 450 μL of ice-cold homogenizing buffer are added, which consists of NETF buffer (100 mM NaCl, 50 mM Tris-Cl, pH 7.4, 5 mM EDTA, 50 mM NaF) containing 1 mM Na$_3$VO$_4$, 1% (v/v) Nonidet P-40, and protease inhibitors (Complete, Boehringer Mannheim Biochemical, Indianapolis, IN). Homogenization duration should be brief (<10 s) to prevent heating of the sample, but can be repeated, after a delay, if necessary.

3.5.2. Denaturing Homogenization for Immunoprecipitation

Slices or tissue pieces are solubilized by sonication in a solution of 1% (w/v) SDS in water containing 1 mM Na$_3$VO$_4$ at 100°C and placed for 5 min in a boiling water bath. For example, 300 μL of solution can be used to homogenize three brain microdisks of 0.5-mm diameter, 0.5-mm thick. For immunoprecipitation, 100 μL of 5X concentrated NETF buffer containing protease inhibitors is added, as well as 100 μL of 15% (v/v) Nonidet P-40. This gives a fivefold final concentration excess of Nonidet P-40 over SDS, and is sufficient to "neutralize" SDS and allow immunoprecipitation in many instances. However, this procedure works only for some antibodies, and should be tested cautiously before being applied. For instance, this procedure was used successfully to immunoprecipitate the NMDA receptor subunits NR2A or NR2B from striatal microdisks (Menegoz et al., 1995).

3.5.3. Preparation of Protein A
or Protein G Sepharose for Immunoprecipitation

Protein A or protein G Sepharose beads are supplemented with 1/10 vol of Sephacryl (Pharmacia-LKB) and washed twice in 10 vol of NETF buffer containing 1% (v/v) Nonidet P-40. They are resuspended in 1 vol of NETF buffer containing 25 mg/mL BSA. The addition of Sephacryl beads improves the consistency of the pellets during washing steps following immunoprecipitation. Protein A or G Sepharose prepared this way can be stored at 4°C for several days.

3.5.4. Preparation of Antibodies for Immunoprecipitation

Crude rabbit antisera can be used for immunoprecipitation, but often contain phosphatase activities that are not fully inhibited. To overcome this problem, for each sample to be precipitated, 20 µL of antiserum are preincubated with 50 µL (settled volume) of protein A Sepharose in 500 µL of NETF buffer containing 1% (v/v) Nonidet P-40 at 4°C, for 30 min, in a rotary incubator. The beads are washed twice in 1 mL of NETF buffer containing 1% (v/v) Nonidet P-40. Note that it is not advisable to prepare a bulk quantity of protein A Sepharose–antibody complex and to aliquot it, since it is extremely difficult to measure accurately small volumes of the Sepharose slurry. Purified antibodies may not require the washing step. For most mouse MAb subclasses, protein G Sepharose is used instead of protein A Sepharose.

3.5.5. Immunoprecipitation

Samples prepared as described above (Sections 3.5.1. or 3.5.2.) are precleared from any residual insoluble material by centrifugation for 30 min at maximum speed in a bench centrifuge at 2°C, after the addition of a small amount of inert resin (Sephacryl) as a carrier. Supernatants are transferred to the tubes containing the antibodies preadsorbed on protein A Sepharose and incubated with constant rotation at 4°C for 2 h. Sepharose beads are then collected by a brief centrifugation in a bench centrifuge. Supernatants are saved and beads washed at 4°C twice in 1 mL ice-cold NETF containing 1% (v/v) Nonidet P-40 and 0.1 mM Na_3VO_4, and once without detergent. For analysis of the immunoprecipitated proteins by immunoblotting, usually with anti-P-Tyr antibodies, the pellets are resuspended in 75 μL of 2X sample buffer and placed in a boiling water bath for 5 min. After a brief centrifugation, the supernatants are loaded on SDS-polyacrylamide gels, as described above. The specificity of immunoprecipitation should be verified by carrying out the same procedure using preimmune or unrelated serum instead of immune serum.

3.6. In Vitro Protein Kinase Assays Using Immunoprecipitates

In vitro protein kinase assays are used to monitor the autophosphorylation of immunoprecipitated PTKs or their activity toward exogenous substrates. It should be noted that assay buffers used for measuring PTK activity usually contain Mn^{2+}, which forms a precipitate in the presence of Na_3VO_4. Therefore, Sepharose beads bearing the immune complexes are washed once with 1 mL of NETF without Na_3VO_4 after immunoprecipitation. The beads are then washed once with 1 mL of kinase buffer (50 mM HEPES, pH 7.4, 10 mM $MnCl_2$). After centrifugation, the supernatant is carefully aspirated, 30 μL of kinase buffer with or without exogenous substrates are added, and the beads are resuspended. Phosphorylation is started by adding 5 μCi of [γ-^{32}P]-ATP (3000 Ci/mmol) in 10 μL of kinase buffer with careful vortexing. Tubes are incubated for various times at 25°C, with frequent resuspension of the beads. Reactions are stopped by the addition of 50 μL of 2X sample buffer and placed in a boiling water bath for 5 min. After a brief centrifugation, the supernatants are analyzed by SDS/PAGE and autoradiography. Various exogenous substrates can be used in this type of assay. Enolase (Sigma) is used in the case of Src family PTKs (Feder and Bishop, 1990). Enolase (1 vol) is dialyzed for 16 h at 4°C against four changes (300 vol each) of distilled water and stored in aliquots (1.5 mg/mL) at –20°C. Immediately before the assay, enolase is denatured by incubation with 1 vol of 50 mM acetic acid for 5 min at 30°C and neutralized by the addition of 1/2 vol of 1M HEPES, pH 7.4. When sufficient amounts of kinase are present in the immunoprecipitate, it is advisable to add unlabeled ATP to the [γ-^{32}P]-ATP solution in order to reach a final concentration of 5 µM, a concentration above the K_m of the enzyme for ATP (e.g., K_m = 2.2 µM for enolase phosphorylation and 0.62 µM for autophosphorylation using Src purified from platelets [Feder and Bishop, 1990]).

References

Baker, L. P. and Peng, H. B. (1993) Tyrosine phosphorylation and acetylcholine receptor cluster formation in cultured *Xenopus* muscle cells. *J. Cell Biol.* **120**, 185–195.

Barbacid, M. (1993) Nerve growth factor: a tale of two receptors. *Oncogene* **8**, 2033–2042.

Barford, D., Flint, A. J., and Tonks, N. K. (1994) Crystal structure of human protein tyrosine phosphatase 1B. *Science* **263**, 1397–1404.

Barnekow, A., Jahn, R., and Schartl, M. (1990) Synaptophysin: a substrate for the protein tyrosine kinase pp60$^{c\text{-}src}$ in intact synaptic vesicles. *Oncogene* **5**, 1019–1024.

Boguski, M. S. and McCormick, F. (1993) Proteins regulating Ras and its relatives. *Nature* **366**, 643–654.

Bolen, J. B. (1993) Nonreceptor tyrosine protein kinases. *Oncogene* **8**, 2025–2031.

Brady-Kalnay, S. M. and Tonks, N. K. (1994) Protein tyrosine phosphatases: from structure to function. *Trends Cell Biol.* **4**, 73–76.

Burgaya, F., Menegon, A., Menegoz, M., Valtorta, F., and Girault, J.-A. (1995) Focal adhesion kinase in rat central nervous system. *Eur. J. Neurosci.* **7**, 1810–1821.

Chao, M. V. (1992) Neurotrophin receptors: a window into neuronal differentiation. *Neuron* **9**, 583–593.

Cooper, J. A. and Howell, B. (1993) The when and how of Src regulation. *Cell* **73**, 1051–1054.

Derkinderen, P., Toutant, M., Burgaya, F., Le Bert, M., Siciliano, J. C., de Franciscis, V., Gelman, M., and Girault, J.-A. (1996) Regulation of a neuronal form of focal adhesion kinase by anandamide. *Science* **273**, 1719–1722.

Erickson, P. F., Minier, L. N., and Lasher, R. S. (1982) Quantitative electrophoretic transfer of polypeptides from SDS polyacrylamide gels to nitrocellulose sheets: a method for their re-use in immunoautoradiographic detection of antigens. *J. Immunol. Methods* **51**, 241–249.

Feder, D. and Bishop, J. M. (1990) Purification and enzymatic characterization of pp60^c-src from human platelets. *J. Biol. Chem.* **265**, 8205–8211.

Girault, J.-A., Siciliano, J. C., Robel, L., and Hervé, D. (1992) Stimulation of protein tyrosine phosphorylation in rat striatum following lesion of dopamine neurons and chronic neuroleptic treatment. *Proc. Natl. Acad. Sci. USA* **89**, 2769–2773.

Grant, S. G. N., O'Dell, T. J., Karl, K. A., Stein, P. L., Soriano, P., and Kandel, E. R. (1992) Impaired long-term potentiation, spatial learning, and hippocampal development in fyn mutant mice. *Science* **258**, 1903–1910.

Halpain, S. and Girault, J.-A. (1995) The use of brain slices to study protein phosphatase regulation and function. *Neuroprotocols* **6**, 46–55.

Halpain, S., Girault, J.-A., and Greengard, P. (1990) Activation of NMDA receptors induces dephosphorylation of DARPP-32 in rat striatal slices. *Nature* **343**, 369–372.

Hanks, S. K. and Hunter, T. (1995) Protein kinases 6: the eukaryotic protein kinase superfamily: kinase (catalytic) domain structure and classification. *FASEB J.* **9**, 576–596.

Haycock, J. W. and Haycock, D. A. (1991) Tyrosine hydroxylase in rat brain dopaminergic nerve terminals. Multiple site phosphorylation in vivo and in synaptosomes. *J. Biol. Chem.* **266**, 5650–5657.

Hervé, D., Trovero, F., Blanc, G., Thierry, A. M., Glowinski, J., and Tassin, J. P. (1989) Nondopaminergic prefrontocortical efferent fibers modulate D1 receptor denervation supersensitivity in specific regions of the rat striatum. *J. Neurosci.* **9**, 3699–3708.

Hopfield, J. F., Tank, D. W., Greengard, P., and Huganir, R. L. (1988) Functional modulation of the nicotinic acetylcholine receptor by tyrosine phosphorylation. *Nature* **336**, 677–680.

Hu, B. R. and Wieloch, T. (1993) Changes in tyrosine phosphorylation in neocortex following transient cerebral ischaemia. *Neuroreport* **4**, 219–222.

Huang, X.-Y., Morielli, A. D., and Peralta, E. G. (1993) Tyrosine kinase-dependent suppression of a potassium channel by the G protein-coupled m1 muscarinic acetylcholine receptor. *Cell* **75**, 1145–1156.

Hubbard, S. R., Wei, L., Ellis, L., and Hendrickson, W. A. (1994) Crystal structure of the tyrosine kinase domain of the human insulin receptor. *Nature* **372**, 746–754.

Huganir, R. L., Miles, K., and Greengard, P. (1984) Phosphorylation of the nicotinic acetylcholine receptor by an endogenous tyrosine-specific protein kinase. *Proc. Natl. Acad. Sci. USA* **81**, 6968–6972.

Hunter, T. and Cooper, J. A. (1985) Protein-tyrosine kinases. *Ann. Rev. Biochem.* **54**, 897–930.

Ip, N. Y. and Yancopoulos, G. D. (1994) Neurotrophic factor receptors: just like other growth factor and cytokine receptors. *Curr. Opinion Neurobiol.* **4**, 400–405.

Johnson, D. A., Gautsch, J. W., Sportsman, J. R., and Elder, J. H. (1984) Improved technique utilizing nonfat dry milk for analysis of proteins and nucleic acids transferred to nitrocellulose. *Gene Anal. Technol.* **1**, 3–8.

Johnson, L. N. and Barford, D. (1993) The effects of phosphorylation on the structure and function of proteins. *Ann. Rev. Biophys. Biomol. Struct.* **22**, 199–232.

Jope, R. S., Johnson, G. V. W., and Baird, M. S. (1991) Seizure-induced protein tyrosine phosphorylation in rat brain regions. *Epilepsia* **32**, 755–760.

Lau, L. F. and Huganir, R. L. (1995) Differential tyrosine phosphorylation of N-methyl-D-aspartate receptor subunits. *J. Biol. Chem.* **270**, 20,036–20,041.

Lee, Y. H., Ryu, S. H., Suh, P.-G., Park, J.-B., Ahn, Y.-M., and Kim, Y. S. (1993) Tyrosine phosphorylation of PLC-gamma$_1$ induced by electroconvulsive shock in rat hippocampus. *Biochem. Biophys. Res. Commun.* **194**, 665–670.

Lev, S., Moreno, H., Martinez, R., Canoll, P., Peles, E., Musacchio, J. M., Plowman, G. D., Rudy, B., and Schlessinger, J. (1995) Protein tyrosine kinase PYK2 involved in CA^{2+}-induced regulation of ion channel and MAP kinase functions. *Nature* **376**, 737–745.

Mahadevan, L. C. and Cano, E. (1995) Parallel signal processing among mammalian MAPKs. *Trends Biochem. Sci.* **20**, 117–122.

Marshall, C. J. (1994) MAP kinase kinase kinase, MAP kinase kinase and MAP kinase. *Curr. Opinion Genet. Dev.* **4**, 82–89.

Mauro, L. J. and Dixon, J. E. (1994) "Zip codes" direct intracellular protein tyrosine phosphatases to the correct cellular "address." *Trends Biochem. Sci.* **19**, 151–155.

Menegoz, M., Lau, L.-F., Hervé, D., Huganir, R. L., and Girault, J.-A. (1995) Tyrosine phosphorylation of NMDA receptor in rat striatum: effects of 6-OH-dopamine lesions. *Neuroreport* **7**, 125–128.

Moon, I. S., Apperson, M. L., and Kennedy, M. B. (1994) The major tyrosine-phosphorylated protein in the postsynaptic density fraction is N-methyl-D-aspartate receptor subunit 2B. *Proc. Natl. Acad. Sci. USA* **91**, 3954–3958.

Moss, S. J., Gorrie, G. H., Amato, A., and Smart, T. G. (1995) Modulation of GABA A receptors by tyrosine phosphorylation. *Nature* **377**, 344–348.

Nishibe, S., Wahl, M. I., Hernández-Sotomayor, S. M. T., Tonks, N. K., Rhee, S. G., and Carpenter, G. (1990) Increase of the catalytic activity of phospholipase C-gamma$_1$ by tyrosine phosphorylation. *Science* **250**, 1253–1256.

O'Callaghan, J. P., Lavin, K. L., Chess, Q., and Clouet, D. H. (1983) A method for dissection of discrete regions of rat brain following microwave irradiation. *Brain Res. Bull.* **11**, 31–42.

Pawson, T. (1995) Protein modules and signalling networks. *Nature* **373,** 573–580.

Qu, Z. and Huganir, R. L. (1994) Comparison of innervation and agrin-induced tyrosine phosphorylation of the nicotinic acetylcholine receptor. *J. Neurosci.* **14,** 6834–6841.

Qu, Z., Moritz, E., and Huganir, R. L. (1990) Regulation of tyrosine phosphorylation of the nicotinic acetylcholine receptor at the rat neuromuscular junction. *Neuron* **4,** 367–378.

Riddihough, G. (1994) More meanders and sandwiches. *Nature Struct. Biol.* **1,** 755–757.

Rosenblum, K., Schul, R., Meiri, N., Hadari, Y. R., Zick, Y., and Dudai, Y. (1995) Modulation of protein tyrosine phosphorylation in rat insular cortex after conditioned taste aversion training. *Proc. Natl. Acad. Sci. USA* **92,** 1157–1161.

Sasaki, H., Nagura, K., Ishino, M., Tobioka, H., Kotani, K., and Sasaki, T. (1995) Cloning and characterization of cell adhesion kinase β, a novel protein-tyrosine kinase of the focal adhesion kinase subfamily. *J. Biol. Chem.* **270,** 21,206–21,219.

Schaller, M. D. and Parsons, J. T. (1994) Focal adhesion kinase and associated proteins. *Curr. Opinion Cell Biol.* **6,** 705–710.

Schlessinger, J. (1993) How receptor tyrosine kinases activate Ras. *Trends Biochem. Sci.* **18,** 273–275.

Schlessinger, J. and Ullrich, A. (1992) Growth factor signaling by receptor tyrosine kinases. *Neuron* **9,** 383–391.

Siciliano, J. C., Menegoz, M., Chamak, B., and Girault, J.-A. (1992) Anti-phosphotyrosine antibodies for studying protein phosphorylation in neural cells: applications to brain slices and cultured cells. *Neuroprotocol* **1,** 185–192.

Siciliano, J. C., Gelman, M., and Girault, J.-A. (1994) Depolarization and neurotransmitters increase neuronal protein tyrosine phosphorylation. *J. Neurochem.* **62,** 950–959.

Smith, P. K., Krohn, R. I., Hermanson, G. T., Mallia, A. K., Gartner, F. H., Provenzano, M. D., Fusimoto, E. K., Gocke, N. M., Olson, B. J., and Klenk, D. C. (1985) Measurement of protein using bicinchoninic acid. *Anal. Biochem.* **150,** 76–85.

Stone, R. L. and Dixon, J. E. (1994) Protein-tyrosine phosphatases. *J. Biol. Chem.* **269,** 31,323–31,326.

Su, X.-D., Agango, E. G., Taddei, N., Stefani, M., Ramponi, G., and Nordlund, P. (1994) Crystallisation of a low molecular weight phosphotyrosine protein phosphatase from bovine liver. *FEBS Lett.* **343,** 107,108.

Taniguchi, T. (1995) Cytokine signaling through nonreceptor protein tyrosine kinases. *Science* **268,** 251–260.

Taylor, S. S., Knighton, D. R., Zheng, J., Sowadski, J. M., Gibbs, C. S., and Zoller, M. J. (1993) A template for the protein kinase family. *Trends Biochem. Sci.* **18,** 84–89.

Temple, S. and Qian, X. (1995) bFGF, neurotrophins, and the control of cortical neurogenesis. *Neuron* **15,** 249–252.

Tessier-Lavigne, M. (1995) Eph receptor tyrosine kinases, axon repulsion, and the development of topographic maps. *Cell* **82,** 345–348.

Valenzuela, C. F., Machu, T. K., McKernan, R. M., Whiting, P., Vanrenterghen, B. B., McManaman, J. L., Brozowski, S. J., Smith, G. B., Olsen, R. W., and Harris, R. A. (1995) Tyrosine kinase phosphorylation of GABA$_A$ receptors. *Mol. Brain Res.* **31,** 165–172.

Van der Geer, P. and Pawson, T. (1995) The PTB domain: a new protein module, implicated in signal transduction. *Trends Biochem. Sci.* **20,** 277–280.

Van der Geer, P., Hunter, T., and Lindberg, R. A. (1994) Receptor protein-tyrosine kinases and their signal transduction pathways. *Ann. Rev. Cell Biol.* **10,** 251–337.

Walton, K. M. and Dixon, J. E. (1993) Protein tyrosine phosphatases. *Ann. Rev. Biochem.* **62,** 101–120.

Wang, Y. T. and Salter, M. W. (1994) Regulation of NMDA receptors by tyrosine kinases and phosphatases. *Nature* **369,** 233–235.

Woodrow, S., Bissoon, N., and Gurd, J. W. (1992) Depolarization-dependent tyrosine phosphorylation in rat brain synaptosomes. *J. Neurochem.* **59,** 857–862.

Identification of Posttranslational Modification Sites by Site-Directed Mutagenesis

James A. Bibb and Edgar F. da Cruz e Silva

1. Introduction

Following translation, many proteins undergo further modifications that can dramatically affect both their physical properties and biological function (Wold and Moldave, 1984; Freedman and Hawkins, 1985; Harding and Crabbe, 1992). These posttranslational modifications are essential to the vitality of all eukaryotic cells, including neurons. Techniques that identify amino acid residues in a given protein that are modified and assess the effect of eliminating a specific modification site on a protein's function, both in vitro or in the context of cellular expression, are useful in studying posttranslational modifications. The more traditional biochemical and immunological methods of studying posttranslational modification are discussed elsewhere in this book. However, the resources required for these approaches may not always be available or may yield equivocal results. Thus, the powerful tools of molecular biology may provide a viable alternative for identifying sites of posttranslational modification.

Originally pioneered by Smith (1985), site-directed mutagenesis has been made feasible by the advances in recombinant DNA technology. This application has been the subject of intense interest, and developmental progress in recent years has been increasingly rapid. The power of this application is that a single precise base pair in a vast stretch

From: *Neuromethods, Vol. 30: Regulatory Protein Modification: Techniques and Protocols*
Ed: H. C. Hemmings, Jr. Humana Press Inc.

of DNA may be specifically changed, resulting in a strand of DNA encoding a protein that is altered only at a specific site. Individual amino acids of a given protein may be deleted, added, or changed to other amino acids. The mutated protein may then be expressed, analyzed, and compared to its normal or wild-type counterpart in a number of different contexts. Site-directed mutagenesis no longer represents a difficult and elaborate project one hesitates to embark on; the wide variety of different methodologies that can be performed easily and economically (Smith, 1985; Shortle and Botstein, 1985; McPherson, 1991; Wu, 1993) have made it almost routine in most laboratories.

In this chapter, we will attempt to describe a few current approaches to site-directed mutagenesis with possible applications to the study of posttranslational modifications, keeping in mind innovation, ease of use, and requirements for specialized resources. A brief explanation of each technique will be presented, including sources of material, methodology, and, where possible, examples of relevant use. A detailed description of all the methods involved is beyond the scope of this chapter, and the readers are directed to the original articles and a number of widely used references (for example, *see* Sambrook et al., 1989; Ausubel et al., 1995). In this way, it is hoped that the reader will gain a better appreciation of the usefulness of this approach and may more easily decide on a method for studying a particular posttranslational modification event. In addition, an appendix at the end of this chapter provides sources of the various commercial products cited.

2. General Considerations

2.1. Choice of Target Amino Acids

Most kinds of posttranslational modifications are restricted to specific amino acid residues that occur within the context of consensus sequences for usage by a modifying enzyme. PROSITE is a set of protein motifs used by a variety of DNA/protein software analysis packages (e.g., Geneworks,

GCG) that can often aid in the selection of target amino acids to mutate. Although several mutations are often as easy to introduce as one, in the interest of efficiency, one should consider all information that might help to exclude putative sites. It is common to change the target amino acid into one that cannot undergo the particular posttranslational modification under study. An example would be to change a serine, suspected to be phosphorylated by a protein kinase, into an alanine. Additionally, by changing the serine to a glutamic acid, thus imparting a negative charge at that position, the effect of constitutive phosphorylation might be mimicked. A prerequisite to site-directed mutagenesis is the availability of a DNA clone (usually cDNA) encoding the protein of interest. One must then develop a mutagenesis strategy based on the nucleotide changes required in order to alter the corresponding translation codon, the sequence surrounding the target nucleotides, and the constraints of the technique.

2.2. Choosing an Oligonucleotide

The common feature of all site-directed mutagenesis strategies is the use of a mutagenic oligonucleotide containing the desired nucleotide changes. Therefore, one must give careful consideration to the choice and design of the mutagenic oligonucleotide. Its length, location of mismatches, annealing temperature, secondary structure, restriction site position, and cost are all important considerations. Oligonucleotides are usually about 16–25 nucleotides in length. A general rule is to use 8–13 perfectly matched nucleotides on either side of a single mutation. However, more may be required for multiple mismatches or when large deletions or insertions are desired. The oligonucleotide melting temperature must be high enough to achieve specificity and overcome the destabilizing effects of any mismatches. In general, the melting temperature should be above 42°C (it may be approximated by calculating 2°C for every A:T base pair and 4°C for every G:C base pair, and subtracting 6–8°C for each mismatch). In designing a primer to mutate a gene coding for a protein of interest, it is often possible to include or remove unique restriction sites in close proximity to the desired mutation without affecting the amino acid sequence (i.e., silent mutation). This may be useful in subsequent steps and in the design of future strategies. Today, oligonucleotides are readily available from commercial sources and are quite inexpensive. Automated oligonucleotide synthesizers are often owned by individual laboratories with a constant demand. Various computer programs for analyzing DNA sequences are available (such as DNA Strider, Geneworks, Oligo Primer, and Genecraft) that greatly facilitate the design of mutagenesis, cloning, and subcloning strategies (Marck, 1988; Rychlik and Rhoads, 1989; Piechocki and Hines, 1994).

2.3. Performing Mutagenesis

In the more traditional methods, the mutagenic oligonucleotide directs the site-specific mutation and serves as a primer for in vitro DNA synthesis once annealed to a closed circular ssDNA template encoding the protein of interest, as well as the rest of the bacterial plasmid, bacteriophage, or hybrid phagemid molecule (Sambrook et al., 1989). An oligonucleotide complementary to the region to be altered, except for the required internal mismatch(es), is hybridized to a single-strand copy of the DNA. A complementary strand is then synthesized by DNA polymerase using the oligonucleotide as primer. DNA ligase is used to seal the new strand to the 5'-end of the oligonucleotide. The double-stranded DNA, homologous except for the intended mutation, is then used to transform *Escherichia coli*, where both strands propagate, resulting in two classes of progeny, the parental and those carrying the oligonucleotide-directed mutation. As will be discussed, a number of methods have been developed that greatly improve the efficiency of the basic method. The improvements rely on the creation of an asymmetry between the two strands of the heteroduplex, permitting selection against the wild-type strand. All of these methods are based on an in vivo amplification step in the bacterial host. Another approach is to amplify the mutation exponentially in vitro utilizing polymerase chain reaction (PCR) methodologies.

2.4. Choice of Vector

Originally, site-directed mutagenesis was performed exclusively on single-stranded templates, but several protocols are now available for double-stranded mutagenesis. We will consider here the most commonly used vectors (single-stranded bacteriophages, plasmids, and phagemids).

2.4.1. Bacteriophage

The filamentous bacteriophage specific to *E. coli*, which include M13, f1, and fd, have a genomic single-stranded closed circular molecule of DNA approx 6400 nucleotides in length. When used as cloning vectors, these bacteriophages have the unique advantage of generating large quantities of dsDNA molecules that carry the sequence of the foreign DNA in one strand. This ssDNA readily serves as the template for oligonucleotide site-directed mutagenesis (Sambrook et al., 1989). The life cycle of these viruses includes uptake of the (+)-stranded genome, replication of the complementary (–)-strand to produce a double-stranded form, transcription of mRNA and translation of protein, and replication of the (–)-strand genome for packaging and release of bacteriophage particles into the media. Because single-stranded DNA is a poor substrate for most restriction endonucleases and ligases, foreign DNA inserts are usually cloned in vitro into the double-stranded replicative form of the bacteriophage. This closed circular dsDNA can easily be purified from infected cells, manipulated essentially in the same way as plasmid DNA, and then reintroduced into bacteria by standard transformation procedures (Sambrook et al., 1989). The recombinant dsDNA then re-enters the replication cycle, eventually generating progeny bacteriophage particles that contain only one of the two strands of the foreign DNA. The other strand of the DNA, the (–)-strand, is never packaged.

2.4.2. Plasmids

More recently, plasmids have also been used as templates. Plasmids are double-stranded closed circular DNA molecules ranging in size from 1 to more than 200 kb. They are found in numerous bacterial species where they behave as accessory genetic units that replicate and are inherited independently of the bacterial chromosome. They rely on enzymes and proteins encoded by the host's genome for replication and transcription. Frequently, plasmids encode genes that provide an advantage to their bacterial hosts, such as antibiotic resistance or enzymes that degrade complex organic compounds (Sambrook et al., 1989). Plasmids were developed as useful cloning vectors by the introduction of DNA encoding selectable markers into plasmids possessing the ability to replicate, their reduction in size by elimination of unnecessary sequences, and the introduction of convenient unique restriction sites (polylinkers). Thus, they became easy to manipulate in vitro and to insert into (transform) bacteria, easy to purify selectively in large quantities, and gained the capacity to accept fragments of foreign DNA generated by cleavage with a wide range of restriction enzymes. Cloning in DNA plasmid vectors involves cleaving the double-stranded plasmid with restriction enzymes and joining the foreign DNA through an enzyme-catalyzed ligation in vitro. The resulting recombinant plasmid is used to transform bacteria.

2.4.3. Phagemids

Phagemids are plasmids containing an origin of replication derived from a filamentous bacteriophage. These vectors, which combine the desirable features of both plasmids and filamentous bacteriophage, are plasmids with an origin of replication, a selectable marker for antibiotic resistance, and a copy of the major intergenic region of a filamentous bacteriophage. They can be propagated as plasmids, or when cells harboring these plasmids are infected with a suitable filamentous bacteriophage, the mode of replication changes so that a single-stranded form is produced, packaged, and secreted as bacteriophage particles. These single-stranded forms may undergo additional mutagenesis or screening by DNA sequencing. Examples of commonly used phagemids are the Bluescript vectors from Stratagene.

3. Specific Methods

3.1. Uracil Incorporation

One method, illustrated in Fig. 1, that provides a very strong selection against the nonmutagenized strand of dsDNA was developed by Kunkel and coworkers. This method involves the utilization of *E. coli* strains which are deficient in dUTPase (*dut*) and uracil N-glycosylase (*ung*) (Kunkel, 1985; Sambrook et al., 1989; Yuckenberg et al., 1991). DNA synthesis in these bacteria carries a number of uracils substituted for thymine (U-DNA). When this uracil-containing DNA is introduced into wild type bacteria, it is selectively degraded. The labile U-DNA can be produced in a single-stranded form using M13 or phagemid vectors, and then used as a template for in vitro DNA synthesis using an oligonucleotide directing the required mutation. The resulting DNA is transformed into a wild-type host that degrades the U-DNA, leaving only the mutagenized in vitro synthesized non-U-DNA strand to replicate. Greater than 50% of the progeny should contain the mutation that can be confirmed by DNA sequencing. Positive clones may then be purified as dsDNA forms, and the mutated DNA subcloned into an expression vector for further analysis. This method has been reported to yield mutant products at frequencies in excess of 90% (Kunkel, 1985, 1987). Bio-Rad produces the Muta-Gene kit based on this method, which can be used with either M13 or phagemid vectors.

3.2. Thionucleotide Incorporation

The phosphorothioate-based oligonucleotide-directed mutagenesis method developed by Eckstein and colleagues is based on the observation that certain restriction endonucleases (e.g., *Nci*I) are incapable of hydrolyzing phosphorothioate internucleotidic linkages (Taylor et al., 1989; Sayers et al., 1992; Olsen et al., 1993). Thus, dsDNA containing phosphorothioate linkages in one strand may only be nicked in the nonsubstituted strand. The mutagenic oligonucleotide primer is annealed to the (+)-strand of a ssDNA molecule.

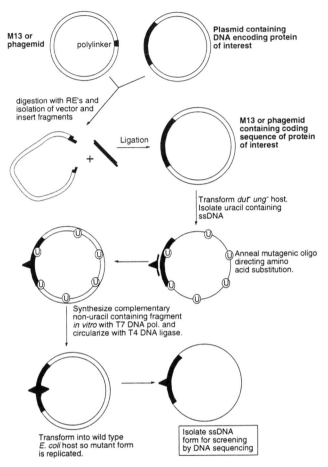

Fig. 1. Uracil incorporation site-directed mutagenesis method. Methodological steps are graphically depicted and accompanied by brief descriptions of each step. dsDNA is shown as a double-lined circle. Polylinker containing the subcloning insertion site for the first two steps, and DNA encoding the protein of interest are shown in black. The mutation is shown as a spike in the mutagenic oligonucleotide or DNA strand. Uracil substitutions of thymidine-containing nucleotides are represented by the circled Us. Text describing the final step appears in a box at the bottom. RE, restriction enzyme; pol., polymerase.

As illustrated in Fig. 2, the primer is extended via a polymerization reaction in which one of the natural deoxynucleoside triphosphates is replaced by the corresponding deoxynu-

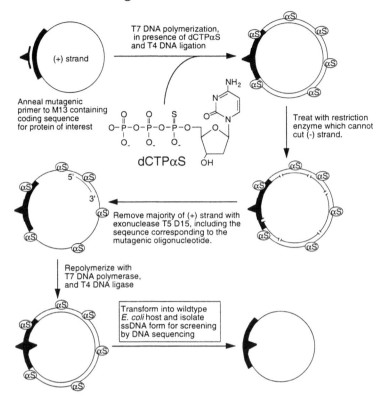

Fig. 2. Phosphorothioate-based oligonucleotide-directed mutagenesis method. The iconic format is similar to that of Fig. 1. The structure of dCTPαS is shown as it is being added to the mutagenic oligonucleotide-primed DNA polymerization step. Its presence in the in vitro synthesized strand is indicated by circled αSs. Nicks introduced by incomplete restriction enzyme digestion are indicated by gaps in the inner strand of the heteroduplex DNA. The orientation of DNA that was not degraded by exonuclease is indicated by the 5'- and 3'-label. Text describing the final step appears in a box.

cleoside 5'-*O*-(1-thiotriphosphate) (dNTPαS). Thus, the phosphorothioate groups are incorporated exclusively into the (–)-strand of the newly synthesized DNA. Several restriction enzymes are selectively unable to cleave the (–)-strand carrying the phosphorothioate and the mutated bases, but will nick the wild-type (+)-strand. Following restriction enzyme treatment, an exonuclease is used to partially digest away the wild-type strand opposite the mismatch introduced by the oligonucleotide. DNA polymerization then replaces the nucleotides in the (+)-strand using the mutant (–)-strand as a template so that the mutation is now present in both strands. This dsDNA is used to transform bacteria and the progeny may be screened by DNA sequencing. The advantages of this method are that mutations are introduced into a double-stranded molecule in vitro, no special host strain is required, the procedure is expedient when using T7 DNA polymerase, and the method has been adapted to both single-stranded bacteriophages and double-stranded vectors (Olsen et al., 1993). Mutational frequencies in excess of 90% have been obtained with this technique (Sayers et al., 1992). Amersham markets the Sculptor kit, which is a 1-d, one-tube, simplified protocol kit for site-directed mutagenesis using this thionucleotide incorporation technique.

3.3. Methyl-DNA Method

The methyl-DNA method, like the previously described U-DNA and thionucleotide methods, also employs the strategy of deoxynucleotide analog incorporation as the basis for selectivity (Vandeyar et al., 1988; Batt et al., 1993). The difference is that 5-methyldeoxycytodinetriphosphate (d5'-MeCTP) replaces dCTP. The incorporation of d5'-MeCTP renders the newly synthesized mutant strand resistant to a number of restriction enzymes, including *Msp*I and *Sau*3A. These enzymes nick the nonmethylated wild-type strand of a hemimethylated dsDNA molecule, so that it may be removed by a subsequent exonuclease step. Stratagene produces the ExSite kit, which utilizes this method and incorporates PCR (discussed below) into the strategy.

3.4. Restriction Site Elimination

Site-directed mutagenesis by unique restriction site elimination is another method that does not require the DNA encoding the protein of interest to be in the context of a ssDNA phage vector (Deng and Nickoloff, 1992; Zhu and Holtz, 1993). Thus, by using plasmid DNA, it is possible to

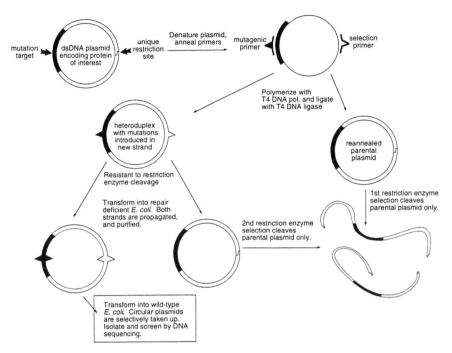

Fig. 3. Restriction site elimination method of site-directed mutagenesis. The positions of the sites that will serve as the targets of the two mutagenic primers are indicated by thick arrows at the top left. The oligonucleotide that is used to introduce the mutation in the DNA encoding the protein of interest is indicated by a solid spike, whereas the selection oligonucleotide that destroys the unique restriction site is an unfilled spike. Linearized plasmid that is inefficient for transformation of bacteria is shown as open ribbons on the right with the central segment encoding the protein of interest indicated in black. Text describing the final step appears in a box, pol., polymerase.

avoid at least one subcloning step. As illustrated in Fig. 3, two oligonucleotide primers are simultaneously annealed to one strand of the denatured plasmid. One primer directs the mutagenesis of the protein, whereas the other changes one nonessential unique restriction site so that it is no longer cleavable. The two mutations are incorporated simultaneously during the in vitro DNA synthesis step. The resulting mutant plasmids are resistant to the restriction enzyme selection step, whereas any parental plasmid produced by reannealing is linearized. The uncleaved heteroduplex plasmids containing the mutations in one strand transform into repair-deficient *E. coli* much more efficiently than the linearized DNA (Conley and Saunders, 1984). Both strands of the circular plasmid are propagated in the bacteria, but a second screening of isolated plasmid removes any parental plasmid contaminant. A second transformation into wild-type *E. coli* further amplifies the mutation-bearing plasmid. Plasmid DNA is then isolated and may be screened by DNA sequencing for positives. The mutation frequency of this method is reported to exceed 70% (Zhu and Holtz, 1993). The use of double-stranded plasmid in this method eliminates some of the requirements of a phage-based system. The strategy can also be applied in reverse with gain of a restriction site serving as a selection marker (Carter, 1991; Kaslow and Rowlings, 1993). The Transformer kit commercially available from Clontech, the Chameleon kit from Stratagene, and the U.S.E. kit from Pharmacia include all necessary materials for performing this technique.

3.5. Screening on the Basis of Antibiotic Resistance

An adaptation of the double-primer technique is based on the activation of antibiotic resistance. In this method, one oligonucleotide primer is used to activate an antibiotic resistance gene, whereas the other primer introduces the desired mutation in the sequence encoding the protein of interest. Characteristically, one of the oligonucleotides introduces a frame shift in an inactivated antibiotic resistance gene so that the proper reading frame is restored to the corresponding mRNA transcript. A separate antibiotic resistance gene may simultaneously be inactivated by a third oligonucleotide (triple-primer mutagenesis), allowing a subsequent mutation to be introduced by repeating the procedure. The methodology is essentially the same as that described for the restriction site elimination procedure. It is compatible with the use of double-stranded plasmids, but without the restriction enzyme screening step. Instead, desired plasmid constructs are selected on the basis of antibiotic resistance, a phenotype imparted on the host strain by the mutated plasmid. The Altered Sites kit, which uses this technique, is available from Promega and claims up to 90% yields of the desired mutant products.

3.6. Gapped-Heteroduplex Formation Method

A less frequently performed, but still effective, method is the use of gapped-heteroduplex formation. This method is applicable to plasmid constructs and is based on the combination of two plasmid single-strands of different origin to form a gapped plasmid (Inouye and Inouye, 1991; Hofer and Kühlein, 1993). In one reaction, the plasmid encoding the protein of interest is cleaved by restriction enzymes in such a way that the target segment for mutagenesis is removed and the nontarget DNA is isolated using agarose gel electrophoresis (Fig. 4) (Sambrook et al., 1989; Ausubel et al., 1995). In a second reaction the plasmid is cleaved at a site outside the target region. These two reaction products are then mixed, denatured, and allowed to reanneal in the presence of the mutagenic oligonucleotide to form a three-component heteroduplex. On in vitro polymerization and ligation, the mutagenic oligonucleotide is incorporated into one strand of the plasmid. In addition to the desired amino acid change, the mutating oligonucleotide may be engineered to introduce a new restriction enzyme cleavage site via a silent mutation, which may be used as the basis for physically separating the mutant plasmid from parental molecules present after bacterial transformation and plasmid isolation. This method has been reported to yield mutational efficiencies of up to 60% (Hofer and Kühlein, 1993).

3.7. PCR-Based Methods of Site-Directed Mutagenesis

3.7.1. General

Since its inception (Saiki et al., 1985), PCR has had a tremendous impact on virtually every aspect of biological and biochemical research. Extensive literature exists regarding its general use for the amplification of DNA and its use in

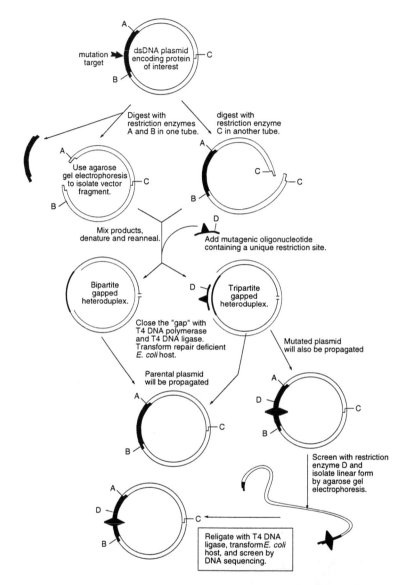

Fig. 4. Gapped-heteroduplex method of site-directed mutagenesis. Four unique restriction sites are indicated by the letters A, B, C, and D. Heteroduplexes formed by two parental strands alone **(left)** and two parental strands plus one mutagenic oligonucleotide **(right)** are shown in the center. ssDNA regions encoding the protein of interest are indicated by a thick line. Text describing the final step appears in a box.

neurobiology (*see,* for example, Saiki et al., 1985; Innis et al., 1994; Dieffenbach, 1995; McPherson, 1995). Likewise, there are numerous descriptions of its use in site-directed mutagenesis procedures (Horton and Pease, 1991; Zhao and Padmanabhan, 1993; Horton et al., 1993; Weiner et al., 1994; Hughes and Andrews, 1996). The basic premise of PCR is the use of flanking oligonucleotide primers to amplify a target DNA sequence exponentially in vitro in a thermal cycling reaction catalyzed by a heat-stable DNA polymerase. Products of the first polymerization reaction serve as templates for the next, and so on. The result is that the product DNA is present in vast excess over the initial template material and the mutation is present in both strands of the double-stranded product. Clearly, this technique may have important applications in site-directed mutagenesis. The primers used in the PCR reaction may accommodate single or multiple base pair mismatches just as with other techniques. The difference is that the mutation is amplified in vitro. Although a screening step may be desirable, the power of exponential propagation of the mutant DNA sequence often makes this unnecessary. One disadvantage of PCR-based techniques is the possible introduction of secondary mutations. The heat-stable DNA polymerases normally used for this technique (e.g., *Taq* polymerase) lack proofreading activity, but high fidelity enzymes are now marketed by several manufacturers (e.g., Stratagene's *Pfu* polymerase), which have much lower rates of error.

3.7.2. PCR SOEing

One method of site-directed mutagenesis that incorporates a clever PCR strategy is called splicing by overlapping extension (PCR SOEing) (Horton and Pease, 1991; Horton et al., 1993). It is usually performed using plasmid dsDNA and, like most PCR strategies, requires a subcloning step to insert the final product in the proper vector for expression and analysis of the protein. This method takes into account the common circumstance that the site that one desires to mutate does not often occur close to a restriction site. As illustrated in Fig. 5, two separate PCRs are conducted with two separate sets of oligonucleotides. Middle oligonucleotides may be designed to overlap in the region of the directed mutation. End oligonucleotides are designed that either correspond to the position of existing restriction sites or introduce new restriction sites. This results in the synthesis of two mutated double-stranded fragments overlapping in the region of the mutation and with restriction sites at their respective nonoverlapping ends. The products of the first reactions are mixed and a second PCR is conducted to generate the full-length insert fragment containing the oligonucleotide-directed mutation. The resulting large fragment is prepared for insertion by trimming the ends with the appropriate restriction enzymes, treatment with a phosphatase to prevent concatamerization, and purification by agarose gel electrophoresis if necessary (Sambrook et al., 1989; Ausubel et al., 1995). Sequence analysis can be performed to confirm the presence of the mutation, but the efficiency of this method would appear to be >90% (unpublished personal observations of both authors). The critical points of this technique are the careful design of the required oligonucleotides required and the choice of proper PCR conditions. To a certain extent, ideal conditions may be determined empirically, since the efficiency of the technique may be estimated by visual observation of the reaction products following agarose gel electrophoresis and ethidium bromide staining, and the experiment is easily repeated several times in 1 d.

Another application for which this technique is particularly suitable is the construction of chimeric proteins. The middle primers can be easily engineered to generate precisely designed chimeric molecules. The power of this approach resides in its independence of pre-existing restriction sites. Any domain or portion of one protein can in this fashion be joined (or deleted) onto any other domain or portion from any other protein. Readers are directed to the original articles for further details of this application (Horton and Pease, 1991; Horton et al., 1993).

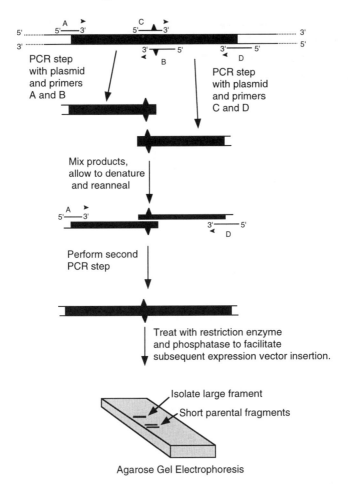

Fig. 5. Use of PCR SOEing technique of site-directed mutagenesis. Unique oligonucleotide primers are indicated by short lines labeled A, B, C, and D. Orientation is indicated by 5'- and 3'-labels and direction of primer extension by polymerase is indicated by small arrowheads. Solid spikes indicate mutations that are directed by oligonucleotides B and C. Physical separation and visualization of the different fragments in an agarose gel are depicted in the final step.

3.8. Screening via Automated Sequencing

Given the high efficiency of most site-directed mutagenesis techniques now commonly used, screening strategies to differentiate mutated from wild-type clones (usually by oligonucleotide hybridization methods) are not usually required. Thus, one usually embarks on the analysis of the DNA sequence of a limited number of the clones generated with the certainty that several are likely to carry the desired mutation. Furthermore, DNA sequencing provides the most definitive proof that a desired mutation has been introduced into a target molecule. Although the dideoxyribonucleotide sequencing method is by now a standard laboratory technique, the uninitiated may still hesitate from undertaking such a project. Also, if only a few mutations are required, the cost of apparatus, reagents, and time spent perfecting technical skills may be exorbitant. An alternative which is gaining in popularity, and bears mentioning here is the use of automated sequencing technology (Hunkapiller, 1991; Ju et al., 1995; Seto et al., 1995). An automated sequencing apparatus uses an argon-ion laser focused near the bottom of a polyacrylamide gel to excite fluorescence from the four ddNTPs (A, T, C, and G) tagged with different fluorescent dyes. Alternatively, the primer may be labeled with four different dyes. Each dye emits light at a different wavelength when excited by the laser. This device is capable of reading longer sequences, requires no radionuclides, and is often available as part of an institution's core facilities. Sequencing reactions are conducted in a single tube and the sequence is read automatically and provided in a computer format for analysis and manipulation. Perkin Elmer provides additional information as part of their product literature on automated sequencing.

4. Expression Systems

4.1. Bacterial Expression

Once one has mutated an amino acid residue suspected of being the site of posttranslational modification, the critical step is to express the protein in a manner so that the effects of the mutation may be analyzed. The effects of the mutation may be assessed in vitro, in which case purity is sometimes a concern, or in the context of the cellular milieu. In the case of the former, bacterial expression is often used. If the protein is too large or insoluble, then one can try to express just a portion of the protein. One of the most successfully used systems for expression in bacteria is the pET vector expression system (Moffatt et al., 1984; Steen et al., 1986; Studier and Moffatt, 1986; Studier et al., 1990; Dubendorff and Studier, 1991). Bacterial expression may be coupled with an affinity-purification strategy, such as the GST-fusion (Smith and Johnson, 1988) or the 6X-histidine fusion (Crowe et al., 1994) affinity-purification systems. Both involve induction of bacteria during log-phase growth, expression, and one-step purification from the bacterial lysate. The wild-type and mutated protein may be subcloned, expressed, and purified simultaneously, and the effect of the mutation may be directly assessed. pET vectors and materials for bacterial expression and affinity-purification are sold by Novagene, 6X-His fusion expression systems are available from Qiagen, and GST-fusion expression systems are available from Pharmacia.

4.1.1. Amber Suppression

An interesting alternative to the general site-directed mutagenesis expression approach discussed above is the amber suppression mutation method (Kleina et al., 1990; Kleina and Miller, 1990;Normanly et al., 1990). This technique still requires that the codon of the amino acid residue of interest be changed, but instead of changing the codon to read for an alanine or some other specific amino acid residue, it is changed to the amber stop codon (TAG). The new construct may then be used to transform a number of different mutant bacterial hosts that have undergone tRNA engineering, so that in each the amber termination codon directs the incorporation of a different amino acid into the protein. In this way, using a single mutant in which the codon of the amino acid under study is mutated to the amber stop codon, one gains the ability to test the effect of substituting the target amino acid by amino acids with hydrophobic, charged, polar, or neutral properties. Translation extracts from such bacterial strains are also commercially available, so that small amounts of the proteins may be synthesized by in vitro translation and the products directly tested or selectively metabolically labeled. The Interchange kit based on this methodology is commercially available from Promega.

4.2. Eukaryotic Expression

The effect of a site-directed mutation which alters a specific posttranslational processing event may often best be assessed in vivo. The most common approach to in vivo studies is to express the mutated protein in eukaryotic cells grown in tissue culture. Before embarking on the expression of heterologous proteins in cultured cells, it is advisable to become thoroughly familiarized with the various parameters that may affect gene expression either directly or indirectly. There are promoters and enhancers, splice signals and polyadenylation signals, and a variety of other cis- and trans-acting elements to consider. To some extent, as with other aspects of modern molecular biology, commercial enterprise has had considerable success in the marketing of predesigned expression vectors. However, in spite of their success, no kit marketed today will substitute for the lack of basic biological knowledge. The reader is particularly advised that many of the elements involved exert their effects in concert, and in a cell-type specific manner. Although it is often convenient to let commercially available vectors dictate your choices, this may not always be ideal for your project. Kriegler (1990) has produced an excellent laboratory manual that discusses the most relevant aspects of gene transfer and expression.

There is a plethora of both viral and cellular promoters and enhancers in use today, with more being characterized almost daily. Please note that although they are often described as "strong" or "weak," they do not function as totally independent elements. Optimal expression depends on the exact selection and arrangement of expression elements, as well as the cellular context. One aspect of eukaryotic expression that is significantly simpler than with bacterial expression is that one need not worry too much about reading frames, as long as the gene construct is cloned downstream of a eukaryotic promoter element. However, it is unrealistic to expect recombinant protein production in eukaryotic cells to approach that obtained in bacteria. On the other hand, one may test directly the effect of mutations in an in vivo cellular context. The use of different radioactive precursors in the culture medium allows the labeling of proteins undergoing a specific type of posttranslational modification. Our considerations here are generally applicable to all eukaryotic expression systems, but especially those involving the use of immortalized mammalian cells grown in tissue culture. Readers are directed to more specialized publications for details of baculovirus mediated gene expression in insect cells (King and Possee, 1992), gene expression in plant cells (Lamb, 1990; NATO, 1991), and expression of foreign proteins in yeast cells (including the recently developed *Pichia* expression system available from Invitrogen).

The general strategy involves cloning the mutated cDNA in a suitable expression vector, gene transfer into a suitable cell line, and subsequent analysis of the expressed protein (with or without prior purification). Thus, one also needs a thorough knowledge of basic cell-culture techniques, from maintenance of a sterile cell culture environment to proper observation and characterization of the cells used. Most cell lines require growth factors not present in chemically defined medium, and bovine serum is usually the source used. However, no two lots of serum are the same, and since some cell lines are particularly sensitive to the serum source used, it usually pays to buy the highest quality serum that your laboratory budget allows. For the introduction of DNA into tissue-cultured cells, there are three general methods in use; those that form DNA precipitates that are internalized by the cells (e.g., the calcium phosphate transfection method), those that create DNA-containing complexes that can be internalized by the cells (e.g., lipofection-type transfection methods), and those that produce pores in the cell membrane of sufficient size for the DNA to enter (e.g., electroporation transfection methods) (Murray, 1991). The methods used to monitor gene expression include both DNA- and RNA-based methods, as well as protein based methods. The latter include monitoring protein expression by techniques, such as immunoblotting, immunoprecipitation, immunofluorescence analysis, and direct bioassays, when appropriate. Each technique has advantages and disadvantages, and most can be used in conjunction with in vivo labeling approaches. Immunoprecipitation combined with radiolabeling allows protein synthesis to be monitored and, if a pulse-chase is used, processing and turnover as well. Immunoblotting techniques are better suited for quantitating the amount of a given protein under a particular set of conditions. Immunofluorescence can provide direct information on the subcellular localization of a protein, whereas immunoblotting combined with subcellular fractionation provides indirect information. Eukaryotic expression can also be combined with the use of specific inhibitors and/or activators of cellular enzymes responsible for particular posttranslational modifications. For example, N-linked glycosylation can be blocked with tunicamycin, which blocks the formation of the dolichol phosphate-linked oligosaccharide donor; deoxynojirimycin, deoxymannojirimycin, and swainsonine are other inhibitors of carbohydrate processing (*see,* for example, Bernhardt et al., 1994; Chapter 11). Palmityolated and myristoylated proteins can be labeled with the addition of radiolabeled palmitic acid or myristic acid to the growth medium (*see* Chapter 9). Finally, there is an overwhelming number of protein kinase inhibitors and activators, and protein phosphatase inhibitors that can be used to study protein phosphorylation (for an example, *see* Bibb et al., 1994; Chapter 4).

5. Examples

In recent years, articles dealing with posttranslational modifications have become replete with examples of the use of site-directed mutagenesis approaches, as described here. Although not attempting to present a comprehensive review of the literature, some illustrative examples are discussed below, based largely on the main field of interest of the authors (protein phosphorylation), which should also serve as a useful guide for other types of modifications.

5.1. Uracil Incorporation

The intracellular cytoplasmic domain of the Alzheimer β-amyloid precursor protein (APP) has been reported to be phosphorylated by Ca^{2+}/calmodulin-dependent protein kinase II (CaMKII) at Thr-654 and by protein kinase C (PKC) at Ser-655 (Gandy et al., 1988; Suzuki et al., 1992). Phorbol esters, known to act via activation of PKC, were also shown to stimulate secretion of APP. Therefore, it was of some interest to determine whether direct phosphorylation of APP by PKC or another protein kinase might play a role in the control of APP secretion. Site-directed mutagenesis was used to change both Thr-654 and Ser-655 to alanine, and the mutant proteins were expressed in mammalian cells in culture. These unphosphorylatable mutant APP proteins were processed and secreted in exactly the same manner as their wild-type counterpart (da Cruz e Silva et al., 1993). Either alone or together in the same construct, changing these residues to alanine was found to have no effect on APP processing in vivo. Furthermore, mimicking constitutive phosphorylation by producing molecules in which these residues were mutated to either glutamate or aspartate had no significant effect on processing and secretion in comparison to the wild-type protein.

5.2. Restriction Site Elimination

The activities of many neuronal cell proteins are regulated through their phosphorylation states. Their activities may be affected not only through protein kinase-mediated phosphorylation, but also via dephosphorylation by protein phosphatases. For example, protein phosphatase 1 (PP1) may affect neuronal excitability by dephosphorylating the Na^+/K^+,-ATPase, as well as Na^+ and Ca^{2+} channels (reviewed in Hemmings et al., 1995). Recently, the phosphatase activity of mammalian PP1 was shown to be negatively regulated through phosphorylation of Thr-320 by a cyclin-dependent kinase, cdc2 kinase (Dohadwala et al., 1994). Site-directed mutagenesis using the restriction site elimination technique

Bibb and da Cruz e Silva

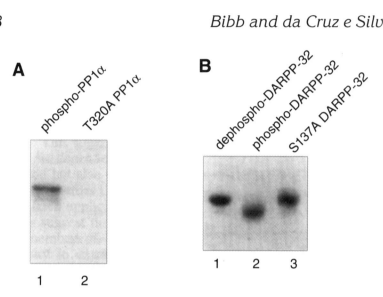

Fig. 6. Identification of a postttranslational modification site using site-directed mutagenesis. **(A)** Changing threonine at amino acid position 320 of PP1α to alanine abolishes phosphorylation by cdc2 kinase. Wild-type (lane 1) and mutant (lane 2) PP1α subjected to cdc2 phosphorylation with [γ-^{32}P]-ATP followed by SDS-PAGE and autoradiography. These results are reprinted with permission of the original publisher (Dohadwala et al., 1994). **(B)** Effect of changing serine at amino acid position 137 of DARPP-32 to alanine on phosphorylation by CKI. Unphosphorylated wild-type DARPP-32 (lane 1), CKI phosphorylated wild-type DARPP32 (lane 2), and CKI phosphorylated mutant DARPP-32 (lane 3) subjected to SDS-PAGE and stained with Coomassie brilliant blue. The mutation abolishes the CKI phosphoylation-induced shift in electrophoretic mobility. These results are reprinted with the permission of the original publisher (Desdouits et al., 1995b).

was used to confirm unequivocally that Thr-320 of PP1 was the residue phosphorylated by cdc2 kinase. Figure 6A shows that conversion of Thr-320 of PP1 into an alanine completely abolished in vitro phosphorylation by cdc2 kinase (Dohadwala et al., 1994).

5.3. Thionucleotide Incorporation

DARPP-32 is a neuronal-specific protein expressed in the basal ganglia. It becomes a potent inhibitor of PP1 on phosphorylation of Thr-34 in response to agonists of the dopamine signaling cascade system that activate cAMP-dependent protein kinase (PKA). Thr-34 is then dephosphorylated in response to signals that raise intracellular Ca^{2+} and activate calcineurin (the Ca^{2+}/calmodulin-dependent protein phosphatase). It has recently been reported that phosphorylation of DARPP-32 at Ser-137 by casein kinase I (CKI) enhances dephosphorylation of phospho-Thr-34 by calcineurin (Desdouits et al., 1995b; reviewed in Hemmings et al., 1995). The site of DARPP-32 phosphorylation by CKI was identified using the thionucleotide incorporation method of site-directed mutagenesis (Desdouits et al., 1995a). As shown in Fig. 6B, phosphorylation of wild-type DARPP-32 by CKI results in a slight downward shift in mobility (compare lanes 1 and 2 of Fig. 6B). However, this effect was abolished when Ser-137 was replaced by alanine and the mutant protein was used as a substrate for CKI in vitro (Fig 6B, lane 3). This result has been confirmed by direct sequencing of HPLC-purified phosphopeptides obtained from DARPP-32 phosphorylated by CKI (Desdouits et al., 1995a).

5.4. Antibody Resistance

CTP-phosphocholine cytidylyltransferase is an enzyme involved in control of lipid biosynthesis, converting phosphocholine to CDP-choline. Its intracellular location is affected by phosphorylation at a number of serine residues located in its carboxyl-terminus. The in vivo effect of converting these serines to alanines or to negatively charged glutamates has recently been assessed using the antibody resistance screening method of site-directed mutagenesis (Wang and Kent, 1995). Changing the serines to unphosphorylatable alanines resulted in the redistribution of the wild-type enzyme from the cytoplasm to a membrane-associated fraction in transfected CHO cell lysates. Mutation of the serines to negatively charged glutamates reversed this effect, mimicking the cytoplasmic localization imparted by the wild-type phosphoserine residues (Wand and Kent, 1995).

5.5. PCR

Ca²⁺-calmodulin-dependent protein kinases (CaM kinases) are known to mediate some of the effects of elevated intracellular Ca^{2+} on cellular functions, including neurotransmitter synthesis and release, regulation of translational rates, transcription of specific genes, and cell-cycle events (Means, 1994). Recently it has been reported that an activating kinase regulates the activity of CaM kinases. Thr-196 of CaM kinase IV was identified as the residue phosphorylated in vitro by the activating CaM kinase Ia using a PCR-based method of site-directed mutagenesis (Selbert et al., 1995).

6. Conclusions

The power of site-directed mutagenesis is being increasingly harnessed to probe structure–function relationships of proteins in general. Recently developed methods have so increased the efficiency of this technique that screening for mutants is almost unnecessary, and easy-to-use kits are now marketed, so that even this once difficult technique is at the disposal of most laboratories equipped with basic molecular biology equipment. Thus, it is no surprise that site-directed mutagenesis is being applied to identify sites of posttranslational modification and to discern their function. In general, the strategy consists of the elimination of one or more previously identified putative sites, or their substitution by another amino acid that cannot undergo the modification under study. Increasingly sophisticated strategies have also been developed, such as mimicking a particular posttranslational modification by another amino acid (e.g., glutamate and aspartate have been used to mimic a phosphorylated serine or threonine residue). In a further refinement of this strategy, phosphorylated serine residues can be changed to threonine and vice versa (Kennelly, 1994).

We have reviewed some of the current and most commonly used methods of site-directed mutagenesis, which can find application not only in the study of posttranslational modification of proteins (e.g., phosphorylation, glycosylation, methylation, and so forth), but also for other mutagenic purposes. Selected examples are given, but many other examples are available in the general biochemistry literature. Without endorsing a particular method, readers are encouraged to consider the requirements of the various techniques described before embarking on a particular project. The identification of posttranslational modification sites by site-directed mutagenesis represents a viable alternative to the more traditional approaches that rely primarily on classical biochemical and/or immunological techniques. The tools and materials required for the site-directed mutagenesis approach are relatively simple, inexpensive, and widely available, whereas other techniques often require more sophisticated equipment and materials.

Appendix

Sources of Material Cited

Sculptor™ IVM Site-directed ZMutagenesis System, Product #RPN 1526, Amersham Life Science Inc., 2636 Clearbrook Dr., Arlington Heights, IL 60005, Tel: (800) 323-9750, http://www.amersham.com

Muta-Gene® In Vitro *Mutagenesis Kit,* vesion 2, cat. #170–3580/1, Bio-Rad Life Science Group, 200 Alfred Nobel Dr., Hercules, CA 94547, Tel: (800) 424–6723.

Transformer™ Site-Directed Mutagenesis Kit, Cat. #K1600–1, Clontech Inc., 1020 E. Meadow Circle, Palo Alto, CA 94303, Tel: (800) 424–8222, http://www.clontech.com.

Altered Sites® II In Vitro *Mutagenesis System,* Product #Q6210, Q6080 & Q6090

Promega Corporation, 2800 Woods Hollow Rd., Madison, WI 53711, Tel: (800) 356–9526, http://www.promega.com.

Interchange™ In vitro *Amber Suppression Mutagenesis System,* Product #Q6250 & Q6260, also from Promega.

Chameleon™ Double-stranded, Site-directed Mutagenesis Kit, cat. #200509, Stratagene, 11011 N. Torrey Pines Rd., La Jolla, CA 92032, Tel: (800) 424–5444, http://www.stratagene.com.

ExSite™ PCR-based Site-directed Mutagenesis Kit, cat. #200502, also from Stratagene.

U.S.E. Mutagenesis Kit, Pharmacia Biotech Inc., 800 Centennial Ave., P.O. Box 1327, Piscataway, NJ 08855-1327, Tel: (800) 5526–3593, http://www.biotech.pharmacia.se.

Bluscript Vectors, also from Stratagene.

pfu Posymerase for PCR, also from Stratagene.

pET Vector Expression System, Novagene, Inc. 597 Science Dr., Madison, WI, 53711, Tel: (800) 526–7319, http://www.novagene.com.

QIAexpress Ni-NTA/6XHis Expression System, Qiagen Inc., 9600 De Soto Ave., Chatsworth, CA 91311, Tel: (800) 426–8157, http://www.qiagen.com.

GST-fusion Expression System, Also from Pharmacia.

The Pichia Expression Kit, cat. #K1710–01, Invitrogen Corporation, 3985 B Sorrento Valley Blvd., San Diego, CA 92121, Tel: (800) 955-6288, htp://www.invitrogen.com.

Automated DNA Sequencing Apparatus, Perkin Elmer, Corporatation, Applied Biosystems Division, 850 Lincoln Centre Dr., Foster City, CA 94404, Tel: (800) 874–9868, http://www.perkin-elmer.com.

DNA Strider™, by Dr. Christian Marck, Service de Biochimie et de Génétique Moléculaire, Bât. 142 CEA/Saclay, 91191 GIF-SUR-YVETTE Cedex, France.

Oligo Primer Analysis Software®, National Biosciences, Inc., 3650 Annapolis Lane N. #140, Plymouth MN 55447, Tel: (800) 747-4362, http://www.natbio.com.

Geneworks on CD-ROM, Intelligenetics, Inc., 700 East El Camino Real, Mountain View, CA 94040, Tel: (800) 486–7489.

Genecraft, 5226 Caminito Vista Lujo, San Diego, CA 92130, Tel: (619) 554–0775, http://www.genecraft.com.

References

Ausubel, F. M., Kingston, R. E., Moore, D. D., Seidman, J. G., Smith, J. A., and Struhl, K., eds. (1995) *Current Protocols in Molecular Biology on CD-ROM*. Liss, New York.

Batt, C. A., Cho, Y., and Jamieson, A. C. (1993) Selection of oligodeoxynucleotide-directed mutants, in *Methods in Enzymology: Recombi-*

nant DNA, part H, vol. 217 (Wu, R., ed.), Academic, New York, pp. 280–286.

Bernhardt, G., Bibb, J. A., Bradley, J., and Wimmer, E. (1994) Molecular characterization of the cellular receptor for poliovirus. *Virology* **199,** 105–113.

Bibb, J. A., Bernhardt, G., and Wimmer, E. (1994) The human poliovirus receptor PVR alpha is a serine phosphoprotein. *J. Virol.* **68,** 6111–6115.

Carter, P. (1991) Mutagenesis facilitated by the removal or introduction of unique restriction sites, in *Directed Mutagenesis, A Practical Approach* (McPherson, M. J., ed.), Academic, New York, pp. 1–24.

Conley, E. C. and Saunders, J. R. (1984) Recombination-dependent recircularization of linearized pBR322 plasmid following transformation of *Escherichia coli. Mol. Gen. Genet.* **194,** 211–218.

Crowe, J., Döbeli, H., Gentz, R., Hochuli, E., Stüber, D., and Henco, K. (1994) 6XHis-Ni-NTA chromatography as a superior technique in recombinant protein expression/purification, in *Methods in Molecular Biology, vol. 31: Protocols for Gene Analysis* (Harwood, A. J., ed.), Humana, Totowa, NJ, pp. 371–387.

da Cruz e Silva, O. A., Iverfeldt, K., Oltersdorf, T., Sinha, S., Lieberburg, I., Ramabhadran, T. V., Suzuki, T., Sisodia, S. S., Gandy, S., and Greengard, P. (1993) Regulated cleavage of Alzheimer beta-amyloid precursor protein in the absence of the cytoplasmic tail. *Neuroscience* **57,** 873–877.

Deng, W. P. and Nickoloff, J. A. (1992) Site-directed mutagenesis of virtually any plasmid by eliminating a unique site. *Anal. Biochem.* **200,** 81–88.

Desdouits, F., Cohen, D., Nairn, A. C., Greengard, P., and Girault, J.-A. (1995a) Phosphorylation of DARPP-32, a dopamine-and cAMP-regulated posphoprotein, by casein kinase I in vitro and in vivo. *J. Biol. Chem.* **270,** 8772–8778.

Desdouits, F., Siciliano, J. L., Greengard, P., and Girault, J.-A. (1995b) Dopamine- and cAMP-regulated phosphoprotein DARPP-32: phosphorylation of Ser-137 by casein kinase-I inhibits dephosphorylation of Thr-34 by calcineurin. *Proc. Natl. Acad. Sci. USA* **42,** 2682–2685.

Dieffenbach, C. W. (1995) in *PCR Primer.* Cold Spring Harbor Laboratory, Cold Spring Harbor, New York, pp. 581–622.

Dohadwala, M., da Cruz e Silva, E. F., Williams, R. T., Carbonaro-Hall, D. A., Nairn, A. C., Greengard, P., and Berndt, N. (1994) Phosphorylation and inactivation of protein phosphatase 1 by cyclin-dependent kinases. *Proc. Natl. Acad. Sci. USA* **91,** 6408–6412.

Dubendorff, J. W. and Studier, F. W. (1991) Controlling basal expression in an inducible T7 expression system by blocking the target T7 promoter with *lac* repressor. *J. Mol. Biol.* **219,** 45–59.

Freedman, R. B. and Hawkins, H. C. (1985) *The Enzymology of Posttranslational Modification of Proteins,* vol. 2, Academic, New York.

Gandy, S. E., Czernik, A. J., and Greengard, P. (1988) Phosphorylation of Alzheimer disease amyloid precursor protein peptide by protein kinase C and Ca^{2+}/calmodulin-dependent protein kinases II. *Proc. Natl. Acad. Sci. USA* **85,** 6218–6221.

Harding, J. J. and Crabbe, M. J. C. (1992) *Posttranslational Modification of Proteins.* CRC, Ann Arbor, MI.

Hemmings, H. C. J., Nairn, A. C., Bibb, J. A., and Greengard, P. (1995) Signal transduction in the striatum: DARPP-32 a molecular integrator of multiple signalling pathways, in *Molecular and Cellular Mechanisms of Neostriatal Function* (Ariano, M. A. and Surmeier, D. J., eds.), Landes, Austin, TX, pp. 283–297.

Hofer, B. and Kühlein, B. (1993) Site-specific mutagenesis in plasmids: a gapped circle method, in *Methods in Enzymology: Recombinant DNA,* part H, vol. 217 (Wu, R., ed.), Academic, New York, pp. 173–189.

Horton, R. M. and Pease, L. R. (1991) Recombinantion and mutagenesis of DNA sequences using PCR, in *Directed Mutagenesis, A Practical Approach* (McPherson, M. J., ed.), IRL, New York, pp. 217–247.

Horton, R. M., Ho, S. N., Pullen, J. K., Hunt, H. D., Cai, Z., and Pease, L. R. (1993) Gene splicing by overlap extension, in *Methods in Enzymology: Recombinant DNA,* part H, vol. 217 (Wu, R., ed.), Academic, New York, pp. 270–279.

Hughes, M. J. G. and Andrews, D. W. (1996) Creation of deletion, insertion and substitution mutations using a single pair of primers and PCR. *Biotechniques* **20,** 188–196.

Hunkapiller, M. W. (1991) Advances in DNA sequencing technology. *Curr. Opinion Gen. Dev.* **1,** 88–92.

Innis, M. A., Gelfand, D. H., and Sninsky, J. J. (1994) *PCR Strategies.* Academic, New York.

Inouye, S. and Inouye, M. (1991) Site-directed mutagenesis using gapped-heteroduplex plasmid DNA, in *Directed Mutagenesis, A Practical Approach* (McPherson, M. J., ed.), Oxford University, New York, pp. 71–82.

Ju, J., Ruan, C., Fuller, C. W., Glazer, A. N., and Mathies, R. A. (1995) Fluorescence energy transfer dye-labeled primers for DNA sequencing and analysis. *Proc. Natl. Acad. Sci. USA* **92,** 4347–5351.

Kaslow, D. C. and Rawlings, D. J. (1993) Introducing restriction sites into double-stranded plasmid DNA, in *Methods in Enzymology: Recombinant DNA,* part H, vol. 217 (Wu, R., ed.), Academic, New York, pp. 295–301.

Kennelly, P. J. (1994) Identifiecation of sites of serine and threonine phosphorylation via site-directed mutagenesis—site transformation versus site elimination. *Anal. Biochem.* **219,** 384–386.

King, L. A. and Possee, R. D. (1992) *The Baculovirus Expression System: A Laboratory Guide.* Chapman and Hall, London.

Kleina, L. G. and Miller, J. H. (1990) Genetic studies of the lac repressor, XII. Extensive amino acid replacements generated by the use of natural and synthetic nosnsense suppressors. *J. Mol. Biol.* **212,** 295–318.

Kleina, L. G., Masson, J.-M., Normanly, J., Abelson, J., and Miller, J. H. (1990) Construction of *Escherichia coli* amber suppressor tRNA genes, II. Synthesis of additional tRNA genes and improvement of suppressor efficiency. *J. Mol. Biol.* **213,** 705–717.

Kriegler, M. (1990) *Gene Transfer and Expression: A Laboratory Manual.* Freeman, New York.

Kunkel, T. A. (1985) Rapid and efficient site-specific mutagenesis without phenotypic selection. *Proc. Natl. Acad. Sci. USA* **82,** 488–492.

Kunkel, T. A. (1987) Rapid and efficient site-specific mutagenesis without phenotypic selection, in *Methods in Enzymology: Recombinant DNA,* part E, vol. 154 (Roberts, J. D. and Zakour, R. A., eds.), Academic, New York.

Lamb, C. J. (1990) *Plant Gene Transfer.* Liss, New York.

Marck, C. (1988) "DNA Strider": a "C" program for the fast analysis of DNA and protein sequences on the Apple Macintosh family of computers. *Nucleic Acid Res.* **16,** 1829–1836.

McPherson, M. J. (1991) *Directed Mutagenesis, A Practical Approach,* IRL, New York.

McPherson, M. J. (1995) *PCR 2: A Practical Approach,* IRL, New York.

Means, A. R. (1994) Calcium regulation of cellular function. *Adv. Sec. Messeng. Phosphopro. Res.* **30,** 1–416.

Moffatt, B. A., Dunn, J. J., and Studier, F. W. (1984) Nucleotide sequence of the gene for bacteriophage T7 RNA polymerase. *J. Mol. Biol.* **173,** 265–269.

Murray, E. J. (1991) *Gene Transfer and Expression Protocols.* Humana, Clifton, NJ.

NATO Advanced Study Institute on Plant Molecular Biology (1991) *Plant Mol. Biol. 2.* Plenum, New York.

Normanly, J., Kleina, L. G., Masson, J.-M., Abelson, J., and Miller, J. H. (1990) Construction of *Escherichia coli* amber suppressor tRNA genes, III. Determination of tRNA specificity. *J. Mol. Biol.* **213,** 719–726.

Olsen, D. B., Sayers, J. R., and Eckstein, F. (1993) Site-directed mutagenesis of single-stranded and double-stranded DNA by phosphorothioate approach, in *Methods in Enzymology: Recombinant DNA,* part H, vol. 217 (Wu, R., ed.), Academic, New York, pp. 189–217.

Piechocki, M. P. and Hines, R. N. (1994) Oligonucleotide design and optimized protocol for site-directed mutagenesis. *BioTechniques* **16,** 702–707.

Rychlik, W. and Rhoads, R. E. (1989) A computer program for choosing optimal oligonucleotides for filter hybridization, sequencing and in vitro amplification of DNA. *Nucleic Acid Res.* **17,** 8543–8551.

Saiki, R. K., Gelfand, D. H., Stoffel, S., Scharf, S. J., Higuchi, R., Horn, G. T., Mullis, K. B., and Erlich, H. A. (1985) Primer-directed enzymatic amplification of DNA with a thermostable DNA polymerase. *Science* **239,** 487–491.

Sambrook, J., Fritsch, E. F., and Maniatis, T. (1989) *Molecular Cloning, A Laboratory Manual.* 2nd ed. Cold Spring Harbor Laboratory, Cold Spring Harbor, NY.

Sarkar, G. (1995) *PCR in Neuroscience.* Academic, New York.

Sayers, J. R., Krekel, C., and Eckstein, F. (1992) Rapid high-efficiency site-directed mutagenesis by the phosphorothioate approach. *BioTechniques* **13,** 592–546.

Selbert, M. A., Anderson, K. A., Huang, Q.-H., Goldstein, E. G., Means, A. R., and Edelman, A. M. (1995) Phosphorylation and activation of Ca^{2+}-calmodulin-dependent protein kinase IV by Ca^{2+}-calmodulin-dependent protein kinase Ia kinase. *J. Biol. Chem.* **270,** 17,616–17,621.

Seto, D., Seto, J., Deshpande, P., and Hood, L. (1995) DMSO resolves certain compressions and signal dropouts in fluorscent dye labeled primer-based DNA sequencing reactions. *DNA Sequence* **5,** 131–140.

Shortle, D. and Botstein, D. (1985) Strategies and applications of in vitro mutagenesis. *Science* **229,** 1193–1201.

Smith, D. B. and Johnson, K. S. (1988) Single-step purification of polypeptides expressed in *Escherichia coli* as fusions with glutathione S-transferase. *Gene* **67,** 31–40.

Smith, M. (1985) In vitro mutagenesis. *Ann. Rev. Genet.* **19,** 423–462.

Steen, R., Dahlberg, A. E., Lade, B. N., Studier, F. W., and Dunn, J. J. (1986) T7 RNA polymerase directed expression of the *Escherichia coli* rrnB operon. *EMBO J.* **5,** 1099–1103.

Studier, F. W. and Moffatt, B. A. (1986) Use of bacteriophage T7 RNA polymerase to direct selective high-level expression of cloned genes. *J. Mol. Biol.* **189,** 113–130.

Studier, F. W., Rosenberg, A. H., Dunn, J. J., and Dubendorff, J. W. (1990) Use of T7 RNA polymerase to direct expression of cloned genes, in *Methods in Enzymology, Gene Expression Technology,* vol. 185, pp. 60–89.

Suzuki, T., Nairn, A. C., Gandy, S. E., and Greengard, P. (1992) Phosphorylation of Alzheimer amyloid precursor protein by protein kinase C. *Neuroscience* **48,** 755–761.

Taylor, J. W., Ott, J., and Eckstein, F. (1985) The rapid generation of oligonucleotide-directed mutations at high frequency using phosphorothioate-modified DNA. *Nucleic Acid Res.* **13,** 8765–8785.

Vandeyar, M. A., Weiner, M. P., Hutton, C. J., and Batt, C. A. (1988) Identification of a simple and rapid method for the selection of oligodeoxynucleotide directed mutants. *Gene* **65,** 129–133.

Wang, Y. and Kent, C. (1995) Effects of altered phosphorylation sites on the properties of CTP: phosphocholine cytidylyltransferase. *J. Biol. Chem.* **270,** 17,843–17,849.

Weiner, M. P., Costa, G. L., Schoettlin, W., Cline, J., Mathur, E., and Bauer, J. C. (1994) Site-directed mutagenesis of double-stranded DNA by the polymerase chain reaction. *Gene* **151,** 119–123.

Weiner, M. P., Felts, K., Simcox, T., and Braman, J. (1993) Directional method for the site-directed mono- and multi-mutagenesis of double-stranded DNA. *Gene* **126,** 35–41.

Wold, F. and Moldave, K. (1984) *Methods in Enzymology, Posttranslational Modifications,* parts A and B, vols. 106,107. Academic, New York.

Wu, R. (1993) in, *Methods in Enzymology, Recombinant DNA,* part H, vol. 217. Academic, New York.

Yuckenberg, P. D., Witney, F., Geisselsoder, J., and McClary, J. (1991) Site-directed in vitro mutagenesis using uracil-containing DNA and phagemid vectors, in *Directed Mutagenesis, A Practical Approach* (McPherson, M. J., ed.), IRL, New York, pp. 27–48.

Zhao, L.-J. and Padmanabhan, R. (1993) Polymerase chain reaction-based point mutagenesis protocol, in *Methods in Enzymology: Recombinant DNA,* part H, vol. 217 (Wu, R., ed.), Academic, New York, pp. 218–227.

Zhu, L. and Holtz, A. E. (1993) Improved protocol for the Transformer™ site-directed mutagenesis kit. *Clontechniques* **Oct.,** 1–3.

Protein Methylation
in the Nervous System

Darin J. Weber and Philip N. McFadden

1. Introduction

This chapter describes the types of protein methylation that might be confronted in investigations of the nervous system, and summarizes technical approaches that have proven useful in elucidating the chemistries and identities of methylated proteins. Several distinct protein methyltransferase activities have been described in nervous tissues, and there are undoubtedly more to be discovered. It has often been the case that identification of a methylation pathway in the nervous system has led to the discovery of a similar system in some other tissue. An excellent recent example is the characterization in the brain of isoprenylated protein methyltransferases (Ben et al., 1993; Paz et al., 1993; Klein et al., 1994), which have proven to have important activities in many other cells (Clarke, 1992b).

Nervous system protein methyltransferases discovered thus far can be categorized on the basis of their amino acid group specificities (e.g., lysine vs arginine vs aspartic acid). However, issues of whether sequence specificities play a role in methylation patterns are much less clear. There are many other outstanding questions as well: How many proteins does a given enzyme methylate? How abundant and how subtly different are the isozymes of a given category of protein methyltransferase? What biological factors regulate the expression and activities of protein methyltransferases? Also,

From: *Neuromethods, Vol. 30: Regulatory Protein Modification: Techniques and Protocols*
Ed: H. C. Hemmings, Jr. Humana Press Inc.

the most hotly debated topic: What are the functions of protein methyltransferases?

Though all of these questions remain unanswered, both
general and specific observations can be made regarding
changes in the proteins that become methylated. In general,
both N-methylation and carboxyl methylation will cause
small shifts in a protein's pI as a result of changes in charge
distribution within the protein owing to the methyl group.
The consequences of such shifts in pI are currently unknown.
Specifically, in several proteins, including histones, riboso-
mal proteins, myelin, and calmodulin, N-methylation has
been found to modulate activity (Rattan et al., 1992) and in
some cases, protease resistance (Lischwe, 1990). Carboxyl
methylation in prenylated, C-terminal cysteinyl residues on
several GTP binding proteins (Ong et al., 1989; Ota and
Clarke, 1989; Fung et al., 1990; Clarke, 1992b; Lerner et al.,
1992; Giner and Rando, 1994; Ghomashchi et al., 1995; Volker
et al., 1995) has been postulated to have a role in signal transduction. Clearly, both N-linked and carboxyl methylation of
proteins play important roles in the nervous system. However, the precise function of these posttranslational modifications remain to be elucidated.

There is a solid foundation of research methods from
which to study nervous system protein methylation. Most fundamental are the methods used to describe protein methyltransferase reactions on the basis of amino acid specificity. Thus
far, every nervous system protein methyltransferase has been
found to be either a carboxyl O-methyltransferase or an N-
methyltransferase. The carboxyl O-methyltransferases include
enzymes that methyl esterify the carboxyl groups of proteins,
including the side chains of age-altered aspartyl residues, the
C-terminal carboxyl groups of certain proteins, and the side
chain of glutamic acid (exclusively in bacteria). The N-methyl-
transferases include enzymes that methylate nitrogen atoms of
lysine, arginine, histidine, and the amino-terminus. The differing chemical stabilities of carboxyl O-methylated sites and N-
methylated sites can be used diagnostically early in the
characterization of a protein methylation reaction. Which

category of reaction is identified will then dictate the repertoire of approaches to be used for characterization.

A recurring theme of research with protein methyltransferases is that structure leads to function—the identification and characterization of methylated proteins have often led to the discovery of novel forms of cellular regulation. A recent example is the identification of the major methylated protein in brain extracts as protein phosphatase 2A (Lee and Stock, 1993; Xie and Clarke, 1993; Favre et al., 1994; Xie and Clarke, 1994a,b). The finding that the C-terminal leucine of this protein undergoes carboxyl O-methylation is interesting from many perspectives, notably as an interface between methylation and phosphorylation control in the nervous system.

Several excellent reviews of the field of protein methylation have been published. A particularly helpful reference for those embarking on the study of a protein methylation reaction is a book on the subject edited by Paik and Kim (1990). Other very useful reviews on protein methylation and the substrates for protein methyltransferases can be found in Clarke (1992b), Clarke (1993), and Aswad (1995b).

2. Methodological Differences in Studies of Protein N-Methylation and Carboxyl O-Methylation

The single greatest factor in determining the choice of methods for characterizing a protein methylation reaction is the distinction between nonhydrolyzable N-linkages and hydrolyzable carboxyl O-linkages. There are three likely approaches:

1. If a protein is N-methylated, many of the standard techniques of biochemistry (e.g., sodium dodecyl sulfate-polyacrylamide gel electrophoresis [SDS-PAGE]), electrophoretic immunoblotting, and Edman degradation) can generally be employed as tools in establishing the identity of the protein;
2. If the protein is carboxyl O-methylated on its C-terminal carboxyl group, these same standard techniques can usually be used, with a few precautions taken to avoid the hydrolysis of the relatively stable methyl ester linkage; and

3. If the protein is carboxyl *O*-methylated on an altered aspartyl residue, a less commonly used subset of acidic techniques will be required, since the methyl ester linkage will spontaneously hydrolyze under the conditions used for most standard biochemical techniques.

Some of the most commonly found types of methylated proteins are those that fall into category 3 because even though many species of proteins are likely to be *N*-methylated (Najbauer et al., 1992) or carboxyl *O*-methylated at their C-termini (Clarke, 1992b), essentially all cellular proteins have the potential for being methylated on the carboxyl groups of age-altered aspartic acids (O'Conner and Clarke, 1984). The methyltransferase that catalyzes the methylation of altered aspartyl residues is protein (D-aspartyl/L-isoaspartyl) methyltransferase (EC 2.1.1.77). This enzyme is also known as protein isoaspartyl methyltransferase (PIMT) in recognition of the fact that the first synthetic substrate for this enzyme contained an L-isoaspartyl residue (Aswad and Guzzetta, 1995). This enzyme is ubiquitous and notorious for its ability to transfer a methyl group to the substoichiometric age-altered fraction of nearly any protein population. The highly labile methyl esterification catalyzed by PIMT has been shown to be capable of converting altered aspartyl residues in proteins back to their normal L configuration (Johnson et al., 1987b). In vitro studies with calmodulin have shown that formation of isoaspartyl residues in Ca^{2+} binding sites prevents binding and that this binding activity can be restored following methylation with PIMT and subsequent demethylation (Johnson et al., 1987a). The finding that PIMT activity is highest in the brain, a tissue composed mostly of postmitotic cells, is intriguing (Aswad, 1995a) and suggests PIMT may play a role in maintaining the functional state of proteins. Reports of the presence of significant levels of isoaspartyl residues in β-amyloid plaques in the brains of Alzheimer's disease patients lend support to this hypothesis (Payan et al., 1992; Roher et al., 1993; Fabian et al., 1994; Tomiyama et al., 1994; Iversen et al., 1995; Sandip and Duffy,

1995). PIMT activity is therefore responsible for a complex pattern of methyl group incorporation into protein mixtures, such as those found in neuronal cells.

2.1. Hydrolytic Lability of Methyl Groups Incorporated by Protein Carboxyl O-Methyltransferases

Four amino acid sites have been shown to be carboxyl O-methylated by nervous system protein carboxyl O-methyltransferases: the α carboxyl group of L-isoaspartic acid (L-iso-Asp); the β carboxyl group of D-aspartic acid (D-Asp); the C-terminal carboxyl group in isoprenylated L-cysteines; and the C-terminal carboxyl group in L-leucine. The generic distinguishing feature of protein carboxyl O-methylation as opposed to protein N-methylation is the lability of the methyl ester linkage to hydrolyzing chemical conditions. All of the known carboxyl O-methylation linkages are severed by strong acid (e.g., $6M$ HCl, 100°C, 6 h) and strong base (e.g., $1M$ NaOH, 37°C, 6 h) (Terwilliger and Clarke, 1981; Paik and Kim, 1990). Furthermore, the methyl ester linkages to L-iso-Asp and D-Asp are hydrolyzed by mild alkaline conditions as a result of a hydrolysis mechanism that involves intramolecular displacement of the methyl ester by the amide nitrogen of the neighboring amino acid (Murray and Clarke, 1984; Johnson and Aswad, 1985). The product of methyl ester hydrolysis is methanol. Therefore, if proteins containing radiochemical methyl groups yield radiolabeled methanol on hydrolytic treatment, this is a strong positive indication of the presence of a protein carboxyl O-methyl linkage. By varying the pH of the hydrolysis reaction, it is possible to ascertain whether or not the ester is of the extremely base-labile variety involving L-iso-Asp and D-Asp carboxyl O-methylation (Clarke et al., 1988).

The simplest method for determining the presence of radioactive methanol is a vapor-phase assay in which volatile products are captured directly into a scintillation cocktail for radiometric measurements (Murray and Clarke, 1984; Chelsky et al., 1989; Aswad and Guzzetta, 1995). Under some conditions of hydrolysis, the breakdown of methylarginine can also yield a potentially volatile product, methylamine, but methylamine can be kept nonvolatile by performing the vapor-phase assay at low pH.

Because methyl esterified amino acids are unstable under hydrolytic chemical conditions, it is not possible to recover and analyze the free carboxyl *O*-methylated amino acids by routine acid hydrolysis procedures used in amino acid analysis. Other methods are therefore necessary to identify positively the amino acid group that is methyl esterified. The most commonly employed method is to degrade the polypeptide as much as practical with proteases, followed by an analysis of the proteolytic digestion products for free amino acids with the methyl group still present (McFadden and Clarke, 1982; Xie and Clarke, 1993). In the case of prenylated cysteine residues, the C-terminal methylated product in rod outer-segment membrane proteins was detected following proteolytic digestion together with performic acid oxidation to remove the prenyl moiety (Ong et al., 1989; Ota and Clarke, 1989). Similar procedures led to the identification of C-terminally methylated cysteines in brain G-protein γ-subunits (Fung et al., 1990).

2.2. Hydrolytically Stable Protein Methylation

Hydrolytic stability of a methyl group in a protein is strong evidence for *N*-methylation. Extensive reviews are available that cite the many known examples of proteins that are *N*-methylated (Park and Paik, 1990; Clarke, 1992a). Although many of these studies have utilized tissues and cells outside of the nervous system, it is clear that numerous proteins that have been characterized as being *N*-methylated are present in many tissues and are therefore likely candidates for methylated proteins in the nervous system. A means of increasing the sensitivity of detection of *N*-methylated protein substrates in PC12 pheochromocytoma cells involves pretreating the cells for a lengthy period with a pharmacological inhibitor of methylation, and then incubating the now hypomethylated cell extracts with radiolabeled *S*-adenosylmethionine (Najbauer et al., 1991, 1992).

The amino acids that are *N*-methylated include lysine, arginine, and histidine. Each of these modified amino acids can be recovered after acid hydrolysis of the protein (Park and Paik, 1990). For lysine, the single nitrogen atom of the side-chain can be mono-, di-, or trimethylated. Calmodulin (Siegel et al., 1990), actin, myosin, and histones are well-known proteins that are methylated on lysines, and of course, these proteins are abundant in the nervous system. For arginine modification, examples are known of methyl group incorporation into a single nitrogen atom in the guanidinium group of the side-chain (monomethylarginine), methyl groups incorporated into both nitrogen atoms of the side-chain ($N^G,N^{'G}$-dimethylarginine), and two methyl groups incorporated into one of the nitrogen atoms of the side-chain ($N^G,N^{'G}$-dimethylarginine). Arginine methylation has been observed in histones, myelin basic protein, and other proteins (Ghosh et al., 1990). Histidine methylation occurs in actin, myosin, and histones among other proteins (Park and Paik, 1990).

3. Radiochemical Methyl Groups and Their Incorporation into Proteins of the Nervous System

The commercial availability of radiochemically labeled methyl groups as *S*-adenosyl-L-[methyl-^{3}H]-methionine and *S*-adenosyl-L-[methyl^{14}C]-methionine, as well as L-[methyl-^{3}H]- and L-[methyl^{14}C]-methionine is fundamentally responsible for the spate of discoveries of methylated proteins that still continues. It is important to consider the chemical and radiochemical characteristics of these commercially available reagents. It is also important to optimize the assays and incubation conditions through which proteins are radioactively methylated.

3.1. Radiochemical Characteristics of Methyl Group Donor Compounds

As is true of any comparison of molecules that are singly labeled by either tritium or carbon-14, the specific radioactivity of tritium can theoretically be almost 500 times as high as the specific radioactivity of carbon-14 (e.g., 29 Ci/mmol for carrier-free tritium vs 62 mCi/mmol for carrier-free carbon-14). Even higher specific activities of S-adenosyl-L-[methyl-^{3}H]-methionine are available, since up to 3 tritium atoms can be present/methyl group. Thus, for a given number of radiolabeled molecules, it is technically possible to achieve the highest number of incorporated dpm by using tritium labeling. This is not always the overriding concern, however. Depending on the specific application, it may be more important to take advantage of the relatively greater decay energies of β particles emitted by carbon-14. For example, the greater decay energy of carbon-14 can be helpful if the radiolabeled protein is to be detected as an electrophoretic band by autofluorography (Skinner and Griswold, 1983).

Another practical consideration that may weigh in favor of the use of carbon-14 as the radiolabel is the typically greater degree of radiation decomposition of tritiated compounds. Manufacturers of radiolabeled S-adenosylmethionine and methionine generally package their materials with stabilizers against radiation decomposition. However, the rate of radiation decomposition of tritiated S-adenosylmethionine is typically about 1%/mo as compared to 1%/yr for S-adenosylmethione labeled by carbon-14.

Suppliers of [^{14}C-methyl]- and [^{3}H-methyl]-S-adenosylmethionine include American Radiolabeled Chemicals (St. Louis, MO), Amersham (Arlington Heights, IL), ICN, and New England Nuclear (Boston, MA). [^{3}H-methyl]S-adenosylmethionine is sold by Sigma (St. Louis, MO). Both L-[methyl-^{3}H]-methionine and L-[methyl^{14}C]-methionine are sold by each of these manufacturers. The availability of S-adenosylmethionine and methionine, which is radiolabeled in a position other than the methyl group, can be useful in certain control experiments. For example, a radiolabel transfer reaction that is catalyzed by hemoglobin is evidently a methyl transfer reaction, since radiolabeled methyl groups are incorporated from S-adenosylmethionine, but not from the radiolabeled carboxyl moiety (Kimzey and McFadden, 1994).

3.2. Radiolabeling Procedures
in the Methylation of Proteins

The starting point for essentially all studies of protein methylation is to perform the radioactive labeling of proteins. Broadly speaking, there are two ways in which this can be done: in vivo cell metabolism and in vitro reconstitution assays.

3.2.1. Protein Methylation Procedures in Intact Cells

The in vivo approach allows the machinery of living cells to accomplish this radiolabeling in the course of cellular methylation metabolism. The radioactive precursor can be methionine (either L-[methyl-^{3}H]-methionine or L-[methyl-^{14}C]-methionine), since cells are thought not to take up *S*-adenosylmethionine, but are generally capable of taking up methionine and converting a portion of this methionine into *S*-adenosylmethionine. A popular approach is to use cultured neuronal cells in such radiolabeling, including clonal lines, such as PC12 cells (Najbauer et al., 1991; Johnson et al., 1993) and murine neuroblastoma cells (O'Dea et al., 1987). Brain tissue slices have also been used in the study of protein methylation in intact nervous system cells (Wolf and Roth, 1985). The medium in which the cells are incubated should be adjusted in its nonradioactive methionine content, so that the specific radioactivity of the radiolabel is high enough to result in the incorporation of detectable levels of methyl group radioactivity in the proteins of interest. For relatively short-term labeling experiments of less than a few hours, methionine-free culture media can be used with additions of high specific radioactivity methionine (including carrier-free radiolabel). For longer-term incubations, the cells may require higher chemical concentrations of methionine for growth, so the addition of unlabeled methionine will dilute the specific radioactivity and decrease the sensitivity with which radioactively methylated proteins can be detected.

Of course, the direct incorporation of radiolabeled methionine into protein will also occur in most cells through protein translation. In the case of protein carboxyl *O*-methylation, the incorporation of radioactive methyl esters can be detected above the much larger background of protein backbone radiolabeling by isolating the proteins, exposing them to an alkali treatment, and measuring volatile radioactive methanol (Kloog et al., 1983; Chelsky et al., 1989). Another way to work around the high background resulting from backbone radiolabeling is to make use of a protein synthesis inhibitor to block protein translation, with the assumption that at least some aspects of protein methylation will continue. A good example of the latter approach is the use of cycloheximide to block protein translation in HeLa cells during their incubation with [³H-methyl]methionine (Ladino and O'Connor, 1992). A further interesting possibility is that some cells can in fact directly take up radiolabeled S-adenosylmethionine and use it in the methylation of proteins (Barten and O'Dea, 1989).

3.2.2. In Vitro Protein Methylation Procedures

Since nearly every type of living cell contains many types of protein methyltransferases and their substrates, cell extracts and subcellular fractions incubated with radiolabeled S-adenosylmethionine usually results in incorporation of methyl groups into proteins. Techniques of gel electrophoresis are often used in conjunction with such radiolabeling in order to resolve the various methylated species of proteins. Many discoveries of protein methyltransferases and their substrates have been made with this basic approach. For example, guanine nucleotide-dependent carboxyl methylation of proteins was investigated by incubating cell membranes with S-adenosyl-[³H-methyl]-methionine and then measuring how the incorporation of radiolabel into particular electrophoretically separated proteins was affected by including GTP analogs in the incubation mixtures (Backlund and Aksamit, 1988). As another example, the identification of protein phosphatase 2A as a C-terminally methylated substrate (Lee and Stock, 1993; Xie and Clarke, 1993,1994a,b; Favre et al., 1994) ultimately stemmed from long-standing observations of highly methylated proteins formed during incubations of cytoplasm with radiolabeled S-adenosylmethionine (O'Conner and Clarke, 1984).

A variation of the in vitro approach is to use cell extracts as the source of the enzymatic activity and to include either recombinantly expressed proteins or synthetic polypeptide substrates as the methyl acceptor substrate. For example, the methylation of recombinant protein A1 by calf brain methyltransferase led to the identification of a specific arginine residue that is methylated in this protein component of the heterogeneous nuclear ribonucleoprotein complex (Rajpurohit et al., 1994). As another example, pathways in the maturation of C-terminally modified proteins can be studied by in vitro methylation of synthetic polypeptides that resemble protein carboxyl-termini (Stephenson and Clarke, 1990,1992). High pressure liquid chromatography (*see* Section 4.4.2.) has often been used to confirm that polypeptide species being methylated by a cellular extract are indeed the synthetic polypeptide substrates (Murray and Clarke, 1986). To demonstrate that polypeptide radiolabeling in vitro is owing to methyltransferase activity and not some other spurious process, a valuable control experiment is to test whether the radiolabel incorporation is blocked by micromolar concentrations of the end product inhibitor of most methyltransferases, *S*-adenosylhomocysteine. Recent work has shown that RNA methylation can potentially exhibit some of the expected characteristics of protein methylation (including radiolabeled electrophoretic bands in SDS-PAGE) (Hrycyna et al., 1994), thus, tests of the nuclease stability of methylated macromolecules might also be an important control experiment to perform.

3.2.3. Radiolabel Quantitation

The quantitation of radioactivity and the conversion of dpm values to mol quantities requires accurate knowledge of the specific radioactivity of the radiolabel in the protein methylation reaction mixture. This is not too difficult in reactions in vitro if no residual quantities of *S*-adenosylmethionine are introduced from the cell extract or enzyme preparation; in this case, the specific radioactivity should be that of the added radiolabel. If, on the other hand, the protein methylation reaction is performed by the uptake of radiolabeled methionine into cells, the specific radioactivity of the ensuing radioactive S-adenosylmethionine will depend on the amount of nonradioactive S-adenosylmethionine that is present from cellular pools. In this case, it is necessary to measure independently the chemical concentrations of S-adenosylmethionine and its radioactivity to arrive at a specific radioactivity that can be used to convert dpm incorporated into protein into mol quantities. Fractionation procedures for purifying S-adenosylmethionine and determining its concentration can be based on chromatography (Glazer and Peale, 1978; Barber and Clarke, 1983); alternatively, the concentration of S-adenosylmethionine can be measured by standardized enzymatic assays, such as the hydroxyindole O-methyltransferase assay procedure (Baldessarini and Kopin, 1966).

4. Methylation of Altered Aspartyl Residues in Proteins

Of the several types of protein methylation reactions known, the study of the methylation of altered aspartyl sites is most dependent on the use of specialized techniques. These radiolabeled methyl ester moieties are very labile at neutral pH. Thus, methods of characterizing proteins with altered aspartyl residues must employ acidic pH, which has been shown to stabilize the methyl ester moiety.

4.1. Methylation Procedures

As described under Section 3.2., there are two general ways of methylating altered aspartyl residues in cells and tissues. One method utilizes the endogenous PIMT to methylate any proteins containing altered aspartyl residues in the intact cell, and the other employs exogenously added PIMT to methylate altered aspartyl residues in proteins from cell lysates, proteins isolated from specific subcellular compartments, or even highly purified proteins.

4.1.1. In Vivo Methylation of Altered Aspartyl Sites by PIMT

To identify which proteins are substrates for endogenous PIMT under in vivo conditions, cells are incubated with radiolabeled methionine and a suitable protein synthesis inhibitor, if necessary, to prevent backbone radiolabeling (Chelsky et al., 1989). PIMT subsequently uses the radiolabeled *S*-adenosylmethione produced in the cell to methylate proteins containing altered aspartyl residues that are accessible to the enzyme under physiological conditions. Following the period of radiolabeling, cell lysates prepared under acidic conditions can be prepared and the individual proteins resolved and analyzed by the techniques described in Section 4.3.

4.1.2. In Vitro Methylation of Altered Aspartyl Sites

Under physiological conditions, endogenous PIMT is localized primarily in the cytoplasm. Thus, intracellular methylation studies do not provide any information about the occurrence and distribution of many proteins containing altered aspartic acid that are inaccessible to methylation (O'Conner and Clarke, 1983). This may include proteins enclosed in organelles, certain cytoskeletal proteins, and integral membrane proteins with extracellular domains. For these inaccessible proteins, an in vitro methylation assay is required. Cell extracts and subcellular fractions can be prepared in several ways and then subjected to methylation with radiolabeled *S*-adenosylmethionine and purified PIMT (Kloog et al., 1983; O'Conner and Clarke, 1983; Barten and O'Dea, 1989; Ladino and O'Connor, 1992; Johnson et al., 1993). A typical approach to examining the distribution of proteins containing altered aspartic acid in cells involves subfractionating cellular proteins using differential solubility in various aqueous buffers. Suitable subfractionation buffers should avoid extremes of pH and ionic strength to avoid spurious alterations of aspartyl sites. Strong denaturants, such as SDS or urea, should be avoided if they are incompatible with altered aspartyl detection assays. A generalized protocol for detecting altered aspartyl residues in proteins is described below.

Use the following reagents for the altered aspartyl detection assay: 1–2 U purified PIMT; 100 mM radiolabeled S-adenosylmethionine (AdoMet); and 10–20 μg of substrate protein. The assay components are incubated in a final volume of 50 μL in 0.2 M sodium citrate, pH 6.0, at 37°C for 20–30 min. PIMT (1 U = 1 pmol methyl esters/min) can be purified as described (Gilbert et al., 1988; Aswad and Guzzetta, 1995). Acidic pH is used to stabilize any methyl esters formed during the reaction. This assay is broadly applicable to analysis of altered aspartyl residues in any protein-containing solution, ranging from cell lysates to highly purified proteins. Optimal reaction conditions can be determined as described in Johnson and Aswad (1991). If a biochemical workup of native proteins is anticipated as the next step in the analysis, 10–20 μM S-adenosylhomocysteine (AdoHcy) can be added to quench the reaction. Another common means of quenching the reaction is to inactivate the PIMT and stabilize methyl esters by adding a denaturing, acidic electrophoresis sample buffer (*see* Section 4.3.).

4.2. Quantitation of Sites Methylated by PIMT

Methylated proteins obtained by either of the two methods described under Section 4.1. can be subjected to a variety of analyses, depending on the goal of the study. Special conditions must be employed in order to minimize the loss of methyl esters, that are labile at neutral and alkaline pH. For protein fractionation, it is imperative that cold temperatures and acidic conditions be employed to minimize hydrolysis of methyl esters during the isolation procedure. Additionally, depending on the preparation, isolation buffers should include a broader spectrum of protease inhibitors than are used at neutral pH, since proteases with acidic pH optima will be favored under acidic isolation schemes (Weber and McFadden, 1995).

4.2.1. Quantitation of Total Levels
of Altered Aspartyl Residues in Proteins

To measure the total level of abnormal aspartyl residues present in a specific sample, the reaction is quenched by add-

ing an equal volume of a 20% (v/v) solution of trichloroacetic acid (TCA) at the end of the altered aspartyl detection
assay. The acid inactivates the enzyme, stabilizes protein
methyl esters, and precipitates the protein substrates. The
precipitated proteins are then washed twice by resuspension
in 5 vol of 10% TCA and centrifuged for 5 min at 5000g to
pellet the protein. This is done to separate any unincorporated radioactive AdoMet, which is soluble in 10% TCA, from
the methyl esterified proteins in the pellet. The level of altered
aspartyl residues in each sample is quantitated by taking
advantage of the fact that the methyl esters formed during
the altered aspartyl detection assay are easily hydrolyzed by
alkaline pH. Treatment with base results in the release of
radiolabeled methanol, a volatile compound. Each acid-pre-
cipitated pellet is resuspended in 200 µL of 0.2M NaOH for
20 min at room temperature in a tightly capped polypropylene tube. During this time, all the methyl esters will hydrolyze to form methanol. At the end of the hydrolysis period,
1/10 vol of 3M citric acid is quickly added and the tube recap-
ped. The acid stops the hydrolysis reaction and prevents the
volatilization of any incidentally formed methylamine.
Aliquots of the radiolabeled methanol-containing samples
are then transferred to a vial that is placed inside a scintillation vial containing scintillation fluid (Murray and Clarke,
1984; Aswad and Guzzetta, 1995). The methanol in the inner
vial is given several hours to volatilize and partition into the
organic scintillation fluid. The radioactivity is then quantitated in a scintillation counter and the volatilization efficiency
corrected using an internal standard of radioactive methanol.

4.2.2. Quantitation of the Total Level
of Altered Aspartyl Residues in Peptides

Since peptides are not quantitatively precipitated by
TCA, an alternative means to separate excess radioactive
AdoMet from the methylated peptides has been developed
(Fig. 1). Peptides that have been radioactively methylated by
PIMT are loaded onto agarose gels and electrophoresed in a
horizontal miniagarose gel apparatus (BRL, Gaithersburg,

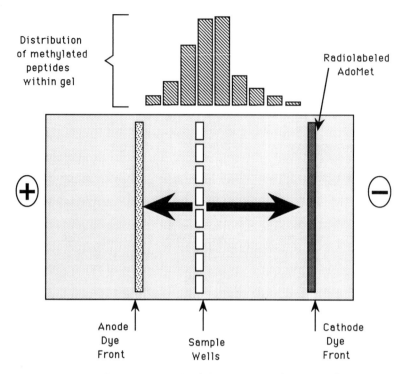

Fig. 1. Diagram of acidic 10% (w/v) agarose gel system for separating methyl esterified proteins or peptides from unincorporated radiolabeled AdoMet remaining at the end of an altered aspartyl detection assay. Aliquots of PIMT reaction mixtures (5 µL) are loaded into the centrally located sample wells and electrophoresed for 30 min at 100 V. The system takes advantage of the small size of radiolabeled AdoMet, which carries two positive charges at the acidic pH used in the system. As a small, doubly charged molecule, it will migrate rapidly toward the cathode, well ahead of any methyl esterified polypeptides, which migrate much more slowly primarily because of size. The low-mol-wt anodic and cathodic dyes included in the loading buffer delineate the maximal distribution of any radiolabeled peptides. A histogram of the typical localization of radiolabeled peptides between the two dye fronts is shown. Radiolabeled AdoMet runs at the leading edge of the cathodic dye on the right side of the diagram.

MD) for 30 min at 100 V. The gels are made by preparing 10% (w/v) ultrapure agarose (BRL) in an Erlenmeyer flask containing a final volume of 100 mL of 50 mM NaH$_2$PO$_4$, pH 2.4. The flask is preweighed and the agarose melted by heating in a microwave. The flask is then reweighed and enough diH$_2$O added to attain the original weight. Although still hot, 20 mL of the viscous solution are cast into a miniagarose gel caster and an 8-well comb is inserted into the center of the gel. After solidification, the comb is removed and the gel immersed in 150 mL of the same buffer used to prepare the gel. The 2X loading buffer consists of 100 mM NaH$_2$PO$_4$, pH 2.4, 5% (v/v) ultrapure glycerol (USB), 0.5 mg bromophenol blue, and 0.5 mg methyl green. The bromophenol blue migrates toward the anode, whereas the methyl green migrates toward the cathode. Excess radiolabeled AdoMet comigrates with the cathode dye, whereas the radiolabeled peptides are distributed between the two dye fronts, with the majority migrating only 1–2 cm beyond the sample wells. Half-centimeter gel slices of each lane are made and the radioactivity is quantified by the volatilization assay as described above for TCA-precipitated proteins. The total number of methyl esters present in each lane is determined from summing over all the gel slices. This method works equally well in separating radiolabeled S-adenosylmethione from high-mol-wt proteins, with the proteins tending to migrate only short distances.

4.3. Electrophoretic Separation of Proteins Containing Altered Aspartyl Residues

The initial stages in the characterization of the distribution of altered aspartyl-containing substrates in cells and tissues are well suited to polyacrylamide gel electrophoresis. This method allows complex mixtures of proteins, such as those fractionated on the basis of differential solubility, to be resolved on the basis of molecular weight, and multiple samples can be resolved simultaneously. Unfortunately, traditional methods of electrophoresis (Laemmli, 1970) cannot be employed, since the system operates at an alkaline pH that causes hydrolysis of any PIMT-methyl esterified proteins. Consequently, a series of systems employing polyacrylamide gel electrophoresis at acidic pH have been utilized in efforts to identify altered aspartyl-containing proteins (Fairbanks and Avruch, 1973; MacFarlane, 1984). The main draw-

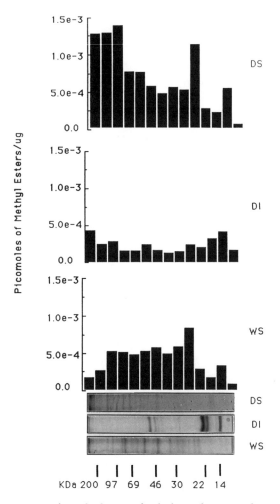

Fig. 2. Separation of methyl esterified altered aspartyl-containing proteins on 12% acidic polyacrylamide gels. Various subcellular fractions
from PC12 cells were assayed for the presence of altered aspartyl-containing proteins. Protein fractions were obtained on the basis of differential solubility in various aqueous buffers. Approximately 1×10^6 cells
underwent a cycle of freeze/thawing in phosphate buffered saline (PBS),
pH 7.4, containing protease inhibitors as described (Weber and
McFadden, 1995). Following centrifugation of the lysate at 10,000g for
10 min at 4°C, proteins recovered in the supernatant were designated
"WS," for water soluble. The insoluble pellet was washed twice by
resuspension in 200 µL of PBS and centrifugation as above. This pellet
was then resuspended in PBS containing 0.1% (v/v) Triton X-100.

backs of these systems are that they either produce broad electrophoretic bands or use cationic detergents that may be incompatible with other protein analysis procedures. Recently, a new electrophoresis system has been developed that employs SDS and an acidic discontinuous buffering system (Weber and McFadden, 1996). This procedure results in sharp electrophoretic bands and would be a good choice for investigators wishing to adhere to SDS as the anionic detergent. An example of the ability of this system to resolve methylated altered aspartyl proteins is shown in Fig. 2.

4.4. Localization of Sites of Altered Aspartyl Residues in Protein Sequences

To determine the total number of altered aspartyl residues in a given protein or a preparation of proteins, it is frequently necessary to fragment native protein(s) into peptides using enzymatic or chemical cleavage techniques in order to disrupt secondary or higher-order structures that prevent sites from being methylated by PIMT. Following a methyla-

---

Following a 30-min extraction period on ice, the supernatant was recovered by centrifugation as above and was designated DS for detergent-soluble. The pellet that was insoluble in the nonionic detergent was washed twice as previously described and resuspended in 200 µL of PBS by brief sonication on ice. The fraction was designated DI for detergent-insoluble. Based on a dye binding assay, an equal amount of protein in each fraction was methylated using the altered aspartyl detection assay described in Section 4. Equal aliquots of methylated proteins were then separated by electrophoresis on a discontinuous, acidic gel system as described (Weber and McFadden, 1996). Following staining of the gel with Coomassie brilliant blue, 0.5-cm gel slices of each sample lane were taken (14 gel slices/lane). Each gel slice was treated with $0.2M$ NaOH to hydrolyze methyl esters. Volatile radioactivity was detected as described in Section 4. **(Top)** Distribution of altered aspartyl-containing proteins in the DS fraction. **(Middle)** distribution of altered aspartyl-containing proteins in the DI fraction. **(Bottom)** distribution of altered aspartyl-containing proteins in the WS fraction. The gel lanes follow the same order as the histograms representing the distribution of methyl esterified proteins. Molecular-weight standards (in kilodaltons) are shown at the bottom of the figure.

tion reaction, any peptides that are methylated can be isolated and the amino acid composition determined.

4.4.1. Protein Fragmentation Methods

Cleavage methods have included trypsin digestion and cyanogen bromide cleavage (Aswad and Guzzetta, 1995). The cleaved proteins can then be radioactively methylated using the altered aspartyl detection assay, and the total amount of methylated residues present quantitated by electrophoresis on 10% agarose gels as described in Section 4.2.2. This method should also be useful for analyzing proteins, such as those of the neuronal cytoskeleton that contain altered aspartyl residues, but that are relatively insoluble under physiological conditions, and therefore incapable of being fully methylated in their intact form.

4.4.2. Separation of Individual Methylated Peptides

Reverse-phase high-performance liquid chromatography (RP-HPLC) of fragmented proteins has proven to be the method of choice for characterizing the location of methylated residues within the amino acid sequence of single, purified proteins. All types of methylated amino acids can be localized in this manner. Even the alkali-sensitive methyl esters formed by PIMT can be localized by this approach, since acidic, high-resolution solvent systems for HPLC are well established. In one common procedure, a purified protein is first radioactively methylated by a purified methyltransferase. Following chemical or enzymatic cleavage, the methylated products are separated by RP-HPLC on a reverse-phase column using a solvent system consisting of 0.1% (v/v) trifluoroacetic acid, water, and acetonitrile (Ota and Clarke, 1990; Aswad and Guzzetta, 1995). Column fractions containing radioactively methylated peptides are then detected and amino acid compositions are used to arrive at the identity of the methylated peptide. The amino acid sequence of a methylated peptide can also be determined using conventional sequence analysis or by mass spectrometry. If only substoichiometric amounts of a protein are methylated, as is generally true of PIMT reactions, the radioactively methylated peptide fragments may be only a small subpopulation of the peptide whose composition is actually measured. In many cases, the presence of the methyl group leads to only a very slight shift in the chromatographic behavior of a peptide, which allows the methylated peptide to be identified by the composition of its coeluting, unmethylated neighbors (Ota and Clarke, 1990; Lindquist et al., 1996).

References

Aswad, D. W. (1995a) Purification and properties of protein L-isoaspartyl methyltransferase, in *Deamidation and Isoaspartate Formation in Peptides and Proteins* (Aswad, D. W., ed.), CRC, Boca Raton, FL, pp. 32–45.

Aswad, D. W. (ed.) (1995b) *Deamidation and Isoaspartate Formation in Peptides and Proteins*, CRC, Boca Raton, FL, 260 p.

Aswad, D. W. and Guzzetta, A. W. (1995) Methods for analysis of deamidation and isoaspartate formation in peptides and proteins, in *Deamidation and Isoaspartate Formation in Peptides and Proteins* (Aswad, D. W., ed.), CRC, Boca Raton, FL, pp. 7–29.

Backlund, P. S. and Aksamit, R. R. (1988) Guanine nucleotide-dependent carboxyl methylation of mammalian membrane proteins. *J. Biol. Chem.* **263,** 15,864–15,867.

Baldessarini, R. J. and Kopin, I. J. (1966) S-Adenosylmethionine in brain and other tissues. *J. Neurochem.* **13,** 769–777.

Barber, J. R. and Clarke, S. (1983) Membrane protein carboxyl methylation increases with human erythrocyte age. Evidence for an increase in the number of methylatable sites. *J. Biol. Chem.* **258,** 1189–1196.

Barten, D. M. and O'Dea, R. F. (1989) Protein carboxylmethyltransferase activity in intact, differentiated neuroblastoma cells: quantitation by S-[³H]adenosylmethionine prelabeling. *J. Neurochem.* **53,** 1156–1165.

Ben, B. G., Paz, A., Marciano, D., Egozi, Y., Haklai, R., and Kloog, Y. (1993) The uniquely distributed isoprenylated protein methyltransferase activity in the rat brain is highly expressed in the cerebellum. *Biochem. Biophys. Res. Commun.* **195,** 282–288.

Chelsky, D., Sobotka, C., and O'Neil, C. L. (1989) Lamin B methylation and assembly into the nuclear envelope. *J. Biol. Chem.* **264,** 7637–7643.

Clarke, S. (1992a) The biological functions of protein methylation, in *Fundamentals of Medical Cell Biology, vol. 3B: Chemistry of the Living Cell,* JAI Press, pp. 413–436.

Clarke, S. (1992b) Protein isoprenylation and methylation at carboxyl terminal cysteine residues. *Ann. Rev. Biochem.* **61,** 355–386.

Clarke, S. (1993) Protein methylation. *Curr. Opinion Cell Biol.* **5,** 977–983.

Clarke, S., Vogel, J. P., Deschenes, R. J., and Stock, J. (1988) Posttranslational modification of the Ha-ras oncogene protein: evidence for a third class of protein carboxyl methyltransferases. *Proc. Natl. Acad. Sci. USA* **85**, 4643–4647.

Fabian, H., Szendrei, G. I., Mantsch, H. H., Greenberg, B. D., and Otvos, L. (1994) Synthetic posttranslationally modified human Ab peptide exhibits a markedly increased tendency to form b-pleated sheets in vitro. *Eur. J. Biochem.* **221**, 959–964.

Fairbanks, G. and Avruch, J. (1973) Four gel systems for electrophoretic fractionation of membrane proteins using ionic detergents. *J. Supramol. Struct.* **1**, 66–75.

Favre, B., Zolnierowicz, S., Turowski, P., and Hemmings, B. A. (1994) The catalytic subunit of protein phosphatase 2A is carboxyl-methylated in vivo. *J. Biol. Chem.* **269**, 16,311–16,317.

Fung, B. K., Yamane, H. K., Ota, I. M., and Clarke, S. (1990) The gamma subunit of brain G-proteins is methyl esterified at a C-terminal cysteine. *FEBS Lett.* **260**, 313–317.

Ghomashchi, F., Zhang, X., Liu, L., and Gelb, M. H. (1995) Binding of prenylated and polybasic peptides to membranes: affinities and intervesicle exchange. *Biochemistry* **34(37)**, 11,910–11,918.

Ghosh, S. K., Syed, S. K., Jung, S., Paik, W. K., and Kim, S. (1990) Substrate specificity for myelin basic protein-specific protein methylase I. *Biochim. Biophys. Acta* **1039**, 142–148.

Gilbert, J. M., Fowler, A., Bleibaum, J., and Clarke, S. (1988) Purification of homologous protein carboxyl methyltransferase isozymes from human and bovine erythrocytes. *Biochemistry* **27**, 5227–5233.

Giner, J. M. and Rando, R. R. (1994) Novel methyltransferase activity modifying the carboxy terminal bis(geranylgeranyl)-Cys-Ala-Cys structure of small GTP-binding proteins. *Biochemistry* **33(50)**, 15,116–15,123

Glazer, R. I. and Peale, A. L. (1978) Measurement of S-adenosyl-L-methionine levels by SP sephadex chromatography. *Anal. Biochem.* **91**, 516–520.

Hrycyna, C. A., Yang, M. C., and Clarke, S. (1994) Protein carboxyl methylation in *Saccharomyces cerevisiae*: evidence for STE14-dependent and STE14-independent pathways. *Biochemistry* **33**, 9806–9812.

Iversen, L. L., Mortishire-Smith, J., Pollack, S. J., and Shearman, M. S. (1995) The toxicity in vitro of β-amyloid protein. *Biochem. J.* **311**, 1–16.

Johnson, B. A. and Aswad, D. W. (1985) Enzymatic protein carboxyl methylation at physiological pH: cyclic imide formation explains rapid methyl turnover. *Biochemistry* **24**, 2581–2586.

Johnson, B. A. and Aswad, D. W. (1991) Optimal conditions for the use of protein L-isoaspartyl methyltransferase in assessing the isoaspartate content of peptides and proteins. *Anal. Biochem.* **192**, 384–391.

Johnson, B. A., Langmack, E. L., and Aswad, D. W. (1987a) Partial repair of deamidation-damaged calmodulin by protein carboxyl methyltransferase. *J. Biol. Chem.* **262,** 12,283–12,287.

Johnson, B. A., Murray, E., Jr., Clarke, S., Glass, D. B., and Aswad, D. W. (1987b) Protein carboxyl methyltransferase facilitates conversion of atypical L-isoaspartyl peptides to normal L-aspartyl peptides. *J. Biol. Chem.* **262,** 5622–5629.

Johnson, B. A., Najbauer, J., and Aswad, D. W. (1993) Accumulation of substrates for protein L-isoaspartyl methyltransferase in adenosine dialdehyde-treated PC12 cells. *J. Biol. Chem.* **268,** 6174–6181.

Kimzey, A. L. and McFadden, P. N. (1994) Spontaneous methylation of hemoglobin by S-adenosylmethionine by a specific and saturable mechanism. *J. Protein Chem.* **13,** 537–546.

Klein, Z., Ben, B. G., Marciano, D., Solomon, R., Altaras, M., and Kloog, Y. (1994) Characterization of the prenylated protein methyltransferase in human endometrial carcinoma. *Biochim. Biophys. Acta* **1226,** 330–336.

Kloog, Y., Axelrod, J., and Spector, I. (1983) Protein carboxyl methylation increases in parallel with differentiation of neuroblastoma cells. *J. Neurochem.* **40,** 522–529.

Ladino, C. A. and O'Connor, C. M. (1992) Methylation of atypical protein aspartyl residues during the stress response of HeLa cells. *J. Cell. Physiol.* **153,** 297–304.

Laemmli, U. K. (1970) Cleavage of structural proteins during the assembly of the head of bacteriophage T4. *Nature* **227,** 680–685.

Lee, J. and Stock, J. (1993) Protein phosphatase 2A catalytic subunit is methyl-esterified at its carboxyl terminus by a novel methyltransferase. *J. Biol. Chem.* **268,** 19,192–19,195.

Lerner, S., Haklai, R., and Kloog, Y. (1992) Isoprenylation and carboxylmethylation in small GTP-binding proteins of pheochromocytoma. (PC-12) cells. *Cell. Mol. Neurobiol.* **12,** 333–351.

Lindquist, J. A., Barofsky, E., and McFadden, P. N. (1996) Determination of two sites of automethylation in bovine erythrocyte (D-aspartyl/L-isoaspartyl) carboxyl methyltransferase. *J. Protein Chem.* **15,** 115–122.

Lischwe, M. A. (1990) Amino acid sequence of arginine methylation sites, in *Protein Methylation* (Paik, W. K. and Kim, S., eds.), CRC, Boca Raton, FL, pp. 98–120.

MacFarlane, D. E. (1984) Inhibitors of cyclic nucleotides phosphodiesterases inhibit protein carboxyl methylation in intact blood platelets. *J. Biol. Chem.* **259,** 1357–1362.

McFadden, P. N. and Clarke, S. (1982) Methylation at D-aspartyl residues in erythrocytes: possible step in the repair of aged membrane proteins. *Proc. Natl. Acad. Sci. USA* **79,** 2460–2464.

Murray, E. J. and Clarke, S. (1984) Synthetic peptide substrates for the erythrocyte protein carboxyl methyltransferase. Detection of a new site of methylation at isomerized L-aspartyl residues. *J. Biol. Chem.* **259,** 10,722–10,732.

Murray, E. D. and Clarke, S. (1986) Metabolism of a synthetic L-isoaspartyl-containing hexapeptide in erythrocyte extracts. *J. Biol. Chem.* **261,** 306–312.

Najbauer, J., Johnson, B. A., and Aswad, D. W. (1991) Amplification and detection of substrates for protein carboxyl methyltransferases in PC12 cells. *Anal. Biochem.* **197,** 412–420.

Najbauer, J., Johnson, B. A., and Aswad, D. W. (1992) Analysis of stable protein methylation in cultured cells. *Arch. Biochem. Biophys.* **293,** 85–92.

O'Conner, C. M. and Clarke, S. (1983) Methylation of erythrocyte membrane proteins at extracellular and intracellular D-aspartyl sites in vitro. *J. Biol. Chem.* **258,** 8485–8492.

O'Conner, C. M. and Clarke, S. (1984) Carboxyl methylation of cytosolic proteins in intact human erythrocytes. Identification of numerous methyl-accepting proteins, including hemoglobin and carbonic anhydrase. *J. Biol. Chem.* **259,** 2570–2578.

O'Dea, R. F., Mirkin, B. L., Hogenkamp, H. P., and Barten, D. M. (1987) Effect of adenosine analogues on protein carboxylmethyltransferase, S-adenosylhomocysteine hydrolase, and ribonucleotide reductase activity in murine neuroblastoma cells. *Cancer Res.* **47,** 3656–3661.

Ong, O. C., Ota, I. M., Clarke, S., and Fung, B. K. (1989) The membrane binding domain of rod cGMP phosphodiesterase is posttranslationally modified by methyl esterification at a C-terminal cysteine. *Proc. Natl. Acad. Sci. USA* **86,** 9238–9242.

Ota, I. M. and Clarke, S. (1989) Enzymatic methylation of 23-29-kDa bovine retinal rod outer segment membrane proteins. Evidence for methyl ester formation at carboxyl terminal cysteinyl residues. *J. Biol. Chem.* **264,** 12,879–12,884.

Ota, I. M. and Clarke, S. (1990) The function and enzymology of protein D-aspartyl/L-isoaspartyl methyltransferases in eukaryotic and prokaryotic cells, in *Protein Methylation* (Paik, W. K. and Kim, S., eds.), Boca Raton, FL, CRC, pp. 179–194.

Paik, W. K. and Kim, S. (eds.) (1990) *Protein Methylation*, CRC, Boca Raton, FL.

Park, I. K. and Paik, W. K. (1990) The occurrence and analysis of methylated amino acids, in *Protein Methylation* (Paik, W. K. and Kim, S., eds.), CRC, Boca Raton, FL, pp. 1–22.

Payan, I. L., Chou, S. J., Fisher, G. H., Man, E. H., Emory, C., and Frey, W. H. D. (1992) Altered aspartate in Alzheimer neurofibrillary tangles. *Neurochem. Res.* **17,** 187–191.

Paz, A., Ben, B. G., Marciano, D., Egozi, Y., Haklai, R., and Kloog, Y. (1993) Prenylated protein methyltransferase of rat cerebellum is developmentally co-expressed with its substrates. *FEBS Lett.* **332,** 215–217.

Rajpurohit, R., Park, J. O., Paik, W. K., and Kim, S. (1994) Enzymatic methylation of recombinant heterogenous nuclear RNP protein A1. Dual substrate specificity for S-adenosylmethionine: histone-arginine N-methyltransferase. *J. Biol. Chem.* **269,** 1075–1082.

Rattan, S. I., Derventzi, A., and Clark, B. F. C. (1992) Protein synthesis, post-translational modifications and aging. *Ann. NY. Acad. Sci.* **633,** 48.

Roher, A. E., Lowenson, J. D., Clarke, S., Woods, A. S., Cotter, R. J., Gowing, E., and Ball, M. (1993) Structural alterations in the peptide backbone of beta-amyloid core protein may account for its deposition and stability in Alzheimer's disease. *J. Biol. Chem.* **268,** 3072–3983.

Sandip, F. L., Vincent, P. L., Neal, T. L., Wright, L. S., Heth, A. A., and Rowe, P. M. (1995) Stabilization of secondary structure of Alzheimer b-protein by aluminum(III) ions and D-Asp substitutions. *Biochem. Biophys. Res. Commun.* **206,** 718–723.

Siegel, F. L., Vincent, P. L., Neal, T. L., Wright, L. S., Heth, A. A., and Rowe, P. M. (1990) Calmodulin and protein methylation, in *Protein Methylation* (Paik, W. K. and Kim, S., eds.), CRC, Boca Raton, FL, pp. 33–58.

Skinner, M. K. and Griswold, M. D. (1983) Fluorographic detection of radioactivity in polyacrylamide gels with 2, 5-diphenyloxazole in acetic acid and its comparison with existing procedures. *Biochem. J.* **209,** 281–284.

Stephenson, R. C. and Clarke, S. (1990) Identification of a C-terminal protein carboxyl methyltransferase in rat liver membranes utilizing a synthetic farnesyl cysteine-containing peptide substrate. *J. Biol. Chem.* **265,** 16,248–16,254.

Stephenson, R. C. and Clarke, S. (1992) Characterization of a rat liver protein carboxyl methyltransferase involved in the maturation of proteins with the -CXXX C-terminal sequence motif. *J. Biol. Chem.* **267,** 13,314–13,319.

Terwilliger, T. C. and Clarke, S. (1981) Methylation of membrane proteins in human erythrocytes: identification and characterization of polypeptides methylated in lysed cells. *J. Biol. Chem.* **256,** 3067–3096.

Weber, D. J. and McFadden, P. N. (1995) A heterogeneous set of urea-insoluble proteins in dividing PC12 pheochromocytoma cells is passed on to at least the generation of great-granddaughter cells. *J. Protein Chem.* **14,** 283–289.

Weber, D. J. and McFadden, P. N. (1996) Identification of proteins modified by protein (D-aspartyl/L-isoaspartyl) carboxyl methyltransferase, in *Protein Protocols Handbook.* (Walker, J. M., ed.), Humana, Totowa, NJ.

Wolf, M. E. and Roth, R. H. (1985) Dopamine autoreceptor stimulation increases protein carboxyl methylation in striatal slices. *J. Neurochem.* **44,** 291–298.

Xie, H. and Clarke, S. (1993) Methyl esterification of C-terminal leucine residues in cytosolic 36-kDa polypeptides of bovine brain. A novel eucaryotic protein carboxyl methylation reaction. *J. Biol. Chem.* **268,** 13,364–13,371.

Xie, H. and Clarke, S. (1994a) An enzymatic activity in bovine brain that catalyzes the reversal of the C-terminal methyl esterification of protein phosphatase 2A. *Biochem. Biophys. Res. Commun.* **203,** 1710–1715.

Xie, H. and Clarke, S. (1994b) Protein phosphatase 2A is reversibly modified by methyl esterification at its C-terminal leucine residue in bovine brain. *J. Biol. Chem.* **269,** 1981–1984.

Long-Chain Fatty Acylation
of Proteins

Sean I. Patterson and J. H. Pate Skene

1. Introduction

Reversible modifications of cellular proteins often serve as key steps in signal transduction and effector pathways, and the investigation of these modifications has been a principal route of access to signaling mechanisms. Ten years ago, for example, tyrosine phosphorylation was an intriguing oddity—a posttranslational modification known to occur in virtually all eukaryotic cells and to be regulated by various manipulations that alter cell growth or differentiation, but with no clear physiological role. From the elucidation of tyrosine kinase pathways emerged a dramatically new view of how cells transduce and integrate signaling cascades controlling cellular functions from growth to death (*see* Chapter 6).

Today, another dynamic posttranslational modification—the covalent attachment of palmitate or other long-chain fatty acids to cysteine residues (Schmidt, 1989; James and Olson, 1990)—is coming to be recognized as an important component of signal transduction pathways. Like tyrosine phosphorylation a decade ago, *S*-palmitoylation today is known to occur in all eukaryotic cells and to be regulated by a variety of signals that alter cell growth, metabolism, or repair (Huang, 1989; James and Olson, 1989; Mouillac et al., 1992; Hess et al., 1993; Wedegaertner and Bourne, 1994; Robinson et al., 1995). Recent work has shown that palmitoylation is a dynamic process that can alter the targeting of proteins

From: *Neuromethods, Vol. 30: Regulatory Protein Modification: Techniques and Protocols*
Ed: H. C. Hemmings, Jr. Humana Press Inc.

to specific membrane domains (Peitzsch and McLaughlin, 1993; Silvius and l'Heureux, 1994; Parton and Simons, 1995), the activation of signal transduction pathways (Casey, 1995; Milligan et al., 1995), and the transforming activity of some oncogenic proteins (Kato et al., 1992).

Long-chain fatty acylation is one of a growing class of lipid modifications of proteins, which includes modifications by phospholipids, isoprenoids, and possibly retinoids as well (Schmidt, 1989; Casey, 1995). Among these modifications, long-chain fatty acylation stands out as the only one with a substantial posttranslational component that includes active cycles of acylation and deacylation throughout the life-time of the protein. Because this dynamic acylation continues after proteins have reached vesicles, plasma membrane, and other cellular compartments, it is segregated spatially from other types of lipid modifications that occur cotranslationally or during early protein processing in the endoplasmic reticulum and Golgi.

In the nervous system, active cycles of protein palmitoylation and depalmitoylation occur not only in neuronal cell bodies, but also in axon terminals (Skene and Virag, 1989; Patterson and Skene, 1994) and myelin membranes (Bizzozero et al., 1987; Bizzozero and Good, 1991). The functional and regulatory roles of this protein palmitoylation are only beginning to be resolved, but available evidence already suggests that dynamic cycles of palmitoylation play critical roles in neural development. Inhibition of ongoing protein palmitoylation in axonal growth cones produces a rapid disruption of axon elongation, implying a fundamental involvement of protein palmitoylation either in the constitutive machinery for axon extension or in signaling pathways for axon guidance (Patterson and Skene, 1994). More recently, characterization of the first enzyme involved in dynamic protein palmitoylation, an extracellular protein depalmitoylating enzyme, revealed that mutations in depalmitoylation may be responsible for one form of neuronal ceroid lipofuscinosis, a severe neurological disorder characterized by early loss of visual function and progressive mental retardation (Camp and Hofmann, 1993; Vesa et al., 1995). Prominent and active palmitoylation of key proteins involved in myelin formation (Bizzozero et al., 1987; Yoshimura et al., 1987; Pedraza et al., 1990) and synaptic transmission (Hess et al., 1992; Gundersen et al., 1995) suggests that investigation of S-palmitoylation and its regulation will provide valuable insights into signaling mechanisms in the mature nervous system brain as well as in development.

S-palmitoylation refers to the posttranslational linkage of long acyl chains to specific cysteine residues on target proteins through a relatively labile thioester bond. The name refers to the thioester linkage and to the observation that the modifying acyl chain is usually palmitate, a saturated 16-carbon fatty acid. Some other long-chain fatty acids may be able to substitute for palmitate when they are available, but because palmitate is the most abundant long-chain fatty acid in cells, it is likely to be the predominant fatty acid for long-chain S-acylation of proteins under most cellular conditions. Although other long-chain fatty acids may occasionally substitute for palmitate, shorter acyl chains are strongly excluded from this reaction. The primary distinction occurs between acyl chain lengths of 16 and 14 carbons (Peitzsch and McLaughlin, 1993; Silvius and l'Heureux, 1994; Vogt et al., 1994). This chain length specificity is one important difference between S-palmitoylation and a separate lipid modification, N-myristoylation. The 14-carbon saturated fatty acid, myristate, is not used for posttranslational S-acylation of cysteine residues, but can be attached cotranslationally in a stable amide linkage to amino terminal glycine residues of appropriate substrate proteins (Grand, 1989; Schmidt, 1989). N-myristoylation is a critical processing steps for some proteins, but unlike S-palmitoylation, N-terminal myristate does not undergo active cycles of posttranslational removal and replacement of the lipid chain, a key requirement of modification reactions used for dynamic regulation of cellular proteins. In addition to its biological importance, the existence of this N-myristoylation reaction is a practical concern when using radiolabeled fatty acids to identify protein substrates for S-palmitoylation, since palmitate may be converted metabolically to myristate.

Although a large number of S-palmitoylated proteins have now been characterized, remarkably little is understood about the enzymes and mechanisms involved in either the attachment or removal of palmitate. Only one depalmitoylating enzyme has been isolated, a secreted palmitoyl-protein thioesterase whose endogenous substrates remain unknown (Camp and Hofmann, 1993). Despite many attempts, no protein palmitoylating enzyme has been isolated. There is, furthermore, no clearly defined recognition sequence for palmitoylation, as in the case of other lipid modifications (Casey, 1995; Milligan et al., 1995). For these reasons, some investigators have suggested that incorporation of long-chain fatty acids may occur by chemical transfer from the activated acyl coenzyme A (Co-A) donor to a receptive thiol group, without involvement of a transferring enzyme. Nonenzymatic chemical acylation has been demonstrated to occur in vitro (O'Brien et al., 1987; Quesnel and Silvius, 1994). Although this matter will not be formally resolved until the purification of a palmitoylating enzyme or demonstration of chemical transfer in vivo, several investigators have described cell-derived protein-acylating activities (Berger and Schmidt, 1984; Bizzozero et al., 1987; Yoshimura et al., 1987; Schmidt and Burns, 1989) that evince properties not easily reconciled with the chemical transfer model. These include specificity for chain length of the acyl Co-A donor, sensitivity to heat and enzymatic degradation, and preferential acylation of two adjacent cysteine residues. Selective acylation of adjacent cysteine residues has been described in two S-palmitoylated proteins, bovine opsin and GAP-43, and may represent a more general trend of dual acylation by the same or different lipids (Ochinnikov et al., 1988; Liu et al., 1993). Such dual acylation may prove to be of functional significance, since in vitro experiments suggest that acylation at two or more sites will tether a protein to a specific membrane, whereas single acylation allows rapid exchange between adjacent membranes (Shahinian and Silvius, 1995).

Protein *S*-palmitoylation is beginning to emerge as an important modification for dynamic regulation of proteins in many cells, including developing and adult neurons and glial cells. In this chapter, we describe methods for labeling and analyzing palmitoylated proteins in neurons, based on approaches originally described in the literature, which we have adapted over several years. Naturally, readers may wish to adapt these procedures, described in detail in the following sections, for their own specific applications. We have therefore included descriptions of some of the problems encountered and some theoretical background to facilitate a painless transfer of these procedures to different systems.

2. Reagents and Equipment

2.1. Sources of Specific Reagents

Most of the reagents described here are easily available from international suppliers, such as Sigma (St. Louis, MO). [³H]Palmitic acid is available from a number of suppliers. We have routinely used the product of DuPont/New England Nuclear (Boston, MA) at a specific activity of 50–60 Ci/mmol, which is economically priced when purchased in 25-mCi lots. [³H]Palmitoyl-CoA is also available commercially, but is prohibitively expensive. In Section 4.1.1., we describe a quick and cheap method for the synthesis of [³H]palmitoyl-CoA from [³H]palmitate. Hydroxylamine·HCl, *N*-ethylmaleimide, and cycloheximide can be obtained from Sigma. Whole tunicamycin and some of its purified homologs are available from Boehringer Mannheim (Indianapolis, IN) and Sigma, or the homologs can be purified by high-pressure liquid chromatography as described (Patterson and Skene, 1995). Organic solvents should be of the highest available purity and at least reagent grade.

2.2. General Equipment Requirements

Equipment required for the techniques described in this chapter include an aspirator line with catchment bottle (for radioactive media), ice buckets, microfuge, centrifugal vacuum desiccator (e.g., Savant Speed-Vac), rotary shaker, vortex, bath sonicator, heating block or boiling water bath, one- and two-dimensional gel apparatus for the separation of proteins, protein electroblotting apparatus, thin layer chromatography tank and plates, autoradiographic film and developing facilities, autoradiographic cassettes, and the facilities for handling and safely disposing of organic solvents and of radioactive solids and liquids.

2.3. Handling of Lipids and Solvents

A major cause of problems in handling lipids is the use of inappropriate containers for storage or transfer. Lipids should routinely be kept in clean glass containers, sealed with some organic-solvent inert cap, such as PTFE, or with Teflon coating. Although the extent of the problem is not clear, palmitate will adhere to the polypropylene containers most commonly used for labeling reactions, and this problem can become severe when working with small amounts or in detergent-free aqueous solutions. Therefore, when drying down fatty acids or lipid extracts for redissolving in aqueous solutions, glass containers are always preferred. Regardless of the type of container, dried fatty acids to be resuspended in aqueous media should be sonicated in a bath sonicator for 10 min after addition of the aqueous buffer to improve recovery. Organic solvents will usually redissolve dried lipids efficiently from polypropylene containers without sonication, although it does not hurt to take precautions. It is difficult to attain quantitative transfer of volatile solvents with air displacement pipets, so we recommend the use of positive displacement pipets for these solvents whenever possible. Never use polystyrene pipets or containers with organic solvents. Many problems with handling lipids and organic solvents arise from contamination with finger grease or detergents from improperly rinsed glassware. Rinsing glassware with ethanol or acetone prior to use and using tubes from containers only accessed with gloves can help prevent these problems.

3. Methodological Considerations

3.1. Approaches for Measuring Different Aspects of the Acylation Cycle

The most commonly used method for measuring protein acylation is the use of a radiolabeled fatty acid tracer (e.g., [³H]palmitate), followed by extraction of the proteins and analysis by gel electrophoresis and quantitative autoradiography. This is a straightforward technique applicable to systems from isolated membranes to tissues in vivo. The most substantial limitation is that the incorporation of label into proteins occurs under nonequilibrium conditions, so that measurement of the incorporated radiolabel cannot be converted to a quantitative estimate of the steady-steady mass or molar ratios of fatty acid incorporated into a particular protein. Nonetheless, this is the technique of choice for demonstrating active protein palmitoylation in a particular tissue, cell type, or membrane system, and for identifying palmitoylated proteins.

Among the problems that can be encountered in using [³H]palmitate incorporation to measure protein acylation are the initial lag as the fatty acid is converted to its metabolically active acyl-coenzyme A derivative, and the sometimes rapid and abrupt saturation of labeling. The latter problem may result from either depletion of radioactive fatty acid or saturation of some intermediate metabolic pool. This saturation effect can confound the comparison of protein palmitoylation observed under different conditions. One approach to solving this problem is to compare the time-course of palmitate incorporation under conditions in which the incorporation of radiolabel into the protein(s) of interest is linear. With this approach, both increased and decreased incorporation of [³H]palmitate can be reliably observed, even at short times occluded by the lag in labeling (*see*, for example, Fig. 3 in Patterson and Skene [1995]) or other transient effects.

Another approach to measuring regulation of palmitoylation is pulse-chase incorporation of [³H]palmitate. This technique has been used successfully to demonstrate receptor-regulated deacylation of specific proteins (James and Olson, 1989; Wedegaertner and Bourne, 1994). In our hands, this technique also suffers from problems arising from saturation when the original labeling is carried out under conditions optimized for maximum labeling. Like other fatty acids, radiolabeled palmitate is rapidly incorporated into a broad range of cellular lipids, creating intracellular pools of radiolabeled palmitate that may confound quantitative analysis of depalmitoylation during a chase period. One solution is to use the cell-permeable thiol-modifying agent *N*-ethylmaleimide to block covalently deacylated cysteine thiols (Hess, personal communication) as described in Section 4.2.1., step 5.

3.2. Criteria for the Identification of Thioester-Linked Long-Chain Acylation

Lipid modification of proteins extends well beyond the incorporation of long-chain fatty acids to cysteine thiol groups. Proteins tethered to the extracellular surface by glycophosphatidylinositol anchors frequently include palmitate as their side chains (*see* Chapter 11). Furthermore, metabolic conversion of the abundant palmitate to the rare cellular fatty acid myristate occurs rapidly, giving rise to overlapping labeling of palmitoylated and myristoylated proteins. Prolonged incubations can even result in labeling of proteins through conversion of the [^{3}H]fatty acid to [^{3}H]acetyl-CoA, and incorporation of the radioactivity into amino acids. It is therefore frequently necessary to demonstrate directly whether labeling of proteins after incubation with [^{3}H]palmitate represents *S*-palmitoylation or another protein modification. This includes determination of the nature of the chemical bond linking the labeled lipid to the protein and identification of the attached lipid moiety.

3.2.1. Linkage Analysis

Fatty acids are known to be attached to proteins through both thioesters (to cysteines) and oxyesters (to serines and threonines), and by amides and oxyesters in phospholipids. Thioesters are particularly sensitive to nucleophilic attack by such agents as hydroxylamine (Schmidt and Lambrecht, 1985), which under neutral conditions, pH 7.0–8.0, can be used as a diagnostic test to distinguish S-palmitoylation of cysteine residues from the other forms of lipid attachment. Under more basic conditions, pH 10.0–12.0, removal of fatty acid is more rapid and quantitative, but oxyesters also become susceptible to cleavage. The common way of analyzing protein–lipid linkage is to separate [^{3}H]palmitate-labeled proteins in two identical samples on two parallel lanes of a polyacrylamide gel (*see* Chapter 3), and then to assess the loss of incorporated radiolabel after incubating one of the gel lanes in neutral hydroxylamine:

1. Disassemble the gel apparatus and fix the gel for 30 min in 25% (v/v) isopropanol in water at room temperature with gentle agitation. Wash the gel in several changes of distilled water (≥10 gel volumes) for 30–60 min (shorter times for lower-percentage acrylamide gels).
2. Divide the lanes and place one in at least 10 gel volumes of 1.0M hydroxylamine·HCl (Sigma) made to pH 7.0–8.0 with KOH. Place the second gel piece in an equal volume of 1.0M Tris-HCl made to pH 7.0–8.0 with KOH. Incubate at room temperature for 4 h with gentle agitation.
3. Wash the gel pieces with water as in step 1.
4. Stain and destain the proteins in the gel, for example, 0.05% (w/v) Coomassie brilliant blue in isopropanol/H$_2$O/acetic acid 25:65:10 to control for differential loading of proteins or loss during incubations.
5. Prepare the gel for autoradiography or fluorography (as described in Section 4.1.3., step 15).

3.2.2. Fatty Acid Identification

Cellular proteins can incorporate a number of different fatty acids, generally segregated into short (12–14 carbon acyl chains) and long (>16 carbon acyl chains) fatty acylation. Techniques exist for the quantitative analysis of endogenous fatty acids within cellular proteins, but they are laborious, require specialized equipment, and for many purposes are excessive. The analysis of incorporated fatty acid is routinely carried out after labeling with radiolabeled myristate or palmitate to identify each kind of modification. However, since cells are capable of converting each label into the other (Olson et al., 1985), formal demonstration that the incorporated radioactivity represents in large part the same fatty acid as added is required. The techniques for fatty acid analysis presume that sufficient radiolabeled protein (about 10-fold the amount used for simple [³H]palmitate incorporation into protein analysis) can be acquired in a reasonably pure form—that is separated from other radiolabeled proteins. Ideally, this can be carried out by polyacrylamide gel electrophoresis, although for rare proteins or when the migration position of the protein on a gel is crowded or ambiguous more exhaustive purification is called for.

1. Separate the [³H]palmitate labeled proteins on an appropriate polyacrylamide gel, remove from the apparatus, and fix in three changes of 25% (v/v) isopropanol in water, with gentle agitation for a total of 3 h. Using prestained markers or a reversible protein stain to identify the migration position of the relevant protein, cut out the part of the gel containing the protein.

2. Dehydrate the gel fragment with three washes in at least 10 vol methanol for 60 min to overnight.

3. Treat the gel fragment with 10 vol 0.1M KOH in methanol for 4 h at room temperature.

4. Transfer the liquid to a separating funnel and extract with 1 vol chloroform and 1 vol H_2O. Collect the lower (organic) phase and re-extract the upper (aqueous) phase with an equal volume of chloroform. Pool the organic extracts and wash three times with equal volumes of chloroform/methanol/H_2O (1:10:10).

5. Dry the washed organic extract in a Speed-Vac or under a nitrogen stream in a glass tube, redissolve in methanol, and spot onto C_{18} reversed-phase TLC plates (Whatman, Hillsboro, OR), with methyl-fatty acid standards (satu-

rated and unsaturated, C_{14}–C_{18}, Sigma) in a parallel lane for comparison. Develop plates twice with acetone/methanol/H_2O (8:2:1), drying thoroughly between the runs.

6. The migration position of the radioactive fatty acid(s) can be identified either by direct autoradiography or by spraying the plate with a 10% (w/v) solution of 2,5-diphenyl oxazole (PPO) in ether followed by fluorography for lower levels of radioactivity. The thoroughly dried plate should be apposed to suitable X-ray film (e.g., Kodak X-OMAT AR) preflashed to an optical density of 0.1 and exposed at –70°C. Marking the plate with fluorescent paint can help to align the fluorogram after development.

7. To visualize the standards, separate the standard lane, spray with a 10% (w/v) solution of phosphomolybdic acid (Sigma) in ethanol, air-dry, and then briefly heat to ≈150°C. The standards should turn a dark, dirty green against a background that varies from yellow to light green.

3.3. Measurements of Lipid Radiolabeling

The metabolic incorporation of radiolabel into proteins occurs through the same intermediate used for the synthesis of lipids, that is, acyl-CoAs. The fraction of label added to intact cells that ends up in protein is very small compared to the incorporation into neutral- and phospholipids. Routine measurement of lipid-labeling confers several advantages. The health and metabolic activity of the biological sample can be monitored, since a robust incorporation of radioactivity into the lipids demonstrates that the cells were capable of taking up the fatty acid, converting it to the activated acyl-CoA in an ATP-dependent step, and finally incorporating the fatty acid into distinct lipid classes. The pattern of lipid-labeling can be quite distinctive and, thus, can be used to assess the purity of the biological preparation, particularly with regard to contamination of cultures by microorganisms. Bacteria appear to take up and incorporate [^{3}H]palmitate into protein and lipid more efficiently than mammalian cells and can thus give misleading results, even when barely detectable by visual inspection of the cultures. If anomalous protein and lipid-labeling patterns appear together, some form of contamination is quite likely.

The relative incorporation of [³H]palmitate into protein and lipid appears to vary with the labeling conditions, and further with the metabolic state of the cells. It therefore cannot be reliably used to normalize results quantitatively between experiments. However, quantitation can often provide useful information when changes in the incorporation of [³H]palmitate are observed between samples within an experiment. Thus, parallel changes in the level of protein and lipid-labeling more likely indicate variation in sample size or an alteration in the uptake or metabolism of fatty acid than direct alterations in the rate of palmitoylation or depalmitoylation.

The starting point for the analysis of cellular lipids described here is the dried organic extract of the labeling reaction, as described in Section 4.1.3. (step 10).

1. Lipid extracts are redissolved in chloroform/methanol (4:1) and spotted onto aluminum-backed silica gel 60 TLC plates (Art. 5553, EM Science) cut to a height of 10 cm.

2. Heat the plates briefly (100–110°C for ≤5 min; do not go over 115°C); after allowing to cool, develop once with butanol/acetic acid/H₂O (5:2:3) in an enclosed pre-equilibrated TLC tank for 8–9 cm.

3. Remove the plates from the TLC tank and allow to air-dry in a fume hood. Using a hair drier can expedite this step.

4. For samples with low amounts of radioactivity, the plates can be permeated with 2,5-diphenyloxazole (PPO) in ether for fluorography. Pour an ice-cold 10% (w/v) solution of PPO in ether over the tilted plate in the direction of chromatography, since some smearing of the neutral lipid and fatty acid will occur; then quickly dry with an air stream or hair drier.

5. Expose the dried TLC plates for 3–10 d at –70°C to X-ray film (Kodak X-OMAT AR) preflashed to an optical density of 0.1 AU. After development, radiolabeled bands

can be quantitated by transmittance optical density and identified by reference to labeled or unlabeled lipid standards. The latter can be transiently visualized by brief exposure to iodine vapor in a sealed tank (in the fume hood). The phosphomolybdate stain described in Section 3.2.2. can also be used, but is not compatible with fluorography.

The incorporation of [^{3}H]palmitate into material comigrating with phosphatidylcholine is often great enough to overwhelm the signal from palmitoyl-CoA. If this is the case, aliquots of lipid extract can be enriched for palmitoyl-CoA by phase-partitioning prior to step 1. Redissolve the dried lipid extracts in 1.1 mL chloroform/methanol/10 mM Tris, pH 7.4 (5:5:1) with brief sonication. Then proceed as described in Section 4.1.1., step 4. The upper aqueous phase obtained from this procedure will be substantially enriched in palmitoyl-CoA. This procedure is very sensitive to the quantity of lipid and the salt content of the sample and may require considerable work to establish reliably on a routine basis.

3.4. The Use of Protein Synthesis Inhibitors and Tunicamycin

In many instances, it is important to distinguish palmitoylation during the initial processing of a protein from ongoing cycles of acylation and deacylation that may subserve more dynamic roles in the regulation of protein localization or activity. Cycloheximide and homologs of tunicamycin may be used to distinguish early processing from the late posttranslational component of S-palmitoylation and to probe the functional roles of ongoing palmitoylation in living cells (Patterson and Skene, 1994,1995). Cycloheximide inhibits protein synthesis and, thus, the cotranslational component of all forms of protein acylation, without directly interfering with S-palmitoylation. Furthermore, primary neuronal cultures and many cell lines can be pretreated with cycloheximide for at least an hour before addition of radiolabeled palmitate without compromising cell viability or metabolism. This effectively isolates the late posttranslational component of palmitoylation and makes it possible to measure addition and turnover of palmitate separate from early protein processing. For the special case of proteins in mitochondria, where cotranslational palmitoylation may continue in the presence of cycloheximide, puromycin or chloramphenicol also may be added.

Tunicamycin, a well-characterized inhibitor of protein glycosylation, also inhibits posttranslational protein palmitoylation and, thus, can be used to probe the function of this modification in intact cells (Patterson and Skene, 1994). Because tunicamycin produces an acute disruption of ongoing cycles of protein palmitoylation, it can be used to investigate the functional roles of ongoing cycles of protein palmitoylation. Experimental procedures and technical considerations in using these agents to investigate protein S-palmitoylation have been described in detail elsewhere (Patterson and Skene, 1995).

4. Incorporation of [^{3}H]Palmitate into Proteins

The techniques for labeling proteins with [^{3}H]palmitate can be loosely divided into two categories based on the form of the radiolabel: use of the immediate precursor of S-palmitoylation, palmitoyl-CoA, or use of palmitic acid and relying on metabolic conversion to the active derivative (Table 1). Conceptually, this can be compared to the use of ^{32}P-labeled ATP or inorganic phosphate, respectively, for the phosphorylation of cellular proteins. In both cases, the energetically activated donor of the protein-modifying group is not cell-permeable and can be used only in those systems in which access to the intracellular compartment is provided, whereas the cell-permeable precursor requires energy-dependent metabolic activation and is incorporated into multiple cellular pools that require separation from the target proteins of interest before analysis.

4.1. S-Palmitoylation with [^{3}H]Palmitoyl-CoA

The labeling of subcellular fractions, such as microsomes or membranes, either in their native state or solubilized with detergent, provides important information about the spatial

Table 1
Comparison of *S*-Palmitoylation with Palmitate or Palmitoyl-CoA

Label	Addition	S-palmitoylation of GAP-43[a]
2 μM [³H]palmitoyl-CoA		100
2 μM [³H]palmitoyl-CoA	ATP/CoA	139 ± 6
2 μM [³H]palmitate		8 ± 6
2 μM [³H]palmitate	ATP/CoA	71 ± 37
2 μM [³H]palmitoyl-CoA	200 μM palmitate	93 ± 9
2 μM [³H]palmitoyl-CoA	200 μM palmitoyl-CoA	7 ± 4

[a]Measured with 10-min incubations with growth cone membranes as described in the text and expressed as a percentage (ISD) of labeling with 2 μM palmitoyl-CoA with no additions.

segregation of palmitoylating reactions in the cell that may reflect the varying functions of the substrates, and can be used to analyze factors impinging on the process in a manner not available in intact cells. Whereas these components can be analyzed separately by the use of protein synthesis inhibitors (Section 3.4.), palmitoylation in other fractions, such as in the terminals of long neuronal axons or in the detergent-resistant complexes that may be specialized for signal transduction are not as easily distinguished.

The patterns of palmitoylated proteins in intact cellular vesicles labeled with [³H]palmitate and membranes derived from these vesicles labeled with [³H]palmitoyl-CoA are remarkably similar (Patterson and Skene, 1994). The palmitoylation of endogenous proteins in membrane and membrane-derived preparations assumes their copurification with the membranes. It is sometimes desirable to use added proteins to follow the process, to control their levels, or if the protein of interest does not copurify with the membranes in the protocol used. Furthermore, addition of a protein that has a free cysteine thiol, but does not undergo *S*-palmitoylation *in situ*, is a useful control for the extent of nonenzymatic palmitoylation under the specific reaction conditions used. Thus, we have used deacylated preparations of the protein GAP-43 and bovine serum albumin (BSA) in conjunction to optimize labeling conditions of the former over the latter. Section 4.1.2. includes a description of the procedure for chemically reducing the thiols to ensure quantitative control over the concentration of free protein thiol in the reaction solutions.

4.1.1. Synthesis of [³H]Palmitoyl-CoA

1. Dry 100–1000 μCi [³H]palmitate (60 Ci/mmol, DuPont NEN) in a glass screw-cap vial (e.g., Kimble Opticlear, 12 × 35 mm), and redissolve in 100 μL of 0.05% (v/v) Triton X-100, 1 mM CoA (Boehringer Mannheim, Mannheim, Germany), 5 mM ATP, 5 mM MgCl$_2$, and 10 mM Tris, pH 7.5, containing 1–10 U/mL long-chain acyl:coenzyme A transferase (Boehringer Mannheim). Vortex gently but thoroughly to dissolve the fatty acid while avoiding foaming. Add the MgCl$_2$ from a 100–1000X stock after dissolving the [³H]palmitate to avoid solubility problems with the fatty acid.

2. Incubate at 37°C for 1–2 h with continuous gentle agitation or occasional gentle vortexing; try to avoid foaming. If preferred, the reaction can be monitored by removing microliter aliquots, diluting in 1.1 mL chloroform/methanol/water (5:5:1), and separating the phases with chloroform and water as described in step 5. Small aliquots of the upper (total ~0.75 mL) and lower (total ~1.13 mL) phases can then be transferred, dried in scintillation vials (important to remove the chloroform that quenches the scintillants), mixed with a scintillation cocktail, and the radioactivity determined in a scintillation counter to calculate the conversion efficiency.

3. To quench the reaction, add 1 mL chloroform/methanol (1:1) and mix thoroughly, preferably with brief bath sonication.

4. Spin down any precipitated material (5 min in a cooled microfuge at maximum speed) and transfer the cleared supernatant to a fresh glass vial.

5. Add 0.5 mL chloroform and 0.275 mL H$_2$O, vortex for ~30 s or bath sonicate, and then centrifuge briefly to obtain a clean separation of the phases.

6. Carefully remove the upper phase, without bringing any interface material or lower phase, and transfer to another glass vial. Dry this material completely in a Speed-Vac (~1–2 h) and redissolve in a small volume of dimethyl sulfoxide (DMSO).

7. Transfer aliquots (2 × 1 μL) to 1.5-mL microfuge tubes containing 0.5 mL methanol and serially dilute 10- and 100-fold. Place the tubes in scintillation vials, add 10 mL scintillant, and determine the radioactivity in a scintillation counter to calculate the recovery.

8. A further aliquot (~1 μL) should be diluted in chloroform/methanol for the analysis of composition by TLC as described in Section 3.3.

Recovery of radioactivity in the upper aqueous phase should be 70–90%. Analysis by TLC should show ≥90% radioactivity comigrating with *S*-palmitoyl-CoA standard (Boehringer Mannheim), and the rest with free palmitic acid close to the solvent front.

An alternative technique for those who do not wish to involve themselves in synthesizing label is to use the metabolic synthesis of [^{3}H]palmitoyl-CoA in the preparation. Membranes (and also detergent-solubilized membranes) contain an endogenous activity that will convert free fatty acids to their CoA derivatives in the presence of ATP, CoA, and MgCl$_2$. For membranes that do not contain a strong long-chain acyl CoA synthetase activity, the purified enzyme can be added. We have used this property to synthesize actively and continuously [^{3}H]palmitoyl-CoA from [^{3}H]palmitate in labeling mixtures by including 5 m*M* ATP and 1 m*M* CoA in the reactions. This method has the advantage of continuously replacing the [^{3}H]palmitoyl-CoA that is consumed by lipid synthesizing pathways. The disadvantage is that the endogenous palmitate will immediately start competing with added radiolabel, rendering the specific activity of the [^{3}H]palmitoyl-CoA unpredictable. Thus, monitoring the level of [^{3}H]palmitoyl-CoA in the reaction will not provide information on the chemical concentration of palmitoyl-CoA. In side-by-side comparisons, we have found that either [³H]palmitoyl-CoA or [³H]palmitate with ATP and CoA at equal concentrations could give more efficient radiolabeling of proteins (Table 1).

4.1.2. Preparation of Reduced Protein Substrates for Palmitoylation

Proteins stored in aqueous solutions can undergo chemical oxidation of free cysteine thiols, which interferes with subsequent quantitation of protein S-palmitoylation. Fortunately, much of this oxidation is inter- or intraprotein formation of cystine bridges (particularly with adjacent cysteines) that can be reversed by treatment with an appropriate reducing agent, such as dithiothreitol (DTT).

1. Dissolve the protein(s) in a buffered aqueous solution suitable for subsequent dilution in the labeling buffer.
2. Adjust to 10 mM DTT by addition from a 1M stock (which can be stored in aliquots at –20°C) and incubate at 55°C for ~15 min.
3. Centrifuge the tube(s) in a microfuge for 5 min at maximum speed. If there is any precipitate, a separate determination of actual protein concentration should be made (this should not be a problem with freshly prepared proteins). Transfer supernatant aliquots to the reaction tubes.
4. If nonreducing conditions are desired for the labeling reaction, proteins should be reduced with DTT in advance, dialyzed against several changes of buffer at 4°C, and can be stored frozen in aliquots. In this case, do not reuse tubes after thawing.

4.1.3. Labeling of Membranes with [³H]Palmitoyl-CoA

Protein palmitoylating activity appears to remain strongly associated with membrane preparations from many sources and probably represents an integral membrane protein(s). Although different conditions are able to modify the efficiency of protein palmitoylation with added radiolabel, overall the membrane-associated activity is robust and tolerant to very different conditions. We have observed effi-

Table 2
Solution for Palmitoylation of Membranes

Component	Final	Comments
NaCl	20 mM	From 11X stock[a]
KCl	100 mM	From 11X stock
CaCl$_2$	0.1 mM	From 11X stock
MgCl$_2$	1 mM	From 11X stock
HEPES, pH 7.1–7.2	20 mM	From 11X stock
BSA (fatty acid-free)	20 µg/mL	Freshly reduced with DTT[b]
DTT	1 mM	From 1M stock[c]
[^{3}H]palmitoyl-CoA	~0.3 µM[d]	From ≥100X DMSO stock

[a]The salts and buffer can be made up as a single 11X stock solution, stable at room temperature.

[b]As described in the text for monitoring nonenzymatic acylation.

[c]Aqueous solution aliquoted and stored at –20°C.

[d]Higher concentrations give progressively worse problems with chemical labeling.

cient labeling under conditions varying from the most simple buffers (2 mM HEPES at pH 7.4) to complex buffers that mimic the intracellular environment (*see* Table 2) or elution conditions from ion-exchange columns.

1. Prepare suspended membranes at a concentration of ~1 mg/mL membrane protein (this can be varied substantially according to availability or need).
2. Add 11X assay buffer (*see* Table 2). Add the appropriate amount of DTT and reduced BSA stock solutions and aliquot the mix into screw-cap reaction tubes.
3. Add GAP-43 (or other protein) substrate if required. Add assay buffer to a final volume of 100 µL, or for time-courses $(n + 1) \times 100$ µL, where n is the number of time-points.
4. Transfer tubes to 37°C and preincubate 10–15 min to allow the temperature to stabilize.
5. Initiate labeling by addition of [^{3}H]palmitoyl-CoA (1–10 µCi/100 µL ≈ 0.25–2.5 µM) from a stock solution in DMSO and mix immediately. Then return the tube to 37°C. Alternatively, [^{3}H]palmitate (1–100 µCi/100 µL), CoA, ATP, and MgCl$_2$ can be used as described in Section 4.1.1.

6. For a time-course, remove an aliquot (100 µL) before addition of label, transfer to a tube containing 1 mL chloroform/methanol (1:1), and add an appropriate amount of radiolabel to the quenched sample for the zero time-point. Transfer further 100-µL aliquots to individual tubes at the desired times and quench with chloroform/methanol (1:1).

7. For single time-point experiments, quench the labeling by addition of 1 mL chloroform/methanol (1:1) to each tube, followed by vortexing. Sonicate the tubes in a bath sonicator for 15 min, and then continue the extraction at room temperature for at least 30 min.

8. If the protein pellet did not break up into a smooth suspension in step 7, we recommend continuing the extraction overnight at –20°C. Precipitation of small amounts of protein can be improved by transfer of the tubes to –20°C for 30 min to overnight. This is not usually necessary in the presence of added protein (e.g., BSA), which acts as carrier protein.

9. Pellet proteins by centrifugation in a microfuge at top speed for 5 min at 4°C. To prevent contamination of the microfuge by radioactivity, it is important that all mixing and centrifugation steps with organic solvents be carried out with tightly fitting screw-cap tubes with some form of seal (e.g., rubber, Teflon, PTFE).

10. Transfer 110-µL aliquots (i.e., one-tenth total volume) of the organic supernatants to a parallel set of glass tubes, and warm to 37°C for ~1 h without caps. Then take to dryness in a Speed-Vac. The predrying period is to reduce the chloroform that boils over at the low pressures of the Speed-Vac. Alternatively, ensure that the vacuum does not drop below 5 mbar.

11. Remove the remaining supernatants from the protein pellets, using a gel loading tip to get the final liquid from the bottom of the tube. Remember that this extract contains most of the added radioactivity and that volatile solvents easily form contaminating aerosols and drip from air-displacement pipets. We perform

this step using a low-pressure aspirator with a "hot" collecting bottle.

12. Resuspend the pellets in 0.5 mL chloroform/methanol (2:1) and transfer to fresh tubes with a further 0.5 mL wash (use positive displacement pipets or aerosol-resistant pipet tips). Break up the pellets with vortexing and sonication as for steps 7 and 8. Centrifuge as for step 9, and discard the radioactive supernatant. Repeat the organic wash with 1 mL chloroform/methanol (2:1) without transferring the pellet to fresh tubes.

13. Allow the pellets to air-dry for ~30 min.

14. Add sample buffer (1% [v/v] SDS, 5 mM EDTA, 10 mM DTT, 20% [v/v] glycerol, 10 mM Tris-HCl, pH 7.0), and heat until the pellet is redissolved.

15. Separate the proteins by one- or two-dimensional gel electrophoresis as appropriate; stain and destain as usual. Prepare the gels for fluorography by incubating in APEX (55% [v/v] acetic acid, 15% [v/v] ethanol, 30% [v/v] xylenes, 1% [w/v] PPO) for 60 min to overnight, followed by at least three 10-min washes in water. Dry the gel and appose to preflashed X-ray film at −70°C. Good exposures depend on the protein of interest, but as a rough guide, they can be obtained within 1 wk to 1 mo.

4.1.4. Labeling with Solubilized Protein Acylating Activity

As mentioned above, a limiting factor in the analysis of protein palmitoylating activities is the presence in membranes of other metabolic pathways that also use [³H]palmitoyl-CoA as a substrate, predominantly lipid synthetic pathways. One approach to dealing with this limitation is the separation of the proteins responsible by any number of techniques currently available, most of which first require solubilization of the proteins. We have previously reported (Patterson and Skene, 1994) the solubilization of a protein palmitoylating activity from the washed membranes of neuronal growth cones using the nonionic detergent Lubrol-PX (currently available under the tradename Thesit, from

Boehringer Mannheim), and a number of other groups have also successfully solubilized similar activities with other detergents. Distinct properties of these crude activities suggest that there may be a family of protein palmitoylating enzymes.

The procedures we have used for protein palmitoylation proteins with detergent-solubilized membranes are essentially the same as for the membranes themselves (Section 4.1.3.). The labeling reactions are carried out in a volume of 100 μL at a protein concentration of 1 mg/mL in the solution described in Table 2. The concentration of [³H]palmitoyl-CoA is the same, as are the time-course, quench, and processing. The major difference is that a smaller aliquot of lipid is used for the analysis by TLC to prevent overloading of the TLC plate owing to the presence of detergent in the organic extract. It should be noted that although the presence of detergent helps to suppress nonenzymatic labeling of proteins (Patterson and Skene, unpublished data), it is worthwhile monitoring this process routinely by addition of reduced BSA, as described in Section 4.1.2.

4.2. S-Palmitoylation with [³H]Palmitate

Labeling with [³H]palmitate shares the problems associated with the metabolism of [³H]palmitoyl-CoA, but has the advantage of not requiring direct access to the intracellular compartment. It is thus useful for cultured cells, tissue slices, or even tissues in vivo, although moving away from dispersed cells dramatically increases the problems of uptake of radiolabel and competition with large endogenous pools of palmitate. Furthermore, the extraction procedures for tissues are more laborious than for cultured cells.

The procedures for labeling require maintenance of the viable cells in a suitable medium with [³H]palmitate. Unlike labeling with [³H]palmitoyl-CoA where concentrations of label greater than low micromolar facilitate nonenzymatic labeling, it is desirable to use high concentrations of extracellular fatty acid to compete as efficiently as possible with the intracellular pools. The limitations encountered are thus physical (the solubility of the fatty acid) or economic. In the presence of physiological concentrations of divalent cations, fatty acids become insoluble between 10 and 50 μM (the exact concentration depends on the composition of the physiological salt solution). The presence of serum proteins can also act as an alternative reservoir for fatty acid that will compete with the cells or tissue (Patterson and Skene, 1994,1995). When encountering failure to label proteins with [^{3}H]palmitate, solubility problems should immediately be considered as a possibility.

4.2.1. Labeling of Continuous Cell Lines and Primary Cultures

This section describes the procedure for labeling intact, cultured cells firmly attached to a substratum, as developed for nerve growth factor-differentiated PC12 cells and cultured adult dorsal root ganglion (DRG) neurons. The amounts and volumes described below are for labeling of subconfluent PC12 cells or 2 DRG ganglia in a 35-mm dish or 6-well plate.

1. Make up the physiological incubation medium. This is normally the same as culture medium, but without serum or other fatty acid binding components. Bicarbonate buffering should be used for incubation in a culture incubator. If incubations are to be carried out without use of a humidified incubator, they should be kept short (≤45 min) to prevent substantial alterations in osmolarity resulting from evaporation. The medium may also include cycloheximide at 10–100 μg/mL in order to inhibit protein synthesis (Section 3.4.). Incubation medium should be pre-equilibrated for temperature and gas content, as appropriate.

2. Dry down the desired amount of [^{3}H]palmitate (plus a little extra for pipeting) in a glass tube in a Speed-Vac. Redissolve in DMSO at 1000X the desired final concentration.

3. Wash the cells with incubation medium. If the cells have been cultured with serum, then at least two washes with an equivalent volume are required. If the cells were cultured serum-free, then in principle the medium need

only be replaced without washing. Take care not to let the cells dry out, which dramatically reduces adhesion and viability. This is particularly easy to do if the cells have been cultured without serum. Return the dishes to the incubator for at least 15 min or at least 30 min if cycloheximide is being used to inhibit protein synthesis.

4. To initiate labeling, add the desired amount of [^{3}H]palmitate in DMSO. When operating near the limit of solubility, it may be preferable to make up the labeling medium separately with sonication in order to ensure good solubility. If this is not feasible, add the label to the medium while swirling, but not so violently that cells are dislodged. With phase-contrast microscopy, precipitated palmitic acid may be clearly seen as a fuzzy contamination that tends to float to the medium surface.

5. If a chase is being performed, remove the medium and either wash two times or replace with medium containing 0.1% (w/v) fatty acid-free BSA ± the relevant stimulant for chasing. At this stage, N-ethylmaleimide (NEM) may also be included in the wash and chase buffer at 0.1–1 mM. If cell adhesion is a problem, the medium may simply be replaced with medium containing NEM. Allow 10 min before adding a chase stimulant. More than 30 min in the presence of NEM can be highly detrimental to cell viability and adhesion. PC12 cells appear to be much more sensitive to NEM treatment than primary cultures of adult sensory neurons, and quite quickly lose adhesion with higher concentrations.

6. To terminate labeling, transfer the dish(es) to ice and wash with ice-cold medium with sufficient EDTA to buffer divalent cations, or preferably with PBS/1 mM EDTA if it causes no cell rupture or loss of cell adhesion. If there is loss of cell adhesion, the original incubation medium should be transferred to microfuge tubes and pelleted (3 min at 1000g at 4°C), the medium aspirated, and the same tubes used to collect the scraped cells from the equivalent wells or dishes (step 7). Furthermore, if BSA is used in a chase, ensure that it is absent from the

wash medium. Immediately after washing, quench the cells by addition of 0.5 mL ice-cold methanol. At this point, either process the dishes immediately, as described below, or store them at –20 or –80°C until ready for processing.

7. Scrape the cells into the methanol with a rubber policeman or a microspatula whose angled end is covered with Tygon tubing (or some other inert tubing). Transfer the suspension to screw-cap tubes (*see* step 6), and pool with a further 0.5-mL methanol wash of each well or dish. The resulting volume is usually ~1 mL and can be assumed to be ~0.75 mL methanol/0.15 mL wash medium. Add 0.75 mL chloroform to make up the extraction solution of chloroform/methanol/water (5:5:1). Mix thoroughly with sonication in a bath sonicator for 15 min.

8. Proceed with the analysis of protein and lipid as described in steps 8–15 of Section 4.1.3.

4.2.2. Labeling of Tissues In Vivo

Similar techniques to those described in the previous section can be used to label proteins in tissue slices and neurons in vivo (our own unpublished data, and data from Hess [1992]). Although there is considerable information available about the pharmacokinetics of [³H]palmitate in neuronal systems in vivo (Kimes et al., 1983,1985; Tabata et al., 1986; Miller et al., 1987; Yamazaki et al., 1989), for labeling of proteins it can briefly be summarized as a problem of delivering a sufficient amount of label to the target region. In vivo, only a fraction of injected radiolabeled palmitate ends up in protein, resulting in a very low specific activity of the palmitate incorporated through *S*-palmitoylation in a comparatively large quantity of protein. There is also a delay in the uptake of the radiolabeled fatty acid as it partitions in many physiological pools, such that longer labeling periods are required.

For the labeling of neural cells in the living rat retina, we (in collaboration with Hess) have injected 2–4 μL vol of a saturated solution of [³H]palmitate in DMSO (estimated at about 1 mCi) into the aqueous humor of the eye of an anesthetized hooded rat using a Hamilton syringe connected through inert tubing to a 28-gage needle inserted 3 mm through the sclera anterior to the retina. The volume was chosen to minimize damage to the neurons through increased pressure in the eye, and the solvent to be miscible with the intraocular solution (to forestall catastrophic precipitation of the fatty acid in the presence of the aqueous humor) and at a level nontoxic to neurons. The rats were permitted to recover from anesthesia and were sacrificed at 4–24 h after the injection. The retinas were dissected out, minced with a razor blade, and homogenized in a glass microhomogenizer in chloroform/methanol/water (5:5:1). Analysis of the protein and lipid was then carried out essentially as described in steps 8–15 of Section 4.1.3. The organic extracts of the homogenized retina with chloroform/methanol (2:1) were monitored for radioactivity by drying and scintillation counting, and the extractions were repeated until the radioactivity extracted had leveled off close to background (three to six washes). The labeling of retinal proteins could be seen at 4 h and peaked after 16 h injection.

5. Perspectives

A central characteristic of posttranslational palmitoylation is the rapid replacement of the fatty acid moieties compared to the degradation of the protein(s) to which they are attached. It is not known which part of this cycle of attachment and removal is under physiological regulation, or whether the primary functional significance is reflected most directly in the rate of acyl chain attachment, removal, or the balance between the acylated and deacylated forms. However, the most commonly used method for measuring palmitoylation is the incorporation of externally added radiolabeled palmitic acid, under conditions in which isotopic equilibrium has not been achieved. Although this is sufficient to demonstrate that the protein(s) in question does indeed incorporate the fatty acid and to proceed some distance in analyzing its regulation, it does not permit the direct determination of any of these three variables required for quantitative analysis of the cycle of palmitoylation *in situ.*

Many of the intracellular substrates for palmitoylation are known to be involved in signal transduction and the control of cytoskeletal and membrane dynamics (Casey, 1995; Milligan et al., 1995). The functional implications of palmitoylation of these proteins are still unclear. However, it has been observed that palmitoylated proteins are enriched in a specialized subcellular membrane fraction known as the detergent-insoluble, glycosphingolipid-enriched complex (Parton and Simons, 1995, and unpublished data). In many cells such domains are enriched in cell-signaling molecules, such as heterotrimeric G-proteins and protein-tyrosine kinases, and in brain-derived synaptosomes they are enriched in protein components of the synaptic membrane fusion machinery (Bennet et al., 1992). Thus, palmitoylation may serve to concentrate components of signal transduction or effector machinery within a small region of the membrane to increase the efficiency of their interaction and promote directed cellular responses in a manner analogous to the phosphorylation-dependent interaction of proteins containing SH2 domains.

The implication of this focus on specialized subdomains of the membrane is that the tools we are currently using to investigate S-palmitoylation may still be too crude to extract useful information about the functioning of the cycle *in situ.* Further technical developments will be required before S-palmitoylation can be measured with the kind of quantitative exactitude currently possible with other protein modifying pathways.

Acknowledgments

Many of the techniques described in this chapter were developed with the help of D. T. Hess (Department of Cell Biology, Vanderbilt University, TN). We would further like to thank him for sharing his own techniques prior to publication.

References

Bennet, M. K., Calakos, N., Kreiner, T., and Scheller, R. H. (1992) Synaptic vesicle membrane proteins interact to form a multimeric complex. *J. Cell Biol.* **116,** 761–775.

Berger, M. and Schmidt, M. F. G. (1984) Cell-free fatty acid acylation of Semliki Forest viral polypeptides with microsomal membranes from eukaryotic cells. *J. Biol. Chem.* **259,** 7245–7252.

Bizzozero, O. A. and Good, L. K. (1991) Rapid metabolism of fatty acids bound to myelin proteolipid protein. *J. Biol. Chem.* **266,** 17,092–17,098.

Bizzozero, O. A., McGarry, J. F., and Lees, M. B. (1987) Acylation of endogenous myelin proteolipid protein with different acyl-CoAs. *J. Biol. Chem.* **262,** 2138–2145.

Camp, L. A. and Hofmann, S. L. (1993) Purification and properties of a palmitoyl-protein thioesterase that cleaves palmitate from H-ras. *J. Biol. Chem.* **268,** 22,566–22,574.

Casey, P. J. (1995) Protein lipidation in cell signaling. *Science* **268,** 221–225.

Grand, R. J. A. (1989) Acylation of viral and eukaryotic proteins. *Biochem. J.* **258,** 625–638.

Gundersen, C. B., Mastrogiacomo, A., and Umbach, J. A. (1995) Cysteine-string proteins as templates for membrane fusion: models of synaptic vesicle exocytosis. *J. Theor. Biol.* **172,** 269–277.

Hess, D. T., Patterson, S. I., Smith, D. S., and Skene, J. H. P. (1993) Neuronal growth cone collapse and inhibition of protein fatty acylation by nitric oxide. *Nature* **366,** 562–565.

Hess, D. T., Slater, T. M., Wilson, M. C., and Skene, J. H. P. (1992) The 25kDa synaptosomal-associated protein SNAP-25 is the major methionine-rich polypeptide in rapid axonal transport and a major substrate for palmitoylation in adult CNS. *J. Neurosci.* **12,** 4634–4641.

Huang, E. M. (1989) Agonist-enhanced palmitoylation of platelet proteins. *Biochim. Biophys. Acta* **1011,** 134–139.

James, G. and Olson, E. N. (1989) Identification of a novel fatty acylated protein that partitions between the plasma membrane and cytosol and is deacylated in response to serum and growth factor stimulation. *J. Biol. Chem.* **264,** 20,998–21,006.

James, G. and Olson, E. N. (1990) Fatty acylated proteins as components of intracellular signaling pathways. *Biochemistry* **29,** 2623–2634.

Kato, K., Der, C. J., and Buss, J. E. (1992) Prenoids and palmitate, lipids that control the biological activity of *ras* proteins. *Sem. Cancer Biol.* **3,** 179–188.

Kimes, A. S., Sweeney, D., and Rapaport, S. I. (1985) Brain palmitate incorporation in awake and anesthetized rats. *Brain Res.* **341,** 164–170.

Kimes, A. S., Sweeney, D., London, E. D., and Rapaport, S. I. (1983) Palmitate incorporation into different brain regions in the awake rat. *Brain Res.* **274,** 291–301.

Liu, Y., Fisher, D. A., and Storm, D. R. (1993) Analysis of the palmitoylation and membrane targeting domain of neuromodulin (GAP-43) by site-specific mutagenesis. *Biochemistry* **32,** 10,714–10,719.

Miller, J. C., Gnaedinger, J. M., and Rapoport, S. I. (1987) Utilization of plasma fatty acid in rat brain: distribution of [¹⁴C]palmitate between oxidative and synthetic pathways. *J. Neurochem.* **49,** 1507–1514.

Milligan, G., Parenti, M., and Magee, A. I. (1995) The dynamic role of palmitoylation in signal transduction. *Trends Biochem. Sci.* **5,** 181–186.

Mouillac, B., Caron, M., Bonin, H., Dennis, M., and Bouvier, M. (1992) Agonist-modulated palmitoylation of β2-adrenergic receptor in Sf9 cells. *J. Biol. Chem.* **267,** 21,733–21,737.

O'Brien, P. J., St. Jules, R. S., Reddy, T. S., Bazan, N. G., and Zatz, M. (1987) Acylation of disc membrane rhodopsin may be nonenzymatic. *J. Biol. Chem.* **262,** 5210–5215.

Ochinnikov, Y. A., Abdulaev, N. G., and Bogachuk, A. S. (1988) Two adjacent cysteine residues in the C-terminal cytoplasmic fragment of bovine rhodopsin are palmitylated. *FEBS Lett.* **230,** 1–5.

Olson, E. N., Towler, D. A., and Glaser, L. (1985) Specificity of fatty acid acylation of cellular proteins. *J. Biol. Chem.* **260,** 3784–3790.

Parton, R. G. and Simons, K. (1995) Digging into caveolae. *Science* **269,** 1398,1399.

Patterson, S. I. and Skene, J. H. P. (1994) Novel inhibitory action of tunicamycin homologs suggests a role for dynamic protein fatty acylation in growth cone-mediated neurite extension. *J. Cell Biol.* **124,** 521–536.

Patterson, S. I. and Skene, J. H. P. (1995) Inhibition of dynamic protein palmitoylation in intact cells with tunicamycin. *Methods Enzymol.* **250,** 284–300.

Pedraza, L., Owens, G. C., Green, L. A. D., and Salzer, J. L. (1990) The myelin-associated glycoproteins: membrane disposition, evidence of a novel disulfide linkage between immunoglobulin-like domains, and posttranslational palmitoylation. *J. Cell Biol.* **111,** 2651–2661.

Peitzsch, R. M. and McLaughlin, S. (1993) Binding of acylated peptides and fatty acids to phospholipid vesicles, pertinence to myristoylated proteins. *Biochemistry* **32,** 10,436–10,442.

Quesnel, S. and Silvius, J. R. (1994) Cysteine-containing peptide sequences exhibit facile uncatalyzed transacylation and acyl-CoA-dependent acylation at the lipid bilayer interface. *Biochemistry* **33,** 13,340–13,348.

Robinson, L. J., Busconi, L., and Michel, T. (1995) Agonist-modulated palmitoylation of endothelial nitric oxide synthase. *J. Biol. Chem.* **270,** 995–998.

Schmidt, M. F. G. (1989) Fatty acylation of proteins. *Biochim. Biophys. Acta* **988,** 411–426.

Schmidt, M. F. G. and Burns, G. R. (1989) Hydrophobic modifications of membrane proteins by palmitoylation in vitro. *Biochem. Soc. Trans.* **17,** 625,626.

Schmidt, M. F. G. and Lambrecht, B. (1985) On the structure of the acyl linkage and the function of fatty acyl chains in the influenza virus haemagglutinin and the glycoproteins of Semliki Forest virus. *J. Gen. Virol.* **66,** 2635–2647.

Shahinian, S. and Silvius, J. R. (1995) Doubly-lipid-modified protein sequence motifs exhibit long-lived anchorage to lipid bilayer membranes. *Biochemistry* **34,** 3813–3822.

Silvius, J. R. and l'Heureux, F. (1994) Fluorimetric evaluation of the affinities of isoprenylated peptides for lipid bilayers. *Biochemistry* **33,** 3014–3022.

Skene, J. H. P. and Virag, I. (1989) Posttranslational membrane attachment and dynamic fatty acylation of a neuronal growth cone protein, GAP-43. *J. Cell Biol.* **108,** 613–624.

Tabata, H., Bell, J. M., Miller, J. C., and Rapaport, S. I. (1986) Incorporation of plasma palmitate into the brain of rat during development. *Dev. Brain Res.* **29,** 1–8.

Vesa, J., Hellsten, E., and Verkruyse, L. A., Camp, L. A., Rapola, J., Santavaori, P., Hofmann, S. L., and Peltonen, L. (1995) Mutations in the palmitoyl-protein thioesterase gene causing infantile neuronal ceroid lipofuscinosis. *Nature* **376,** 584–587.

Vogt, T. C. B., Killian, J. A., and Kruijff, B. D. (1994) Structure and dynamics of the acyl chain of a transmembrane polypeptide. *Biochemistry* **33,** 2063–2070.

Wedegaertner, P. B. and Bourne, H. R. (1994) Activation and depalmitoylation of G_{sa}. *Cell* **77,** 1063–1070.

Yamazaki, S., Noronha, J. G., Bell, J. M., and Rapoport, S. I. (1989) Incorporation of plasma [^{14}C]palmitate into the hypoglossal nucleus following unilateral axotomy of the hypoglossal nerve in adult rat, with and without regeneration. *Brain Res.* **477,** 19–28.

Yoshimura, T., Agrawal, D., and Agrawal, H. C. (1987) Cell-free acylation of rat brain myelin proteolipid protein and DM-20. *Biochem. J.* **246,** 611–617.

Protein ADP-Ribosylation

Keith D. Philibert and Henk Zwiers

1. Introduction

ADP-ribosylation of proteins was originally discovered in 1966 by Chambon et al. (1966), who detected polymers of ADP-ribose attached to protein substrates. Since that time, protein ADP-ribosylation has been identified in a diverse range of species, from bacteria to human, as well as within virtually every cellular compartment (Hilz et al., 1984; Ueda and Hayaishi, 1985; Ogura et al., 1990; Williamson and Moss, 1990; Aktories and Wegner, 1992). In general, ADP-ribosylation is a posttranslational modification in which one or more ADP-ribose moieties are transferred from a nicotinamide adenine dinucleotide (NAD$^+$) donor to an amino acid acceptor. It is important to note that NAD$^+$ is a substrate in this process rather than a cofactor as it is in many enzyme reactions.

Although the reaction has been known for nearly 30 yr, it is only relatively recently that endogenous mammalian ADP-ribosylation has come under investigation. Considerable research has focused on bacterial ADP-ribosylation, which is an important pathogenic mechanism for a variety of microorganisms. Some common bacteria that use ADP-ribosylation in their attack on host cells include diphtheria whose toxin modifies a histidine residue in Elongation Factor 2, and *Clostridium botulinum* C2 and C3, whose toxins ADP-ribosylate an actin arginine and a Rho asparagine, respectively (Van de Kerckhove et al., 1988; Aktories and Wegner, 1992). However, the most well-characterized bacte-

From: *Neuromethods, Vol. 30: Regulatory Protein Modification: Techniques and Protocols*
Ed: H. C. Hemmings, Jr. Humana Press Inc.

rial ADP-ribosylating toxins are from *Vibrio cholerae* and *Bordetella pertussis*. Cholera toxin (CTX) ADP-ribosylates an arginine in the α_s-subunit of heterotrimeric G-proteins, whereas pertussis toxin (PTX) modifies a cysteine in the α_i-subunit. Frequently, CTX and PTX have been utilized as experimental tools in the identification of new members of the G-protein family (Hepler and Gilman, 1992) and the regulation of G-protein activity in signal transduction pathways. In contrast to bacterial ADP-ribosylation, mammalian ADP-ribosylation occurs in two forms: poly-ADP-ribosylation, in which multiple ADP-ribose units are transferred to a single amino acid forming a polymer, and mono-ADP-ribosylation, which will be the main focus of this chapter.

1.1. Poly-ADP-Ribosylation

An exclusively nuclear process, poly-ADP-ribosylation involves the attachment of multiple ADP-ribose moieties through ester bonds to free carboxyl groups of glutamic acid and lysine acceptors (Ogata et al., 1980). Hence, poly-ADP-ribosylation is a form of O-linked glycosylation (Ueda and Hayaishi, 1985). PolyADP-ribosylation is catalyzed by a single poly(ADP-ribose) polymerase enzyme in a three-step process that includes initiation, elongation, and polymer branching (Alvarez-Gonzalez and Mendoza-Alvarez, 1995). To date, this dimeric, 110–130 kDa protein has been identified in virtually every eukaryotic cell type (Ogura et al., 1990). Furthermore, poly(ADP-ribose) polymerase requires DNA for activation (Spina-Purello et al., 1990). The primary substrates for this enzyme include histones, endonucleases, topoisomerases I and II, and DNA polymerases α and β (Althaus and Richter, 1987). Given the nature of the proteins modified, it is not surprising that poly-ADP-ribosylation has been implicated in regulating the repair of DNA strand breaks (Ding et al., 1992). Protocols for poly-ADP-ribosylation studies are quite similar to those discussed below for mono-ADP-ribosylation reactions; experimental details pertaining to poly-ADP-ribosylation of proteins were recently reviewed elsewhere (Shah et al., 1995).

1.2. Mono-ADP-Ribosylation

Mono-ADP-ribosylation, as the name implies, involves the transfer of a single ADP-ribose to a single amino acid acceptor. Mono-ADP-ribosylation is essentially an extra-nuclear process that constitutes >90% of the total cellular ADP-ribosylation (Ueda and Hayaishi, 1985). In this reaction, an ADP-ribosyltransferase (ADPRT) catalyzes the attachment of ADP-ribose to three primary amino acid acceptors and two secondary acceptors (Fig. 1A,B). The primary acceptors include arginine, which is involved in roughly two-thirds of all reported systems, cysteine, and histidine or its diphthamide derivative. ADP-ribose can also be incorporated into asparagine and lysine. Moreover, as Fig. 1 shows, the majority of ADP-ribosylation occurs on the side-chain amino groups of the acceptor amino acid. Hence, mono-ADP-ribosylation is primarily a form of N-linked glycosylation (Ueda and Hayaishi, 1985).

Although still in its infancy, neuronal ADP-ribosylation has been studied and several substrates have been identified, including actin and tubulin (Scaife et al., 1992; Aktories and Wegner, 1992; Palkiewicz et al., 1994), myelin basic protein (Boulias and Moscarello, 1994), heterotrimeric G-proteins (Tamir and Gill, 1988), the neuronal tissue-specific proteins B-50 (also known as GAP-43), BICKS (neurogranin) (Coggins et al., 1993 a,b), and MARCKS (Chao et al., 1994). Unlike most other proteins, the phosphoprotein B-50 undergoes cysteine-linked ADP-ribosylation (Philibert and Zwiers, 1995). When B-50 is incubated with NAD$^+$ and an appropriate source of ADP-ribosylating activity, the nicotinamide group is cleaved and the reactive cysteine thiol group binds to the vacated C1' of the ribose (Fig. 2). The products of the reaction include nicotinamide, a proton, and ADP-ribosylcysteine B-50. As with many other enzyme systems, ADP-ribosylation is subject to product inhibition by nicotinamide (Rankin et al., 1989). The removal of ADP-ribose from the substrate protein is catalyzed by the enzyme ADP-ribosylhydrolase (for a review of the ADP-ribosylation cycle, *see* Okazaki et al., 1995). Both

A   Primary ADP-ribose acceptors

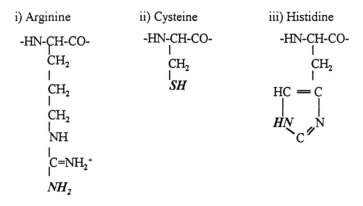

i) Arginine   ii) Cysteine   iii) Histidine

B   Secondary ADP- ribose acceptors

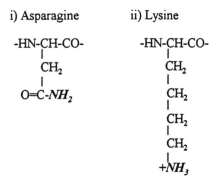

i) Asparagine   ii) Lysine

Fig. 1. Diagram of mono-ADP-ribosylation acceptors in proteins. The three predominant amino acid acceptors of ADP-ribose are shown in **(A)**, whereas the minor amino acid acceptors of ADP-ribosylation are shown in **(B)**. The groups involved in ADP-ribose attachment are shown in boldfaced italics.

ADPRT and ADP-ribosylhydrolase have specificity for the type of amino acid modified.

The aim of this chapter is to give an overview of the protocols commonly used in the study of mammalian endogenous ADP-ribosylation and, in particular, of neuronal ADP-ribosylation. Assay conditions and detection and quan-

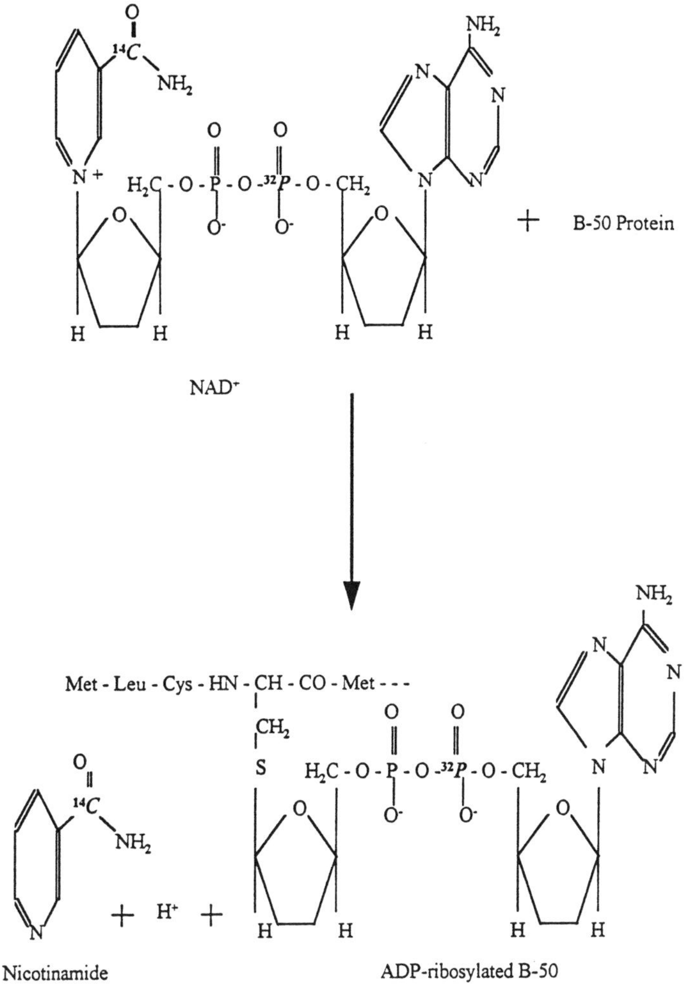

Fig. 2. Chemistry of B-50 ADP-ribosylation. Diagramatic represen-
tation of the ADP-ribosylation reaction showing the structures of the
ADP-ribose donor NAD$^+$ as well as the major products of B-50 ADP-
ribosylation. The common radiolabels are shown in italics.

tification of ADP-ribosylating reactions will be described, as
will the methodology involved in the identification of acceptor amino acids. The regulation of and common purification
schemes for the enzyme systems involved in mono-ADP-
ribosylation will also be considered.

2. ADP-Ribosylation Reactions

2.1. Assay Conditions

In order to follow the ADP-ribosylation reaction, several radiolabeled NAD⁺ precursors are commercially available containing ^{32}P, ^{14}C, or ^{3}H at various sites within the molecule (*see* Fig. 2). The label of choice is [adenylate-^{32}P]NAD⁺ for most applications, although the ^{3}H and ^{14}C labels are used for certain applications, which will be discussed later. Recently, a nonradioactive tracer was developed by Zhang and Snyder (1993) in which NAD⁺ was biotinylated. This allows ADP-ribosylated proteins to be isolated with avidin affinity chromatography or visualized with streptavidin-horseradish peroxidase chemiluminescence. The reaction mixture pH is usually maintained with a 20–100 mM Tris-HCl buffer at pH 7.5. To inhibit ADP-ribosylhydrolase activity, 1 mM 3'-acetylpyridine adenine dinucleotide (APAD) (Tamir and Gill, 1988) and 20 mM isonicotinic acid hydrazide are included. However, since APAD, an NAD⁺ analog, is a minor contributor of ADP-ribose, precise determinations of the stoichiometry of ADP-ribose incorporation are difficult (Tamir and Gill, 1988). Since many ADP-ribosylation substrates and ADP-ribosyltransferases are membrane-associated, 0.5% (v/v) Triton X-100 is added (Duman et al., 1991). To inhibit poly-ADP-ribosylation, 10 mM thymidine is also included (Williams et al., 1992). Several groups have also shown that ADP-ribosylation requires GTP for activation; hence, the nonhydrolyzable GTP analog 5'-guanylylimidodiphosphate (Gpp[NH]p) is added to 0.1 mM (Moss et al., 1980; Watkins et al., 1980; Duman et al., 1991). Dithiothreitol (DTT) at 5–10 mM is another common ingredient, particularly if the reaction is believed to be cysteine-specific, as it is for B-50 (Philibert and Zwiers, 1995). The retrograde neurotransmitter nitric oxide, produced in vitro by decomposition of sodium nitroprusside (0.1–0.5 mM), has been shown to stimulate endogenous ADP-ribosylation (Brune and Lapetina, 1989; Molina y Vedia et al., 1992). DTT appears to be a require-

Table 1
Common Constituents of ADP-Ribosylation Reactions[a]

Tris-HCl, pH 7.3	20–100 mM
3'-Acetylpyridine adenine dinucleotide	1 mM
5'-Guanylylimidodiphosphate	0.1 mM
DTT	5 mM
Triton X-100	0.5% (v/v)
Thymidine	10 mM
Isonicotinic acid hydrazide	20 mM
NAD$^+$	5–100 µM
Sodium nitroprusside	0.1–0.5 mM
MgCl$_2$	10 mM
CaCl$_2$	10 mM

[a]These buffer components and the listed concentrations are variable and represent the reagents most frequently seen in the literature.

ment for stimulation by nitric oxide (Brune and Lapetina, 1989). Protease inhibitors phenylmethylsulfonylfluoride, leupeptin, pepstatin, and EDTA do not appear to influence ADP-ribosylation (Just et al., 1994). The general buffer constituents for ADP-ribosylation reactions are summarized in Table 1. The buffer should be made fresh prior to use and stored in the dark, particularly when sodium nitroprusside is utilized. A further note of caution should be made in regard to vortex mixing, which has been shown to inactivate certain isoforms of purified arginine-specific ADPRTs (Ohno et al., 1994). Hence, low ADP-ribose incorporation may be the result of enzyme inactivation.

When purified enzymes or bacterial toxins are being used, necessary cofactors, such as the small GTP binding proteins known as ADP-ribosylation factors (ARFs), may be absent (Terashima and Shimoyama, 1993). Therefore, in some situations, it is useful to include a crude cell preparation to supply these cofactors. Furthermore, CTX and PTX must be preactivated by incubation in the presence of 1–5 mM DTT and 1–5 mM ATP for 30 min at 37°C (Moss et al., 1983; Terashima and Shimoyama, 1993). ADP-ribosylation assays are customarily performed at 37°C for 30–60 min.

We have observed linear incorporation of ^{32}P-ADP-ribose into brain proteins for up to 30 min, depending on the reaction conditions. This is in stark contrast to protein phosphorylation, for which maximal incorporation is reached within seconds after the addition of the phosphoryl donor [γ-^{32}P]ATP. One possible explanation for this is substrate availability. Neuronal membranes are extremely rich in ATPase activity, which may leave little time for the protein kinase to be active (Wiegant et al., 1978). It is premature to conclude at this time that ADP-ribosylation reactions are slower than phosphorylation reactions. Further resolution of this problem will require detailed characterization of the enzymes involved in the reaction.

Poly-ADP-ribosylation assays should be performed in the presence of DNA (Spina-Purello et al., 1990) and should not include 10 mM thymidine. The type of ADP-ribosylation present in a crude cell preparation can be distinguished by the use of selective inhibitors of poly vs mono-ADP-ribosylation (Rankin et al., 1989). Benzamide and its derivatives, as well as thymidine, produces 50% inhibition of poly-ADP-ribosylation at low-micromolar concentrations, whereas millimolar concentrations are necessary for a similar degree of inhibition of mono-ADP-ribosylation. Poly-ADP-ribosylating activity can also be eliminated by acid inactivation (Doi et al., 1993). In brief, the enzyme preparation is dialyzed for 10 h vs 2 L of 0.2M glycine-HCl, pH 3.5, containing 2 mM 2-mercaptoethanol, 1 mM EDTA, and 0.5 mM EGTA. The enzyme is then dialyzed for an additional 6 h vs 2 L of 0.2M Tris-HCl, pH 8.0, containing 2 mM 2-mercaptoethanol, 1 mM EDTA and 0.5 mM EGTA. ADP-ribosylation can then be assayed as described below.

2.2. Detection and Quantification

2.2.1. Sodium Dodecyl Sulfate-Polyacrylamide Gel Electrophoresis (SDS-PAGE)

As with most posttranslational modification studies, visualization and detection of modified proteins uses SDS-

PAGE as the method of choice. SDS-PAGE is particularly useful for determining incorporation of ADP-ribose into specific proteins, and allows both qualitative and quantitative determinations (Fig. 3). Unfortunately, SDS-PAGE is not of value when small substrate molecules, such as arginine, cysteine methyl esters, or agmatine, are used in the assay. A further complication in the use of SDS-PAGE is caused by unreacted [^{32}P]NAD$^+$ which leaves a dark smear of radioactivity on autoradiograms (Fig. 3). To eliminate the excess label, proteins can be precipitated by 10–20% (w/v) trichloroacetic acid (TCA) (Duman et al., 1991; Philibert and Zwiers, 1995). Commonly, either 100% (w/v) TCA is added to the desired final concentration or the sample is diluted with 1 mL of 20% TCA. After 10–30 min on ice, the samples are centrifuged at 10,000g for 10 min. The supernatant containing the unincorporated radioactivity is discarded and the protein pellets are washed with 2 × 1 mL of 95% (v/v) ethanol, followed by 2 × 1 mL diethyl ether (Piron and McMahon, 1990; Williams et al., 1992) to remove the TCA. The final pellet is resuspended in SDS-PAGE sample buffer and subjected to electrophoresis followed by autoradiography (*see* Chapter 3). By following this precleaning protocol, autoradiograms with a clear background can be obtained. It is important to note that the traditional boiling of SDS-PAGE samples prior to electrophoresis should be approached cautiously, since heat can cleave the phosphodiester bond, releasing the ^{32}P-label (Fig. 4). It remains to be determined whether this finding is limited to B-50 and BICKS, or is a more general phenomenon.

Western blot analysis of ADP-ribosylated proteins separated by SDS-PAGE may pose some difficulties. Problems of detectability have been described after phosphorylation as well as ADP-ribosylation of the enzyme 6-phosphofructo-2-kinase/fructose-2,6-bis-phosphatase (Rosa et al., 1995). The acidic protein B-50 in its unmodified form transfers and binds readily to nitrocellulose membranes. However, ADP-ribosylation appears to diminish the ability of B-50 to bind nitrocellulose (Philibert and Zwiers, unpublished observation). To overcome this problem partially, electrophoretic trans-

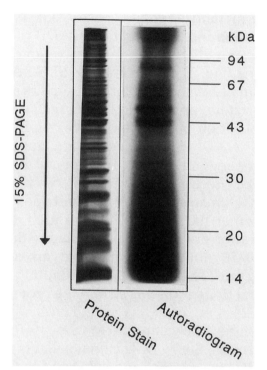

Fig. 3. Analysis of rat brain ADP-ribosylated proteins. Rat brain
homogenate was prepared by homogenization of rat forebrain tissue
(1 g of tissue/25 mL of buffer) in 50 mM Tris-HCl, pH 7.5. The protein
concentration was determined by the method of Bradford (1976) and 50
μL of homogenate (100 μg protein) was mixed with 30 μL of ADP-
ribosylation buffer (RB: 50 mM Tris-HCl, pH 7.5, 1% (v/v) Triton X-100,
2 mM ADAP, 0.2 mM 5'-guanylylimide diphosphate, and 10 mM DTT)
and preincubated for 5 min at 37°C. Labeled NAD$^+$ (20 μL of RB containing 25 μL of [^{32}P]NAD$^+$ and unlabeled NAD at a final concentration in
the assay of 10 μM) was added to the preincubated tissue fraction. The
reaction was allowed to continue for 1 h at 37°C. An aliquot of 10μL was
removed and the reaction was terminated by addition of 5 μL of SDS
sample buffer followed by vigorous mixing. The sample mixture contains 6% (w/v) SDS, 30% (w/v) glycerol, 15% (v/v)2-mercaptoethanol,
and 0.003% (w/v) bromophenol blue in 187 mM Tris-HCl, pH 6.8. The
samples were separated by SDS-PAGE gel (15% acrylamide in the running gel) for 3 h at 45 mA constant current until the bromophenol blue
was just about to leave the gel. At that time, the bulk of the unreacted
radioactivity has already left the gel. The gel was immersed for 10 min
in a solution of 0.1% (w/v) fast green in 10% (v/v) acetic acid 50% (v/v)
methanol and destained for 3 h to visualize the proteins (left lane). The

fers at much lower current and longer times seems to improve recovery of ADP-ribosylated B-50 on the nitrocellulose membrane. Incubation of the protein-loaded membrane with 5% (v/v) glutaraldehyde also shows improved attachment. As yet no antibodies directed specifically against ADP-ribosylated epitopes have been reported. They could be extremely useful in the study of the function of ADP-ribosylation of proteins akin to the use of antiphosphotyrosine antibodies for studies on hormone-receptor signal transduction pathways (*see* Chapters 5 and 6).

An improvement of standard one-dimensional SDS-PAGE is two-dimensional (2D) gel electrophoresis (Fig. 5), in which the first dimension is isoelectric focusing (IEF) followed by SDS-PAGE in the second dimension. The advantage of 2D gel electrophoresis is that the proteins do not need to be subjected to the relatively harsh acidic conditions used during TCA precipitation, since free label is quantitatively removed, resulting in a low background on autoradiograms. When the decision is made to do the analysis on 2D gels, the ADP-ribosylation reaction can be stopped by snap freezing of samples at –80°C followed by lyophilization. Samples (90 µg total protein) are resuspended in 24 mg urea, 25 µL H_2O, 2 µL 10% (v/v) Triton X-100, and 12 µL of IEF sample buffer (25% [w/v] sucrose, 0.015 g/mL ampholines 4–6, and 0.022 g/mL ampholines 3.5–10), and run on a 5% IEF slab gel overnight at 200 V (constant voltage) (Zwiers et al., 1985). By using a slab gel, many samples can be run simultaneously under identical conditions. After overnight electrophoresis at room temperature (in the cold room urea crystallizes in the gel), the voltage is increased to 1000 V for 1 h. The gel is removed from the electrophoresis apparatus, the glass plates cleaned, and the tracks of the samples are marked on both glass plates. Carefully, the glass plates are separated without disturbing

---

gel was wrapped with BioDesign Gel Wrap (Carmel, NY) and dried overnight at room temperature. The dried gel was subjected to autoradiography (Kodak X-Omat film; –80°C for 10 h). The right lane shows labeling of protein bands; however, the background is extremely high.

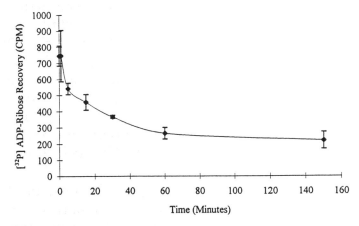

Fig. 4. Thermostability of ADP-ribosylated B-50. ADP-ribosylated B-50 (400 cpm [^{32}P]ADP-ribose/µg B-50) was heated to 95°C for 1, 5, 15, 30, 60, and 150 min followed by separation with SDS-PAGE. Loss of label was quantified by liquid scintillation counting of excised gel pieces.

the position of the gel on the glass plate to which it remains attached. The addition of 2 µg hemoglobin to the sample prior to IEF allows for more accurate cutting of the soft gel. Time is of the essence because of ongoing diffusion of proteins in the gel. After cutting, the gel lanes (1 × 10 cm) are transferred to 10 mL of 1:2 diluted SDS-PAGE sample buffer, and are gently shaken for 10 min to reduce the urea content in the strips and to saturate the proteins with SDS. Using a pair of tweezers, each lane is then placed longitudinally on a flat-surfaced stacking gel of an SDS slab gel (3% stacking gel/11% running gel for normal one-dimension SDS-PAGE). The lane is immersed in molten 1% agarose in 1:2 diluted SDS-PAGE sample buffer, which solidifies in a few minutes. The gel is placed in an electrophoresis apparatus; stacking is for 2 h at 15 mA or until the front reaches the running gel, and subsequently the current is increased to 45 mA until the front reaches the bottom of the gel. The gel is then transferred to a solution containing 0.1% (w/v) fast green in 10% (v/v) acetic acid/50% (v/v) methanol, stained for 10 min, and destained for 2–3 h. If protein spots are barely visible, the gel can be subjected to a more sensitive silver staining procedure (Merril et al., 1981; Fig. 5A). If quantitative data are required from a set of samples, it is necessary and possible to determine the relative amounts of an individual protein by densitometric scanning of the separated spot. This allows for a correction for recovery of the radioactivity measured in the protein of interest (Fig. 5B).

When [carbonyl-^{14}C]NAD$^+$ is used as the precursor for ADP-ribosylation reactions, SDS-PAGE gels must be treated with a scintillant for autoradiography. Following staining and destaining with 10% (v/v) acetic acid/30–50% (v/v) methanol, the gel is incubated in EN^{3}HANCE (DuPont, Wilmington, DE) for 1 h (for 1–1.5 mm thick gels) with gentle agitation. EN^{3}HANCE is discarded, and the gel is equilibrated for a maximum of 30 min in ice-cold dH$_2$O. The gel is then subjected to autoradiography at –80°C for an appropriate exposure time.

2.2.2. Filter Assay

Filter assays are frequently used to measure ADP-ribosylation. In contrast to the relatively slow SDS-PAGE procedure, which provides both qualitative and quantitative data for larger protein substrates, the filter assay is fast and is ideal for quantitatively determining total ADP-ribose incorporation into small substrates. To ensure the accuracy of the filter assay, the procedure should initially be run in parallel with an SDS-PAGE assay. A wide variety of filter assay protocols is available. In the most common procedure, the sample is TCA-precipitated (10–20% [w/v]) as described above. The acid-insoluble material is collected on a Whatman GF/A filter disk and subjected to liquid scintillation counting (Taniguchi et al., 1993; Yamada et al., 1994; Tsuchiya et al., 1995) in the presence of scintillation cocktail. In a variation of this procedure, the acid-insoluble material is applied to a nitrocellulose membrane (0.45 µm) and washed with 15 mL of 6% (w/v) TCA. The membrane is dried and liquid scintillation counted (Just et al., 1995). This assay can be performed with other types of membranes,

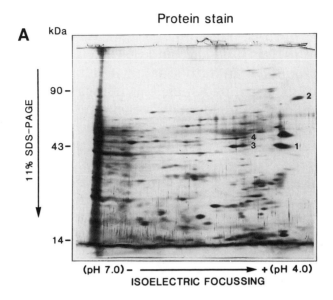

Protein stain

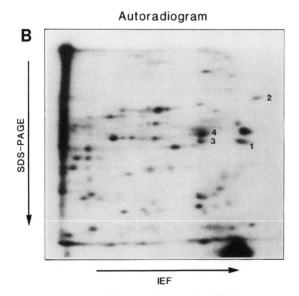

Autoradiogram

Fig. 5. 2-D analysis of ADP-ribosylated proteins. The remaining 90-μL aliquot of the ADP-ribosylation reaction (*see* caption for Fig. 3) was rapidly frozen and lyophilized. The sample was prepared for IEF *(see text)* in a polyacrylamide slab gel. After electrophoresis to equilibrium (overnight at 200 V, followed by 1 h at 1000 V), the relevant lane was excised, washed for 10 min in 3X sample mixture (caption to Fig. 3), and placed although the membranes should be tested for stability to TCA and scintillant prior to use. A third alternative was developed by Nestler et al. (1995), in which the reactions are quenched by the addition of an excess of unlabeled NAD⁺. The samples are applied to a Whatman 3MM filter disk and washed five times for 5–10 min with 10% (v/v) TCA in 10 mM sodium pyrophosphate. The filters are further washed for 10 min in 95% (v/v) ethanol and rinsed in diethyl ether. The membranes are allowed to dry completely before scintillation counting.

2.2.3. Capillary Electrophoresis

Capillary electrophoresis provides an opportunity not only to identify ADP-ribosylated substrates and quantify ADP-ribose incorporation, but also to purify the reaction product. Capillary electrophoresis is particularly useful when amino acids and their derivatives are used as ADP-ribosylation substrates. The ADP-ribosylation reaction is performed as described in Section 2.1. and is quenched by the addition of SDS to a final concentration of 0.1%. A fused silica capillary tube is washed with 0.2N NaOH for 2 min prior to equilibration in 30 mM citrate buffer, pH 2.5, containing 2.5 mM sodium 1-hexanesulfonate for 2 min. The sample is injected into the capillary tube by vacuum at a flow rate of 3.5 nL/s for 3 s. Capillary electrophoresis is performed at 30 kV for 30 min in citrate buffer (Tsuchiya et al., 1995). The ADP-ribosylation products are detected at 254 nm and can be identified by their specific retention times.

2.2.4. High-Performance Liquid Chromatography (HPLC)

As with capillary electrophoresis, HPLC has the advantage of allowing purification of the ADP-ribosylation product

---

on top of an 11% SDS-polyacrylamide slab gel, allowing separation of proteins in the second dimension. The gel was stained for proteins **(A)** using fast green, followed by silver stain. The gel was dried at room temperature (*see* caption to Fig. 3) and subjected to autoradiography **(B)**. Several neuronal proteins that are now known to be subject to ADP-ribosylation are indicated: 1: B-50; 2: MARCKS; 3: Actin; 4: Tubulin.

in addition to identification and quantification. Likewise, HPLC is very useful when very small substrates are utilized in the assay. As with the filter assay, there are numerous HPLC protocols available, but two methods will be discussed here, one involving radioactive tracers and the other fluorescent tracers. Other possible HPLC systems for ADP-ribosylation assays include strong anion-exchange (Jacobson et al., 1990) and affinity chromatography using a dihydroxybornyl column (Jacobson et al., 1990). For radioactive assays, the sample protein(s) is ADP-ribosylated as described above and the reaction is terminated by the addition of trifluoroacetic acid (TFA) to a final concentration of 0.1% (v/v). The sample is applied to a C18 reverse-phase HPLC (rpHPLC) column (e.g., Waters μBondapak) previously equilibrated in buffer containing 0.1% (v/v) TFA and 0.0125% (v/v) glacial acetic acid. The column is washed with equilibration buffer until the baseline stabilizes and the column is eluted with a linear gradient of 0–70% (v/v) acetonitrile containing 0.1% (v/v) TFA (Coggins and Zwiers, 1989; Boulias and Moscarello, 1994). This protocol works well for some larger substrates; however acidic proteins, such as ADP-ribosylated B-50, show lower than expected recoveries probably because of irreversible adsorption to the column.

The fluorescent ADP-ribosylation assay is used for arginine-specific reactions in which L-arginine methyl ester (LAME) is the substrate. This assay is based on *o*-phtalaldehyde (OPA), which reacts specifically with the primary amine present in LAME. LAME is ADP-ribosylated in the presence of unlabeled NAD^+, and the reaction is terminated with 5% (w/v) TCA. Samples are then incubated with OPA/2-mercaptoethanol reagent (Larew et al, 1991) for 1 min at room temperature, and are subsequently injected onto a C-8 rpHPLC column pre-equilibrated with buffer A (50 mM potassium phosphate, pH 4.0). The column is eluted at a flow rate of 1.5 mL/min with a linear gradient of 5–100% buffer B (5% [v/v] THF and 95% [v/v] methanol). The ADP-ribosylated products are detected with a fluorometer with excitation and emission wavelengths of 365 and 418 nm, respectively (Peterson et al., 1990).

Other instrumental techniques are also available for the qualitative and quantitative analysis of ADP-ribosylated substrates. In particular, some groups have had success with nuclear magnetic resonance (NMR) spectroscopy (Cervantes-Laurean et al., 1993,1995). ADP-ribosylated proteins have additional resonances between 120 and 160 ppm in the ^{13}C spectra, which correspond to the carbon atoms of the adenine ring of ADP-ribose. The proton spectra also contain several resonances distinctly associated with ADP-ribose. Mass spectroscopy has also recently been used to examine ADP-ribosylated proteins and to determine the exact site of the modification (Just et al., 1995). An in-depth discussion of the methods for either NMR or mass spectroscopy of ADP-ribosylated proteins is beyond the scope of this chapter.

3. Characterization of ADP-Ribosylated Proteins

When investigating protein ADP-ribosylation, a number of questions must be answered. These include: What is the site of the modification? Is the process enzymatic? Is the process actually ADP-ribosylation rather than, for instance, phosphorylation?

3.1. ADP-Ribosylation vs Phosphorylation

To determine qualitatively that a protein is indeed ADP-ribosylated and not phosphorylated, a number of procedures are available. The protein sample in question is subjected to ADP-ribosylation as described with [^{32}P]NAD$^+$ as substrate. The reaction is terminated with 10% (v/v) TCA followed by centrifugation. The supernatant is discarded and the pellet washed with diethyl ether to remove residual TCA. The pellet material is resuspended in a Tris-MgCl$_2$ buffer and incubated in the presence of snake venom phosphodiesterase (details can be found in Brune and Lapetina, 1989; Piron and McMahon, 1990; Duman et al., 1991; Williams et al., 1992). After incubation, the identity of the released label can be revealed by cellulose polyethyleneimine chromatography according to the procedure of Lehman et al. (1974) using appropriate markers. The radioactivity of 5-mm wide strips is determined with liquid scintillation counting or autoradiography. This system clearly separates NAD$^+$ from ATP, PO$_4$, and AMP (Duman et al., 1991). It is also possible to determine indirectly if the label on protein is phosphodiesterase-sensitive. This can be done by analysis of the labeled protein of interest by SDS-PAGE before and after incubation. Removal of label can be an indication of the presence of a phosphodiester bond. However, additional evidence should prove that this is not the result of the presence of contaminating protease activity, as we observed when we used some preparations of commercially available phosphodiesterase. Phosphodiesterase releases the ^{32}P label as [^{32}P]5'-AMP if the modification is indeed ADP-ribosylation.

As mentioned in Section 1., mono-ADP-ribosylation occurs primarily on arginine, cysteine, and histidine, and occasionally on lysine and asparagine (Fig. 1A,B). The particular amino acid modified in a given protein can be determined indirectly by characterization of the chemical stability of the ADP-ribosyl-protein bond. In addition, selective amino acid blockage and numerous peptide mapping/sequencing, and site-directed mutagenesis protocols have been used in the elucidation of the exact site(s). Some of these procedures will be reviewed here.

3.1.1. Amino Acid Blockage

Procedures are available to modify chemically potential substrate amino acids in proteins. The amino acid cysteine is an ADP-ribose acceptor and can be modified by iodoacetamide under reducing conditions. Using B-50 as an example, the protein (7 μM) is incubated in the presence of 50 mM Tris-HCl, pH 7.4, 280 μM DTT, and 580 μM iodoacetamide overnight at room temperature in the dark (Philibert and Zwiers, 1995). For an iodoacetamide alkylation to be specific for cysteine, the pH of the reaction mixture must be below 8.5, since otherwise both arginine and histidine can be modified. The reaction is quenched by the addition of TFA to 0.1% (v/v) followed by rpHPLC as described above to purify the modified B-50. The repurified protein can then be subjected to ADP-ribosylation. A cysteine-specific ADP-ribosylation reaction will be inhibited under these circumstances. An alternative to the multispecific alkylation by iodoacetamide is the cysteine-specific modification by *N*-ethylmaleimide (NEM). The protein sample of interest is incubated in the presence of 100 μM NEM in 50 mM Tris-HCl, pH 7.5, for 60 min at 21°C (Just et al., 1994). Excess NEM can be removed by dialysis against 50 mM Tris-HCl, pH 7.5, or by HPLC. If pretreatment of the protein of interest by iodoacetamide or NEM leads to reduced ADP-ribosylation, it is highly likely that cysteine is the substrate. For the amino acid arginine, different reagents are available. Modification has been achieved with 1,2-cyclohexanedione as well as with phenylglyoxal (Hebert and Fackrell, 1987).

3.1.2. Chemical Stability Assay

Knowledge of the exact chemical structure of the bonds linking ADP-ribose to the various amino acid acceptors led to the development of a chemical stability assay to determine the site of attachment (Jacobson et al., 1994). The thioether bond in ADP-ribosylcysteine is labile to 1M NaOH and 10 mM mercuric chloride, but is stable to neutral hydroxylamine. Conversely, ADP-ribosylhistidine and ADP-ribosylasparagine are essentially stable to all treatments, whereas ADP-ribosylarginine is labile to NaOH and neutral hydroxylamine, but stable to mercury. Finally, ADP-ribosyllysine is unstable in NaOH and CHES, pH 9.0, but stable in the presence of mercury and of hydroxylamine. To conduct the chemical stability assay, ADP-ribosylated proteins are prepared as described above for the phosphodiesterase assay (Jacobson et al., 1990; Piron and McMahon, 1990; Williams et al., 1992), but are redissolved in 50 μL of dH$_2$O and an equal volume of 1M NH$_2$OH, pH 7.5, 1M NaCl, 10 mM mercuric chloride, 1M NaOH, or 100 mM CHES buffer, pH 9.0 (for more information, *see also* Piron and McMahon, 1990; Williams et al., 1992; Just et al., 1995; Philibert and Zwiers, 1995; Terashima et al., 1995). All incubations are 1 h at 37°C with the exception of those in the presence of NH_2OH, pH 7.5, which last 12 h. The reactions are terminated with TCA and prepared for SDS-PAGE or filter assay as described above to determine the amount of radioactivity bound to the protein after the various stability treatments. The stability profile thus obtained, in combination with results from the amino acid modification experiments described above, usually gives a reasonably accurate indication of the nature of the ADP-ribosylation site(s).

4. The ADP-Ribosylation Reaction

The process of ADP-ribosylation can either be performed enzymatically by an ADPRT or it can be nonenzymatic, in which case it involves a cysteine-specific reaction (McDonald et al., 1992; McDonald and Moss, 1994). The nonenzymatic reaction involves the attachment of ADP-ribose, produced by NADase activity, to a cysteine acceptor resulting in the formation of an ADP-ribosylthiazolidine linkage. This reaction requires a reactive free amino group in the vicinity of the thiol. This type of bond is distinguishable from the enzymatic ADP-ribosylcysteine product by its decreased chemical stability to neutral hydroxylamine (McDonald et al., 1992). If the ADP-ribosylation of a given protein is nonenzymatic, the protein will become labeled when incubated in the presence of ADP-ribose. Although [^{32}P]ADP-ribose is not readily available commercially, its synthesis from [^{32}P]NAD$^+$ is a relatively simple procedure (Zocchi et al., 1993).

4.1. Generation and Purification of [^{32}P]ADP-Ribose

To generate [^{32}P]ADP-ribose, 10^8 cpm of [^{32}P]NAD$^+$ (1000 Ci/mmol) (Amersham, UK) are incubated in the presence of 0.1 U/mL NAD$^+$ glycohydrolase (EC 3.2.2.5, Sigma, St. Louis, MO) in phosphate-buffered saline for 20 min at 37°C (Zocchi et al., 1993). The reaction is quenched and deproteinated by the addition of TCA to 10% (w/v). After centrifugation at 10,000g for 10 min, the pellet is discarded and the supernatant is extracted three times with diethyl ether to remove the TCA. The samples are dried under nitrogen, and the [^{32}P]ADP-ribose is separated from nicotinamide and NAD$^+$ using strong anion-exchange (SAX) HPLC. The sample is applied to an SAX column pre-equilibrated in 20 mM Tris-HCl, pH 7.5, and eluted with a linear gradient of 0–1M NaCl in Tris buffer (Zolkiewska and Moss, 1995). The reaction products are detected by absorption at 260 nm. Alternatively, the NADase reaction can be quenched by a 20-fold dilution with distilled water and the ADP-ribose isolated on a Dowex AG 1-X8 (formate form) column. The ADP-ribose is eluted with a linear gradient of 1–10M formic acid. Fractions containing ADP-ribose are lyophilized and redissolved in ADP-ribosylation assay buffer for assay under the conditions described above for ADP-ribosylation in the absence of NAD$^+$.

4.2. ADP-Ribosyltransferase Activity

As alluded to before, the first mammalian ADPRT was isolated from turkey erythrocytes by Moss et al. (1980). Since then the presence of ADPRTs has been confirmed in virtually every mammalian tissue, and ADPRTs have been isolated from spleen and bone marrow (Tsuchiya et al., 1994,1995), rat liver (Tanigawa et al., 1984), cardiac muscle (McMahon et al., 1993), rabbit skeletal muscle (Zolkiewska et al., 1992), and several other tissues. Although cytosolic and mitochondrial ADPRTs have been discovered, the majority of ADPRTs are membrane-associated. Most ADPRTs appear to be arginine-specific (Peterson et al., 1990), have a mol-wt between 27 and 45 kDa, and bind to DEAE at neutral pH. Interestingly, Saxty and van Heyningen (1995) recently reported the isolation of a cysteine-specific ADP-ribosylating activity. ADPRT purification protocols usually contain a Concanavalin A affinity chromatography step, indicating that many mammalian ADPRTs are glycoproteins.

Endogenous regulation of brain ADPRT activity has received relatively little attention to date. A number of potential physiological effectors have been identified and their effects on ADP-ribosylation have been partially characterized (Saito et al., 1989; Duman et al., 1991; Williams et al., 1992). Over the last few years the role of the Ras superfamily of low-mol-wt GTP binding proteins in signal transduction has been the subject of intense investigations. The ARFs, a subclass of the Ras superfamily (Kahn and Gilman, 1984), have been identified as crucial allosteric activators of ADP-ribosylation, particularly by bacterial toxins, such as cholera toxin (Moss and Vaughan, 1995). The second known effector of ADP-ribosylation is GTP, which is not surprising since the ARFs require binding of GTP for activation. Currently, the best studied regulator of ADP-ribosylation is nitric oxide. Brune and Lapetina (1989) discovered that nitric oxide generated from sodium nitroprusside and 3-morpholinosydnonimine greatly enhanced the incorporation of ADP-ribose into a 39-kDa soluble protein in human platelets. The effect of nitric oxide was demonstrated to be independent of cyclic GMP, the most common second messenger of nitric oxide signaling.

4.3. ADP-Ribosylhydrolase Activity

As with the ADPRTs, the ADP-ribosylhydrolases also seem to be amino acid-specific. Furthermore, ADP-ribosylhydrolase activities have been identified in many different tissue types and species (Takada et al., 1994), indicating the existence of an ADP-ribosylation-deADP-ribosylation cycle. Currently, little is known about in vivo regulation of ADP-ribosyl hydrolysis. However, Mg^{2+} and DTT appear to be required for activation in vitro (Takada et al., 1994). To assay ADP-ribosylhydrolase activity, a radioactive ADP-ribosylated substrate is incubated in the presence of 50 mM Tris-HCl, pH 7.5, containing 10 mM $MgCl_2$, 5 mM DTT, and an appropriate amount of the ADP-ribosyl hydrolyzing activity. After 60 min at 30°C, the reactions are quenched with TCA to 20% (w/v) and the samples are analyzed for the amount of radioactivity as described above in ADP-ribosylation protocols. ADP-ribosylhydrolase activity can be determined by measuring the increase in radioactivity in the TCA supernatant in proportion to the release of [^{32}P]ADP-ribose from the substrate. The TCA pellet is analyzed for a loss of radiolabel in specific protein bands using SDS-PAGE and autoradiography as described above.

5. Functional Implications

In brain and numerous other tissues, arginine-specific ADP-ribosylation of actin and tubulin has been shown to inhibit the assembly of microfilaments and microtubules (Aktories and Wegner, 1992; Scaife et al., 1992). ADP-ribosylation also appears to promote depolymerization of these cytoskeletal elements. The general methods employed in these polymerization studies started with the isolation and ADP-ribosylation of purified actin and tubulin, followed by ADP-ribosylation of the proteins in more complex subcellular fractions. Subsequently, polymerization assays were performed to establish the role of ADP-ribosylation (Scaife et al., 1992; Terashima et al., 1995). A further important regulatory role for ADP-ribosylation in brain (like in other tissues) is in signal transduction pathways, where G-proteins are targets for endogenous ADP-ribosylation (Nestler et al., 1989; Saito et al., 1989; Wu et al., 1992).

Nitric oxide-mediated ADP-ribosylation has recently been implicated in the generation of long-term potentiation (LTP) in the CA1 region of the hippocampus (Schuman et al., 1994). Studies on the effects of ADP-ribosylation on neuronal function typically involve microinjection of specific ADPRT activators or inhibitors into the cells prior to functional assays involving electrophysiology, growth cone collapse, and numerous others.

Poly-ADP-ribosylation also has pronounced effects on cellular function, since it is involved in the repair of DNA strand breaks (Benjamin and Gill, 1980; Shah et al., 1995). Further, a role for poly-ADP-ribosylation has been reported in cell-cycle regulation (Kidwell et al., 1982), oncogenesis (Wang et al., 1995), and programmed cell death or apoptosis (Radons et al., 1994).

Acknowledgments

This work was supported by an operating grant from the Medical Research Council of Canada. H. Z. is a Senior Scholar of the Alberta Heritage Foundation for Medical Research. The authors would like to express their gratitude to Kim McLean for technical support and to Daryl Beers for secretarial assistance.

References

Althaus, F. R. and Richter, C. (1987) ADP-ribosylation of proteins-enzymology and biological significance. *Mol. Biol. Biochem. Biophys.* **37,** 1–125.

Aktories, K. and Wegner, A. (1992) Mechanisms of the cytopathic action of actin-ADP-ribosylating toxins. *Mol. Microbiol.* **6,** 2905–2908.

Alvarez-Gonzalez, R. and Mendoza-Alvarez, H. (1995) Dissection of ADP-ribose polymer synthesis into individual steps of initiation, elongation, and branching. *Biochimie* **77,** 403–407.

Benjamin, R. C. and Gill, D. M. (1980) ADP-ribosylation in mammalian cell ghosts: dependence of poly(ADP-ribose) synthesis on strand breakage in DNA. *J. Biol. Chem.* **255,** 10,493–10,501.

Boulias, C. and Moscarello, M. A. (1994) ADP-ribosylation of human myelin basic protein. *J. Neurochem.* **63,** 351–359.

Bradford, M. M. (1976) A rapid and sensitive method for the quantitation of microgram quantities of protein utilizing the principle of protein-dye binding. *Anal. Biochem.* **72,** 248–254.

Brune, B. and Lapetina, E. G. (1989) Activation of a cytosolic ADP-ribosyltransferase by nitric oxide-generating agents. *J. Biol. Chem.* **264,** 8455–8458.

Cervantes-Laurean, D., Loflin, P. T., Minter, D. E., Jacobson, E. L., and Jacobson, M. K. (1995) Protein modification by ADP-ribose via acid-labile linkages. *J. Biol. Chem.* **270,** 7929–7936.

Cervantes-Laurean, D., Minter, D. E., Jacobson, E. L., and Jacobson, M. K. (1993) Protein glycation by ADP-ribose: studies of model conjugates. *Biochemistry* **32,** 1528–1534.

Chambon, P., Wehe, P., Doly, J. D., Strosser, M. L., and Mandel, P. (1966) On the formation of a novel adenylic compound by enzymatic extracts of liver nuclei. *Biochem. Biophys. Res. Commun.* **25,** 638–643.

Chao, D., Severson, D. L., Zwiers, H., and Hollenberg, M. D. (1994) Radiolabeling of bovine myristoylated alanine-rich protein kinase C substrate (MARCKS) in an ADP-ribosylation reaction. *Biochem. Cell Biol.* **72,** 391–396.

Coggins, P. J., McLean, K., and Zwiers, H. (1993a) Neurogranin, a B-50/GAP-43-immunoreactive C-kinase substrate (BICKS), is ADP-ribosylated. *FEBS Lett.* **335,** 109–113.

Coggins, P. J and Zwiers, H. (1989) Evidence for a single PKC-mediated phosphorylation site in rat brain protein B-50. *J. Neurochem.* **53,** 1895–1901.

Coggins, P. J., McLean, K., Nagy, A., and Zwiers, H. (1993b) ADP-ribosylation of the neuronal phosphoprotein B-50/GAP-43. *J. Neurochem.* **60,** 368–371.

Ding, R., Pommier, Y., Kang, V. H., and Smulson, M. (1992) Depletion of poly(ADP-ribose) polymerase by antisense RNA expression results in a delay in DNA strand break rejoining. *J. Biol. Chem.* **267,** 12,804–12,812.

Doi, S., Tanigawa, Y., Tsuchiya, M., and Shimoyamma, M. (1993) Loss of poly(ADP-ribose) synthetase activity without change in arginine-specific ADP-ribosyltransferase activity by acid-treatment; application of the treatment to simply assay for transferase in the presence of poly(ADP-ribose) synthetase. *Biochim. Biophys. Acta* **1162,** 115–120.

Duman, R. S., Terwilliger, R. Z., and Nestler, C. J. (1991) Endogenous ADP-ribosylation in brain: initial characterization of substrate proteins. *J. Neurochem.* **57,** 2124–2132.

Hebert, T. E. and Fackrell, H. B., (1987) Inhibition of staphylococcal alphatoxin by covalent modification of an arginine residue. *Biochem. Biophys. Acta* **916,** 419–427

Hepler, H. R. and Gilman, A. G. (1992) G proteins. *Trends Biochem. Sci.* **17,** 383–387.

Hilz, H., Koch, R., Fanick, W., Klapproth, K., and Adamietz, P. (1984) Nonenzymatic ADP-ribosylation of specific mitochondrial polypeptides. *Proc. Natl. Acad. Sci. USA* **81,** 3929–3933.

Jacobson, E. L., Cervantes-Laurean, D., and Jacobson, M. K. (1994) Glycation of proteins by ADP-ribose. *Mol. Cell. Biochem.* **138,** 207–212.

Jacobson, M. K., Loflin, P. T., Aboul-Ela, N., Mingmuang, M., Moss, J., and Jobson, E. L. (1990) Modification of plasma membrane protein cystein residues by ADP-ribose in vivo. *J. Biol. Chem.* **265,** 10,825–10,828.

Just, I., Wollenberg, P., Moss, J., and Aktories, K. (1994) Cysteine-specific ADP-ribosylation of actin. *Eur. J. Biochem.* **221,** 1047–1054.

Just, I., Sehr, P., Jung, M., van Damme, J., Puype, M., Vandekerckhove, J., Moss, J., and Aktories, K. (1995) ADP-ribosyltransferase type A from turkey orythrocytes modifies actin at Arg-95 and Arg-372. *Biochemistry* **34,** 326–333.

Kahn, R. A. and Gilman, A. G. (1984) Purification of a protein cofactor required for ADP-ribosylation of the stimulatory regulatory component of adenylate cyclase by cholera toxin. *J. Biol. Chem.* **259,** 6228–6234.

Kidwell, W. R., Nolan, N., and Stone, P. (1982) Variations in Poly(ADP-ribose) and Poly(ADP-ribose) synthetase in synchronously dividing cells, in *ADP Ribosylation Reactions* (Hayaishi, O. and Ueda, K., eds.), Academic, London, pp. 373–388.

Larew, J. S., Peterson, J. E., and Graves, D. J. (1991) Determination of the kinetic mechanism of arginine-specific ADP-ribosytransferases using a high performance liquid chromatographic assay. *J. Biol. Chem.* **266**, 52–57.

Lehman, A. R., Kirk-Bell, S., Shall, S., and Whish, W. J. D. (1974) The relationship between cell growth, macromolecular synthesis and poly ADP-ribose polymerase in lymphoid cells. *Exp. Cell Res.* **83**, 63–72.

McDonald, L. J. and Moss, J. (1994) Enzymatic and nonenzymatic ADP-ribosylation of cysteine. *Mol. Cell Biochem.* **138**, 221–226.

McDonald, L. J., Wainschel, L. A., Oppenheimer, M. J., and Moss, J. (1992) Amino acid-specific ADP-ribosylation: structural characterization and chemical differentiation of ADP-ribose-cysteine adducts formed nonenzymatically and in a pertussis toxin-catalyzed reaction *Biochemistry* **31**, 11,881–11,887.

McMahon, K. K., Piron, K. H., Ha, V. T., and Fullerton, A. T. (1993) Developmental and biochemical characteristics of the cardiac membrane-bound arginine-specific mono-ADP-ribosyltransferase. *Biochem. J.* **293**, 789–793.

Merril, C. R., Goldman, D., Sedman, S. A., and Ebert, M. H. (1981) Ultrasensitive stain for proteins in polyacrylamide gels shows regional variation in cerebrospinal fluid protein. *Science* **211**, 1437,1438.

Molina y Vedia, L., McDonald, B., Reep, B., Brune, B., Di Silvio, M., Billiar, T. R., and Lapetina, E. G. (1992) Nitric oxide-induced S-nitrosylation of glyceraldehyde-3-phosphate dehydrogenase inhibits enzymatic activity and increases endogenous ADP-ribosylation. *J. Biol. Chem.* **267**, 24,929–24,932.

Moss, J. and Vaughan, M. (1995) Structure and function of ARF proteins: activators of cholera toxin and critical components of intracellular vesicular transport processes. *J. Biol. Chem.* **270**, 12,327–12,330.

Moss, J., Stanley, S. J., and Watkins, P. A. (1980) Isolation and properties of an NAD- and guanidine-dependent ADP-ribosyltransferase from turkey erythrocytes. *J. Biol. Chem.* **255**, 5838–5840.

Moss, J., Yost, D. A., and Stanley, S. J. (1983) Amino acid-specific ADP-ribosylation. *J. Biol. Chem.* **258**, 6466–6470.

Nestler, E. J., Erdos, J. J., Terwiliger, R., Duman, R. S., and Tallmen, J. F. (1989) Regulation of G-proteins of chronic morphine in the rat locus coeruleus. *Brain Res.* **476**, 230–239.

Nestler, E. J., Duman, R. S., and Terwilliger, R. Z. (1995) Regulation of endogenous ADP-ribosylation by acute and chronic lithium in rat brain. *J. Neurochem.* **64**, 2319–2324.

Ogata, N., Ueda, K., Kagamiyama, H., and Hayaishi, O. (1980) ADP-ribosylation of histone H1: identification of glutamic acid and residues 2, 14, and the COOH-terminal lysine residues as modification sites. *J. Biol. Chem.* **255,** 7616–7620.

Ogura, T., Takenouchi, N., Yamaguchi, M., Matsukage, A., Sigimura, T., and Esumi, H. (1990) Striking similarity of the distribution patterns of the poly(ADP-ribose) polymerase and DNA polymerase beta among various mouse organs. *Biochem. Biophys. Res. Commun.* **172,** 377.

Ohno, T., Badruzzaman, M., Nishikori, Y., Tsuchiya, M., Jidoi, J., and Shimoyama, M. (1994) Vortex-mixing-induced inactivation of arginine-specific ADP-ribosyltransferase activity and re-activation of the less-active form by dithiothreitol plus NaCl under anaerobic conditions. *Biochem. Mol. Biol. Int.* **32,** 213–220.

Okazaki, I. J., Zolkiewska, A., Takada, T., and Moss, J. (1995) Characterization of mammalian ADP-ribosylation cycles. *Biochimie* **77,** 319–325.

Palkiewicz, P., Zwiers, H., and Lorscheider, F. L. (1994) ADP-ribosylation of brain neuronal proteins is altered by in vitro and in vivo exposure to inorganic mercury. *J. Neurochem.* **62,** 2049–2052.

Peterson, J. E., Larew, J. S.-A., and Graves, D. J. (1990) Purification and partial characterization of arginine-specific ADP-ribosyltransferase from skeletal muscle microsomal membranes. *J. Biol. Chem.* **265,** 17,062–17,069.

Philibert, K. and Zwiers, H. (1995) Evidence for multisite ADP-ribosylation of neuronal phosphoprotein B-50/GAP-43. *Mol. Cell. Biol.* **149/150,** 183–190.

Piron, K. J. and McMahon, K. K. (1990) Localization and partial characterization of ADP-ribosylation products in hearts from adult and neonatal rats. *Biochem. J.* **270,** 591–597.

Radons, J., Heller, B., Burkle, A., Hartmann, B., Rodriguez, M.-L., Kroncke, K.-D., Burkart, V., and Kolb, H. (1994) Nitric oxide toxicity in islet cells involves poly(ADP-ribose) polymerase activation and concomitant NAD$^+$ depletion. *Biochem. Biophys. Res. Commun.* **199,** 1270–1277.

Rankin, P. W., Jacobson, E. L., Benjamin, R. C., Moss, J., and Jacobson, M. K. (1989) Quantitative studies of inhibitors of ADP-ribosylation in vitro and in vivo. *J. Biol. Chem.* **264,** 4312.

Rosa, J. L., Perez, M. X., Ventura, F., Tauler, A., Gil, J., Shimoyama, M., Pilkis, S. J., and Bartrons, R. (1995) Role of the N-terminal region in covalent modification of 6-phosphofructo-2-kinase/fructose-2,6-biphosphatase: comparison of phosphorylation and ADP-ribosylation. *Biochem. J.* **309,** 119–125.

Saito, N., Guitart, X., Hayward, M., Tallman, J. F., Duman, R. S. and Nestler, E. J. (1989) Corticosterone differentially regulates the expression of messenger RNA and protein in rat cerebral cortex. *Proc. Natl. Acad. Sci. USA* **86,** 3609,3910.

Saxty, B. A. and van Heyningen, S. (1995) The purification of a cysteine-dependent NAD⁺ glycohydrolase activity from bovine erythrocytes and evidence that it exhibits a novel ADP-ribosyltransferase activity. *Biochem. J.* **310,** 931–937.

Scaife, R. M., Wilson, L., and Purich, D. L. (1992) Microtubule protein ADP-ribosylation in vitro leads to assembly inhibition and rapid depolymerization. *Biochemistry* **31,** 310–316.

Schuman, E. M., Meffert, M. K., Schulman, H., and Madison, D. B. (1994) An ADP-ribosyltransferase as a potential target for nitric oxide action in hippocampal long-term potentiation. *Proc. Natl. Acad. Sci. USA* **91,** 11,958–11,962.

Shah, G. M., Poirier, D., Duchaine, C., Brochu, G., Desnoyers, S., Lagueux, J., Verreault, A., Hoflack, J.-C., Kirkland, J. B., and Poirier, G. G. (1995) Methods for biochemical study of poly(ADP-Ribose) metabolism in vitro and in vivo. *Anal. Biochem.* **227,** 1–13.

Spina-Purello, V., Avola, R., Condorelli, D. F., Nicoletti, V. G., Insirello, L., Reale, S., Costa, A., Ragusa, N., and Giuffrida Stella, A. M. (1990) ADP-ribosylation of proteins in brain regions of rats during postnatal development. *Int. J. Dev. Neurosci.* **8,** 167–174.

Takada, T., Okazaki, I. J., and Moss, J. (1994) ADP-ribosylarginine-hydrolases. *Mol. Cell. Biochem.* **138,** 119–122.

Tamir, A. and Gill, D. M. (1988) ADP-ribosylation by cholera toxin of membranes derived from brain modifies the interaction of adenylate cyclase with guanine nucleotides and NaF. *J. Neurochem.* **50,** 1791–1797.

Tanigawa, Y., Tsuchiya, M., Imai, Y., and Shimoyama, M. (1984) ADP-ribosyltransferase from hen liver nuclei. *J. Biol. Chem.* **259,** 2022–2029.

Taniguchi, M., Tsuchiya, M., and Simoyama, M. (1993) Comparison of acceptor protein specificities on the formation of ADP-ribose acceptor adducts by arginine-specific ADP-ribosyltransferase from rabbit skeletal muscle sarcoplasmic reticulum with those of the enzyme from chicken peripheral polymorphonuclear cells. *Biochim. Biophys. Acta.* **1161,** 265–271

Terashima, M. and Shimoyama, M. (1993) ADP-ribosylation of A1 peptide of cholera toxin by chicken arginine-specific ADP-ribosyltransferase with a concomitant increase in ADP-ribosyltransferase activity of the peptide. *Biomed. Res.* **14,** 329–335.

Terashima, M., Yamamori, C., and Shimoyama, M. (1995) ADP-ribosylation of Arg28 and Arg206 on the actin molecule by chicken arginine-specific ADP-ribosyltransferase. *Eur. J. Biochem.* **231,** 242–249.

Tsuchiya, M., Hara, N., Yamada, K., Osago, H., and Shimiyama, M. (1994) Cloning and expression of cDNA for arginine-specific ADP-ribosyltransferase from chicken bone marrow cells. *J. Biol. Chem.* **269,** 27,451–27,457.

Tsuchiya, M., Osago, H., and Shimoyama, M. (1995) Assay of arginine-specific adenosine-5'-diphosphate-ribosyltransferase by capillary electrophoresis. *Anal. Biochem.* **224**, 486–489.

Ueda, K. and Hayaishi, O. (1985) ADP-ribosylation. *Ann. Rev. Biochem.* **54**, 73–100.

Van de Kerckhove, J., Schering, B., Bärmann, M., and Aktories, K. (1988) Botulinum C2toxin ADP-ribosylates cytoplasmic β/γ actin in arginine 177. *J. Biol. Chem.* **263**, 696–700.

Wang, Z., Auer, B., Stingl, L., Berghammer, H., Haidacher, D., Schweiger, M., and Wagner, E. F. (1995) Mice lacking ADPRT and poly(ADP-ribosyl)ation develop normally but are susceptible to skin disease. *Genes Dev.* **9**, 509–520.

Watkins, P. A., Moss, J., and Vaughan, M. (1980) Effects of GTP on choleragen catalyzed ADP-ribosylation of membrane and soluble proteins. *J. Biol. Chem.* **255**, 3959–3963.

Wiegant, V. M., Zwiers, H., Schotman, P., and Gispen, W. H. (1978) Endogenous phosphorylation of rat rain synaptosomal plasma membranes in vitro: some methodological aspects. *Neurochem. Res.* **3**, 443–453.

Williams, M. B., Li, X., Gu, X., and Jope, R. S. (1992) Modulation of endogenous ADP-ribosylation in rat brain. *Brain Res.* **592**, 49–56.

Williamson, K. C. and Moss, J. (1990) Mono-ADP-ribosyltransferase and ADP-ribosylarginine hydrolases: a mono-ADP-ribosylation cycle in animal cells, in *ADP-Ribosylating Toxins and G. Proteins: Insights into Signal Transduction* (Moss J. and Vaughan M. eds.), American Society for Microbiology, Washington, DC, pp. 493–510.

Wu, K., Nigam, S. K., LeDoux, M., Huang, Y.-Y., Aoki, C., and Siekevitz, P. (1992) Occurrence of the subunits of G-proteins in cerebral cortex synaptic membrane and postsynaptic density fractions: modulation of ADP-ribosylation by Ca2+/calmodulin. *Proc. Natl. Acad. Sci. USA* **89**, 8686–8690.

Yamada, K., Tsuchiya, M., Nishikori, Y., and Shimoyama, M. (1994) Automodification of arginine-specific ADP-ribosyltransferase purified from chicken peripheral heterophils and alteration of the transferase activity. *Arch. Biochem. Biophys.* **308**, 31–36.

Zhang, J. and Snyder, S. H. (1993) Purification of a nitric oxide-stimulated ADP-ribosylated protein using biotinylated nicotinamide adenine dinucleotide. *Biochemistry* **32**, 2228–2233.

Zocchi, E., Guida, L., Franco, L., Silvestro, L., Guerrini, M., Benatti, U., and De Flora, A. (1993) Free ADP-ribose in human erythrocytes: pathways of intra-erythrocytic conversion and nonenzymatic binding to membrane proteins. *Biochem. J.* **295**, 121–130.

Zolkiewska, A. and Moss, J. (1995) Processing of ADP-ribosylated integrin in skeletal muscle-myotubes. *J. Biol. Chem.* **271**, 9227–9233.

Zolkiewska, A., Nightingale, M. S., and Moss, J. (1992) Molecular characterization of NAD: arginine ADP-ribosyltransferase from rabbit skeletal muscle. *Proc. Natl. Acad. Sci. USA* **89,** 11,352–11,356.

Zwiers, H., Verhaagen, J., van Dongen, C. J., de Graan, P. N. E., and Gispen, W. H. (1985) Resolution of rat brain synaptic phosphoprotein B-50 into multiple forms by two-dimensional electrophoresis: evidence for multisite phosphorylation. *J. Neurochem.* **44,** 1083–1090.

Glycosylation and Glycosylphosphatidylinositol Membrane Anchors

Anthony J. Turner, Edward T. Parkin, and Nigel M. Hooper

1. Introduction

Glycosylation is one of the most common covalent modifications of proteins. It is not normally a reversible phenomenon and does not play a role in regulation of protein activity. Both glycoproteins and glycolipids are especially abundant in the plasma membrane of eukaryotic cells where the oligosaccharide chains face the extracellular space. In general, the carbohydrate moieties of glycoproteins play little or no role in the biological functions of the proteins and, in the case of enzymes, generally do not participate in catalytic activity. However, because of the hydrophilicity of the sugars, covalently attached carbohydrates maintain the solubility of glycoproteins and ensure the correct folding of the extracellular domains. Glycans also can protect peptide chains from proteolysis, and provide for intercellular recognition and adhesion. Many proteins of importance to neural function, e.g., enzymes, receptors, ion channels, cell adhesion, and axonal guidance molecules, are glycoproteins in nature.

A specialized form of covalent modification involving oligosaccharides is the attachment of a glycolipid tail to the C-terminus of certain plasma membrane proteins that are commonly referred to as "GPI-anchored," "glypiated" or

From: *Neuromethods, Vol. 30: Regulatory Protein Modification: Techniques and Protocols*
Ed: H. C. Hemmings, Jr. Humana Press Inc.

"PIG-tailed" proteins. This glycosyl-phosphatidylinositol (GPI) "anchor" provides a relatively common and functionally important mechanism for the stable association of eukaryotic proteins with the membrane. A diverse group of proteins are attached to the plasma membrane in this way, including some ectoenzymes, cell-adhesion proteins, differentiation antigens, receptors, and tumor markers. A number of these proteins have special significance for the nervous system, including neural cell adhesion molecules, the Thy-1 antigen, the prion protein, the ciliary neurotrophic factor (CNTF) receptor, and ecto-5'-nucleotidase. This chapter will focus particularly on methods for identification and characterization of GPI-anchored proteins and the membrane domains in which they localize, with particular attention to the nervous system. More detailed information on analysis of carbohydrates and glycoproteins can be found in Chaplin and Kennedy (1994) and Fukuda and Kobata (1993). For specific reference to GPI-anchored proteins the reader is directed to Turner (1990, 1994) and Hooper and Turner (1992).

2. Protein Glycosylation

2.1. Identification of Glycoproteins

Sugar residues in glycoproteins are termed *N*-linked if they are attached to the amide nitrogen of asparagine or *O*-linked if attached to the hydroxyl oxygen of serine or threonine. *N*-linked oligosaccharides are found in many membrane and secreted proteins. Deduction of the amino acid sequence of a protein from cDNA cloning studies can predict *N*-linked glycosylation sites by virtue of the consensus sequence for *N*-linked glycans (-Asn-Xaa-Ser/Thr-), although not all the available sites may be modified in the mature protein. *O*-linked glycans are typically found in Pro/Ser/Thr-rich regions of the protein.

Lectin affinity chromatography provides a valuable and specific purification step for glycoproteins while also providing concentration, although high purification factors are rare. In the case of membrane glycoproteins, such purification must be carried out in the presence of detergents. It is useful to screen a range of immobilized lectins for binding the protein of interest, although wheat germ agglutinin, concanavalin A, and lentil lectin are commonly employed. Elution is effected by the appropriate sugar. Protocols for such procedures are found in Carlsson (1993) and in manufacturers' information sheets. A range of other conventional purification steps are also required to achieve a homogeneous product. Highly glycosylated proteins are usually stained poorly by conventional procedures, such as Coomassie brilliant blue or silver staining, however, the periodic Schiff reagent is effective for staining glycoproteins after sodium dodecyl sulfate-polyacrylamide gel electrophoresis (SDS-PAGE), producing red bands. Blotting techniques can also be used for detection of glycoproteins after SDS-PAGE by using labeled lectins.

2.2. Protein Deglycosylation

The same protein when isolated from different tissues commonly shows small differences in M_r as a result of differences in glycosylation. For example, porcine endopeptidase-24.11 (E-24.11; "enkephalinase," neprilysin, EC 3.4.24.11) migrates as a polypeptide of M_r 89,000 when purified from kidney, but 94,000 from intestine, and 87,000 from brain (Relton et al., 1983). Another example is provided by angiotensin-converting enzyme (ACE; peptidyl dipeptidase A, EC 3.4.15.1), which migrates as a polypeptide of M_r 180,000 when isolated from porcine lung or kidney, but an additional band of M_r 170,000 is detected in porcine brain (striatum) (Hooper and Turner, 1987). The larger species represents the endothelial isoform, whereas the smaller species is restricted to neurons where it may play a role in inactivation of peptide transmitters (Williams et al., 1991). The difference in size between these two forms of ACE is attributable principally to differences in glycosylation, since deglycosylation of the purified brain protein caused it to comigrate with the kidney protein on SDS gels (Fig. 1) (Hooper and Turner, 1987). Glycoproteins, even when isolated from a single tissue, commonly exhibit microheterogeneity on SDS-PAGE as a result of minor differences in glycosylation.

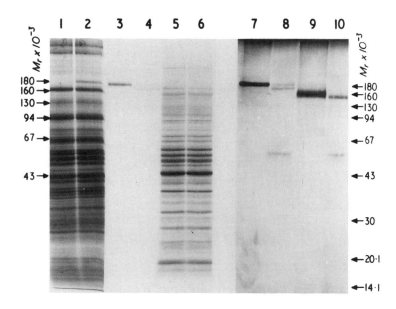

Fig 1. SDS-PAGE and deglycosylation of pig kidney and brain ACE. Lanes 1 and 2, pig kidney membrane fraction solubilized with Triton X-100 after (lane 1) and before (lane 2) affinity chromatography on lisinopril-2.8 nm-Sepharose (200 μg protein); lane 3, pig kidney ACE purified by affinity chromatography (2 μg); lane 4, pig brain (striatal) ACE purified by affinity chromatography (6 μg); lanes 5 and 6, pig striatal membrane fraction solubilized with Triton X-100 before (lane 5) and after (lane 6) affinity chromatography on lisinopril-2.8 nm-Sepharose (100 μg protein); lane 7, pig kidney ACE untreated (0.75 μg); lane 8, pig striatal ACE untreated (6 μg); lane 9, pig kidney ACE (1 mg) treated with N-glycanase (10–20 U/mL incubated with ACE for 24 h at 37°C in 0.25M sodium phosphate buffer, pH 8.6, containing 10 mM 1,10-phenanthroline in a final volume of 30–50 mL. The reaction was terminated by the addition of 50 μL of electrophoresis buffer and boiled for 4 min; lane 10, pig striatal ACE (6 μg) treated with N-glycanase. Lanes 1–6 were stained with Coomassie blue; lanes 7–10 were silver-stained. The two species of striatal ACE (M_r 170,000 and 180,000) migrate identically (M_r 150,000) after deglycosylation (lane 10), indicating the differences are principally owing to differences in N-linked glycosylation (from Hooper and Turner [1987] with permission of Portland Press, London).

Commercial kits are now available both for the sensitive detection of glycoproteins after gel electrophoresis and for partial or complete enzymatic deglycosylation (Oxford Glycosystems, Oxford, UK). Chemical deglycosylation, e.g., with trifluoromethanesulfonic acid (TFMS), is a less expensive alternative to enzymatic deglycosylation, but invariably results in denaturation of the protein. TFMS was originally introduced by Edge et al. (1981), who showed it to be efficient in cleaving O-glycosyl bonds, whereas N-glycosyl bonds between asparagine and N-acetylglucosamine remained intact. Thus, apart from these residues, glycoproteins can be effectively stripped of carbohydrate.

For deglycosylation by TFMS, protein (typically 10–50 μg) is precipitated twice with ice-cold acetone and dried under N_2. The protein pellet is resuspended in anisole (25 μL) and TFMS (50 μL) is added. N_2 is bubbled through the solution and the mixture left on ice for 2.5 h. The reaction is terminated by the addition of 250 μL of pyridine/water (4:1, v/v) in 25 μL portions and the reaction mixture is then quenched by cooling in an acetone/solid CO_2 slurry. After a final precipitation with acetone, 30 μL of 3*M* Tris-HCl, pH 7.5, are added, followed by 50 μL of electrophoresis buffer. After boiling, the protein can then be analyzed by SDS-PAGE.

Chemical cleavage of O-glycans can also be achieved by the action of alkali in the presence of borohydride. For N-linked glycans, hydrazinolysis or alkaline cleavage may be used (*see* Montreuil et al., 1994). A number of glycosidases are useful for removal of carbohydrate structures from glycoproteins. N-linked glycans can generally be removed from a protein by using N-glycanase (N-glycosidase F), which trims the glycan chain back to the aspartic acid residue at the glycosylation site. O-glycanase (endo-α-N-acetylgalactosaminidase; O-glycosidase) will trim O-linked sugar chains attached to serine or threonine, although any sialic acid groups must first be removed with neuraminidases. Endo-β-N-acetylglucosaminidase H (Endo H) cleaves high-mannose-type N-linked chains and is commonly used in biosynthetic studies to follow the intracellular transport of nascent glycoproteins. Biosynthetic studies of intestinal microvillar proteins (reviewed in Danielsen et al., 1987) provide an excellent example of the methodology.

3. Structure and Functions of GPI Membrane Anchors

GPI membrane anchors are present in organisms at most stages of eukaryotic evolution, including protozoa, yeast, slime molds, invertebrates, and vertebrates, and are found on a diverse range of proteins. Although over 150 examples of GPI-anchored proteins have been described, very few GPI anchor structures have been characterized in detail. Among those proteins whose GPI anchor structures have been determined are rat brain Thy-1 antigen (Homans et al., 1988), human erythrocyte acetylcholinesterase (Deeg et al., 1992), hamster brain scrapie prion protein (Stahl et al., 1992), bovine liver 5'-nucleotidase (Taguchi et al., 1994), and mouse skeletal muscle neural cell-adhesion molecule (N-CAM) (Mukasa et al., 1995) (Fig. 2). Recently, we determined the complete structure of the GPI anchor on porcine kidney membrane dipeptidase, and the first interspecies comparison of the GPI glycan core structure on the same protein (Brewis et al., 1995) (Fig. 2). The GPI anchor on both human and porcine membrane dipeptidase was found to consist of the conserved core structure: ethanolamine phosphate-6Manα1-2Manα1-6Manα1-4GlcNH$_2$α1-6*myo*-inositol-1-PO$_4$-lipid. Attached to this conserved core are variable side-chains (*see* Fig. 2), which may be protein- and/or tissue-specific. The function, if any, of these side-chain modifications in general remains obscure, although the α-galactose branch in some forms of the trypanosome variant surface glycoprotein may be involved in the dense packing of the protective surface coat (Homans et al., 1989).

The nascent form of a GPI-anchored protein contains a hydrophobic cleavable N-terminal signal sequence that directs the protein into the endoplasmic reticulum, and a second hydrophobic peptide at the C-terminus that directs addition of the GPI anchor (reviewed in Englund, 1993; Stevens, 1995; Udenfriend and Kodukula, 1995). N-terminal to this second hydrophobic peptide and separated from it by a hydrophilic

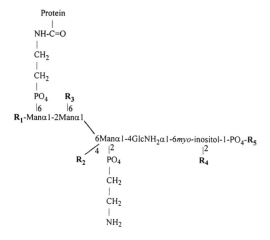

Fig. 2. Structures of mammalian GPI anchors. Various side-chain modifications of carbohydrate, including additional ethanolamine phosphate (EtNP), and/or palmitate (R_1–R_4) are found as indicated. In some of the proteins certain residues may only be present in a proportion of the GPI anchors (indicated by ±). OH indicates that no modification is thought to be present. R_5 = lipid moiety present. n.d. = not determined.

Protein	R_1	R_2	R_3	R_4	R_5
Rat brain Thy-1	±Manα1-2	±GalNAcβ1-4	OH	OH	Alkylacylglycerol
Hamster brain scrapie prion protein	±Man	±NANA±Hex-HexNAc	OH	OH	n.d.
Mouse skeletal muscle NCAM	±Manα1-2	±GalNAcβ1-4	n.d.	OH	n.d.
Human erythrocyte acetylcholinesterase	OH	OH	±EtNP	Palmitate	Alkylacylglycerol
Human kidney membrane dipeptidase	±Manα1-2	±Galβ1-3GalNAcβ1-4 ±GalNAcβ1-4	n.d.	OH	n.d.
Porcine kidney membrane dipeptidase	OH	±Galβ1-3GalNAcβ1-4 ±GalNAcβ1-4 ±NANA±GalNAcβ1-4	±EtNP	OH	Diacylglycerol
Bovine liver 5'-nucleotidase	±Man	±HexNAc	±EtNP	OH	n.d.

spacer region is the site of cleavage and anchor attachment. Cleavage of the polypeptide chain occurs on the C-terminal side of an internal amino acid residue (ω) lying in a particular consensus sequence ω, ω + 1, ω + 2, with the concomitant addition of a preformed GPI anchor to the newly exposed COOH group of the ω residue. Analysis of native GPI-anchored protein sequences and extensive site-directed mutagenesis (*see* Chapter 7) of these residues in alkaline phosphatase and decay-accelerating factor has shown that ω is restricted to the amino acids with small side chains (Ala, Gly, Ser, Asp, Asn, or Cys), whereas ω + 1 can be any residue except for Pro and Trp, and ω + 2 is Gly, Ala, or occasionally Ser.

The main function of a GPI moiety is to anchor the protein in the lipid bilayer, and in this respect, a GPI anchor can be considered as an alternative to a single membrane-spanning polypeptide anchor. However, a number of additional functions have been proposed for GPI anchors, which will be described here briefly. The reader is referred to a number of excellent reviews that cover these topics in more detail (Cross, 1990; Ferguson, 1991,1992b; McConville and Ferguson, 1993). First, cleavage of a GPI anchor by either a phospholipase C or D would allow for rapid down-regulation of the protein from the cell surface and/or for the generation of a soluble form that could act at sites distant from the original point of membrane attachment. In this respect, a soluble form of 5'-nucleotidase has been identified in bovine cerebral cortex, which appears to be derived from the membrane anchored form as a result of cleavage of the GPI anchor by a phospholipase C (Vogel et al., 1992). Second, a GPI anchor may take up less space in the membrane than a transmembrane polypeptide anchor, a factor that may be critically important for the dense packing of some 10 million GPI-anchored variant surface glycoproteins in the protective surface coat of trypanosomes, and for the dense clustering of GPI-anchored proteins in membrane microdomains. Third, owing to the lack of interaction of the anchor with cytoskeletal proteins, GPI-anchored proteins may generally be more laterally mobile in the membrane than transmembrane polypeptide-anchored proteins. Fourth, a GPI anchor has been shown to act as an intracellular sorting signal, directing the protein to the apical membrane of polarized epithelial cells (Rodriguez-Boulan and Powell, 1992) or to the axonal membrane in hippocampal neurons (Dotti et al., 1991). However, the cellular distribution of GPI-anchored proteins within neurons is complex and depends on cell type, differentiation state, and the protein itself (Faivre-Sarrailh and Rougon, 1993). Fifth, some

Table 1
Some Examples of GPI-Anchored Proteins in the Nervous System

Cell adhesion/axon growth	Enzymes
and guidance	Ecto-5'-nucleotidase
N-CAM 120	Acetylcholinesterase
Thy-1	(insects, *Torpedo*)
F3/F11 (Contactin)	Alkaline phosphatase
TAG-1 (Axonin-1)	
BIG-1	Receptors
Fasciclin I	CNTF Receptor (α-subunit)
	Folate receptor (choroid plexus)
Other	
Prion protein	
CD59	

of the actions of insulin and a number of other hormones, growth factors, and cytokines appear to be mediated by inositol phosphoglycans that are chemically, structurally, and immunologically related to, and may be derived from, GPI anchors on cell surface proteins (Romero and Larner, 1993; Saltiel, 1994). Sixth, studies on the folate receptor indicate that certain GPI-anchored proteins may cluster in caveolae and be involved in the novel endocytic process of potocytosis (Rothberg et al., 1990; Anderson, 1993). Finally, GPI-anchored proteins, such as Thy-1, may be involved in transmembrane signaling through interaction with nonreceptor protein-tyrosine kinases (Brown, 1993).

4. GPI-Anchored Proteins in the Nervous System

A comprehensive list of GPI-anchored proteins can be found in Low (1989). Examples with particular relevance to the nervous system are listed in Table 1. Of GPI-anchored enzymes in the mammalian nervous system, ecto-5'-nucleotidase (EC 3.1.3.5) is the best characterized (Zimmermann, 1992). Its functional role is in the hydrolysis of extracellular AMP, which represents an intermediate step in the turnover of purine nucleotides, specifically ATP, released as part of the neurotransmitter cycle. Ecto-5'-nucleotidase also binds to laminin and fibronectin, and can carry the cell-adhesion epitope L2/ HNK-1, which suggests it may play an additional role in cell adhesion or interactions with the extracellular matrix. The enzyme has been detected immunocytochemically on endothelial and glial cells in brain, as well as in some neurons, particularly during development (Maienschein and Zimmermann, 1996). Although some ectopeptidases are GPI-anchored, e.g., membrane dipeptidase (EC 3.4.13.19) and aminopeptidase P (EC 3.4.11.9), those that function in the inactivation of peptide neurotransmitters in the brain (E-24.11, ACE, aminopeptidase N) are all transmembrane-polypeptide-anchored, as is acetylcholinesterase (EC 3.1.1.7) in mammalian brain, although a GPI-anchored form of this enzyme is found in mammalian erythrocytes, insect heads, and *Torpedo* electric organ (Rosenberry et al., 1990). The functional significance of the distinct membrane-anchoring domains of acetylcholinesterases in different tissues is unknown.

Other GPI-anchored proteins of relevance to the nervous system occur in differently spliced forms possessing either a GPI or a transmembrane anchor. Of particular importance is the family of N-CAMs, which are expressed at various stages of neural differentiation, as well as in the adult. Three major forms of N-CAM are produced by alternative splicing: N-CAM 180 (M_r 180,000) and N-CAM 140 are anchored by a single hydrophobic transmembrane region and differ in the length of their cytoplasmic domains. N-CAM 120 is GPI-anchored (He et al., 1987). Differentiating nerve, glial, and muscle cells all express N-CAMs. The strength of cell–cell adhesion by N-CAMs during differentiation is modified by differential glycosylation, especially sialylation of the extracellular regions. As can be seen in Table 1, an increasing number of GPI-anchored molecules are being identified in the nervous system that appear to play a role in cell–cell adhesion and axonal growth and guidance (Walsh and Doherty, 1991). Many of them, including *Drosophila* fasciclin I, and the mammalian glycoproteins N-CAM, F3, TAG-1, and Thy-1, are members of the immunoglobulin gene superfamily. Thy-1 is a developmentally regulated cell surface glycoprotein expressed at high levels both in the brain and in the immune system (Morris, 1992). It has been well characterized structurally (Morris, 1985), but its functional role in developmental processes is less clear, although it has been suggested to interact with astrocytes to restrict axonal growth (Tiveron et al., 1992). In adult nervous tissue, it appears to be a surface component of most neurons with the major exception of primary olfactory neurons.

CNTF was discovered and purified through its ability to promote survival and neurite outgrowth of ciliary neurons. Subsequently, it has been shown to act as a survival factor for motor neurons, although it is unrelated to other neurotrophins, such as nerve growth factor (Stahl et al., 1994). CNTF is a member of a cytokine family that also includes interleukin-6, leukemia inhibitory factor, and oncostatin M. The CNTF receptor complex shares receptor components with other members of this cytokine family, but also possesses a CNTF-specific α-subunit, which is GPI-anchored, a surprising finding for a signal-transducing receptor (Davis et al., 1991). Whether CNTFα is released through a regulated cleavage of the GPI anchor is currently unknown.

Prion diseases exist in both genetic and infectious forms. The GPI-anchored prion protein is the transmissible agent causing scrapie in sheep, bovine spongiform encephalopathy in cows, and a number of neurodegenerative diseases in humans, e.g., Creutzfeld-Jakob and Gerstmann-Straüssler-Scheinker diseases (Prusiner, 1991; Stahl et al., 1992,1994). The prion protein is present in normal brain as a single membrane-spanning glycoprotein that is targeted to the cell surface of neurons via its GPI anchor. Conformational or structural changes in the prion protein lead to its conversion to the infective or "scrapie" isoform, which is protease-resistant and can aggregate to form amyloid plaques. The normal physiological function of the prion protein is unknown. Although the GPI anchor of the prion protein exists in several different glycoforms, some that are sialylated (Stahl et al., 1992), the glycolipid itself is not believed to be involved in the infectivity of prions.

5. Techniques for the Identification of GPI-Anchored Proteins

The most definitive means of identifying the presence of a GPI anchor on a protein is by direct chemical analysis of one or more of the anchor components, i.e., inositol, glucosamine, ethanolamine, or mannose. However, this requires both relatively large amounts of highly purified protein and sophisticated equipment (Ferguson, 1992a). A number of techniques, however, are available to enable one to identify the presence of a GPI anchor on a protein of interest that do not require purified protein and that use little material (*see* Hooper, 1992,1993 for further details). With any of the techniques described herein, it is advisable to include an appropriate GPI-anchored protein to act as a positive control for the method. Alkaline phosphatase is one of the easiest GPI-anchored proteins to assay, and is widely distributed in mammalian systems. A simple spectrophotometric assay for this enzyme using *p*-nitrophenyl phosphate as substrate is available (Hooper, 1992).

5.1. Phospholipase Release

The susceptibility of a protein to release from the membrane by bacterial phosphatidylinositol-specific phospholipase C (PI-PLC) is the most commonly used and simplest criterion for demonstrating the presence of a GPI anchor. PI-PLCs from a variety of sources are available, often in recombinant form, from a number of suppliers (Boehringer Mannheim [Mannheim, Germany], Calbiochem [La Jolla, CA], Pensinsula Laboratories [Belmont, CA], and Sigma [St. Louis, MO]). A relatively inexpensive source of PI-PLC that we use for large scale work is the commercial preparation of *Bacillus cereus* phospholipase C from Sigma (catalog no. P6135) or Fluka (catalog no. 79484), which contains a varying amount of a contaminating PI-PLC activity (Hooper, 1992). The standard protocol (Hooper et al., 1987) is to incubate membranes or cells in the presence of increasing amounts of bacterial PI-PLC, and then monitor the dose-dependent loss of the protein from the membrane and/or its appearance in the supernatant following centrifugation (Fig. 3). Alternatively, temperature-induced phase separation in Triton X-114 (Bordier, 1981) can be used to distinguish between the uncleaved amphipathic form and the cleaved hydrophilic form of a GPI-anchored protein (Hooper, 1992). A combination of PI-PLC digestion and phase separation in Triton X-114 has been widely used to identify GPI-anchored proteins in membrane extracts (Lisanti et al., 1988). When the membrane preparation is subjected to phase separation in Triton X-114, integral membrane proteins are recovered in the detergent-rich phase, and peripheral membrane proteins in the aqueous phase. The detergent-rich phase is then incubated with bacterial PI-PLC and subjected to a second phase separation. The GPI-anchored proteins are then recovered in the second aqueous phase, whereas transmembrane polypeptide-anchored proteins remain in the detergent-rich phase.

The results obtained using even the most highly purified bacterial PI-PLCs must be interpreted with some caution. If prolonged incubations are used, there is the possibility that phospholipase or protease activities in the sample preparation could nonspecifically release the protein of interest. Using a fixed incubation time and varying amounts of bacterial PI-PLC can usually guard against this, since a dose-dependent release should be observed for GPI-anchored proteins (*see* Fig. 3). Misleading results can also arise if the protein of interest is not itself GPI-anchored, but is tightly associated with a protein that is GPI-anchored and therefore appears to be susceptible to release by PI-PLC. An example is lipoprotein lipase, which associates with the GPI-anchored heparan sulfate proteoglycan (Chajek-Shaul et al., 1989). Also, the lack of release by PI-PLC does not necessarily rule out the presence of a GPI anchor on a protein, since several GPI-anchored proteins are now known to be resistant to the action of PI-PLC owing to direct acylation of the inositol ring (Ferguson, 1992c; Wong and Low, 1992), for example, human erythrocyte acetylcholinesterase. Prior removal of this additional acyl group by treatment with mild base (1.0M hydroxy-

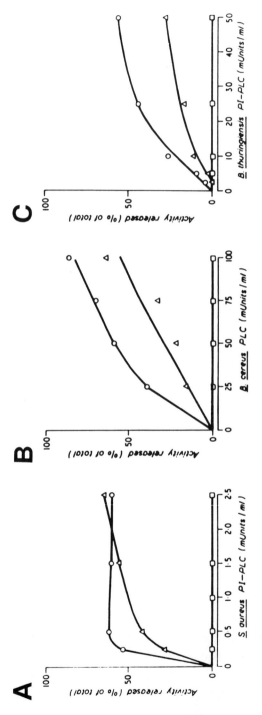

Fig. 3. Release of proteins from membranes by bacterial PI-PLC. Porcine kidney microsomal membranes were incubated at 37°C with (A) *Staphylococcus aureus* PI-PLC for 30 min; (B) *Bacillus cereus* phospholipase C (Sigma catalog no. P6135) for 2 h; (C) *Bacillus thuringiensis* PI-PLC for 30 min. Enzyme activities were determined in the total incubation mixture after treatment with PI-PLC, but before centrifugation, and in the supernatant after centrifugation at 31,000g for 90 min, and the released activity expressed as a percentage of the total activity in the original incubation mixture. GPI-anchored proteins: ○, alkaline phosphatase; △, membrane dipeptidase. Transmembrane polypeptide anchored proteins: □, endopeptidase-24.11 and aminopeptidase N (from Hooper et al. [1987] with permission of Portland Press, London).

lamine, 0.1M triethylamine, pH 10.0–12.0 for 2–48 h at 4°C or 50 mM NaOH in 20% [v/v] 1-propanol for 60 min at room temperature) then renders the protein susceptible to cleavage by PI-PLC (Guther et al., 1994). It should also be noted that some GPI-anchored proteins differ in their susceptibility to release by PI-PLC depending on the source of the phospholipase or the species or cell type under investigation.

5.2. Identification
of the Crossreacting Determinant (CRD)

Polyclonal antisera raised to the PI-PLC cleaved form of one GPI-anchored protein often crossreact with other unrelated GPI-anchored proteins. The CRD involved in this recognition is cryptic in the membrane-bound or amphipathic form of a GPI-anchored protein, and is only exposed when the protein is converted into a hydrophilic form on cleavage of its GPI anchor by the action of a phospholipase C (Fig. 4A,B). Several anti-CRD antisera are available, the best characterized of which are those generated against the trypanosome variant surface glycoprotein, porcine or human membrane dipeptidase, or porcine aminopeptidase P (Zamze et al., 1988; Hooper et al., 1991; Broomfield and Hooper, 1993). The major epitope recognized by these anti-CRD antisera is inositol-1,2-cyclic monophosphate, which is generated by phospholipase C cleavage of the anchor (Fig. 4B). Recognition of a protein, following phospholipase C cleavage, by an anti-CRD antiserum either by Western blot analysis or enzyme-linked immunoradiometric assay (Hooper et al., 1991) is good, and often confirmatory, evidence for the presence of a GPI anchor. As with all these procedures, it is advisable to carry out the appropriate controls. Treatment of the protein prior to analysis with 1M HCl for 30 min at room temperature selectively decyclizes the inositol-1,2-cyclic phosphate ring (Fig. 4C), whereas treatment with nitrous acid (fresh 0.25M NaNO$_2$, 0.25M Na acetate, pH 4.0, for 3 h at room temperature) deaminates the glucosamine residue, hydrolyzes the glucosamine–inositol bond (Fig. 4D), and thereby abolishes the crossreactivity.

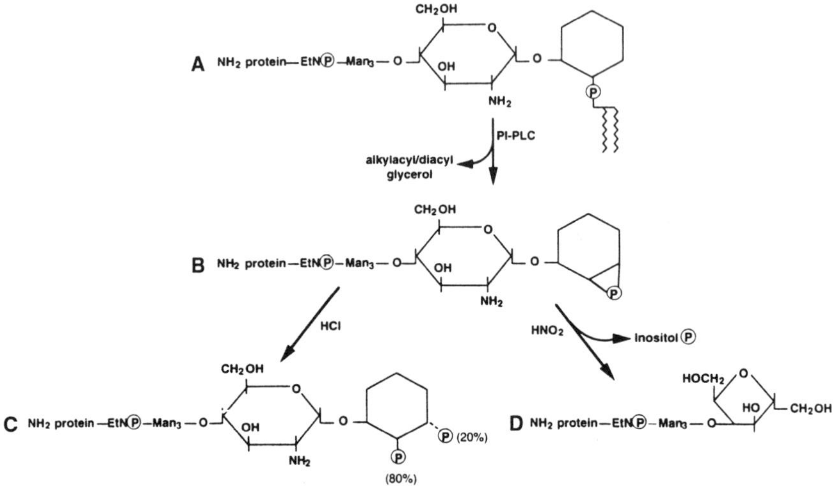

Fig. 4. Effects of 1M HCl and nitrous acid on the CRD. Cleavage of a
GPI anchor **(A)** with bacterial PI-PLC leads to the formation of inositol-
1,2-cyclic phosphate, the CRD **(B)**, which is the major epitope involved
in the crossreactivity of GPI-anchored proteins with anti-CRD antisera.
Treatment of this structure with 1M HCl results in the formation of 80%
inositol 1-phosphate and 20% inositol 2-phosphate **(C)**. Treatment with
nitrous acid deaminates the glucosamine residue converting it into 2,5-
anhydromannose and releases the inositol-1,2-cyclic phosphate. Both (C)
and **(D)** are unreactive with anti-CRD antisera.

5.3. Detergent Solubilization
and Phase Separation in Triton X-114

On examining the ability of a range of detergents to solubilize GPI-anchored proteins from porcine kidney microvillar membranes, we observed that those detergents with a high
critical micellar concentration, such as *n*-octyl-β-D-gluco-
pyranoside and 3-([3-cholamidopropyl] dimethylammonio)-
1-propane-sulfonate (CHAPS), solubilized substantial (>50%)
amounts of such proteins (Hooper and Turner, 1988a,b). In
contrast, those detergents with a low critical micellar concentration, such as Triton X-100, Triton X-114, and Nonidet
P-40, were relatively ineffective at releasing substantial
amounts of GPI-anchored proteins. Those proteins with a
single membrane-spanning polypeptide anchor were solu-

bilized efficiently by all of the detergents examined. Thus, the pattern of detergent solubilization of a protein, in particular the ratio of amount solubilized by *n*-octyl-β-D-glucopyranoside to that solubilized by Triton X-100, can be used to distinguish between a GPI and a polypeptide membrane anchor.

Based on the observation that those detergents with a low critical micellar concentration are relatively ineffective at solubilizing GPI-anchored proteins, we modified a technique of differential solubilization and temperature-induced phase separation in Triton X-114 to distinguish between those proteins anchored by a GPI moiety and those anchored by a single membrane-spanning polypeptide (Hooper and Bashir, 1991). Following incubation of a membrane sample with 2% (v/v) Triton X-114 at 0°C, GPI-anchored proteins can be isolated by low speed centrifugation in the detergent-insoluble pellet (Fig. 5). The supernatant can then be further fractionated by phase separation at 30°C into a detergent-rich phase containing transmembrane polypeptide-anchored proteins and an aqueous phase containing peripheral membrane proteins, or the hydrophilic forms of GPI-anchored proteins following phospholipase cleavage. One advantage of this method is that it can be used to identify a GPI anchor on those proteins that are resistant to phospholipase C cleavage owing to acylation of the inositol ring (Hooper and Bashir, 1991).

5.4. Metabolic Labeling Studies

If the protein of interest is expressed at a sufficiently high level in a cultured cell line, the cells can be metabolically labeled with components of the GPI anchor structure, e.g., [3H] or [14C]ethanolamine, [3H]inositol, [3H]mannose, [3H]glucosamine, [3H] or [14C]fatty acids (Masterson and Magee, 1992). The protein of interest is then isolated by immunoprecipitation and analyzed by SDS-PAGE followed by fluorography. The location of the fatty acid label can be confirmed by digestion with PI-PLC or cleavage with nitrous acid, and the location of the inositol label by nitrous acid treatment. However, there is no selective degradation method for con-

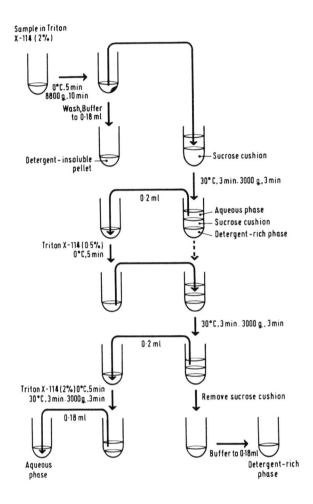

Fig. 5. Differential solubilization and temperature-induced phase separation in Triton X-114. GPI-anchored proteins are recovered in the detergent-insoluble pellet, transmembrane polypeptide-anchored proteins in the detergent-rich phase, and peripheral membrane proteins in the aqueous phase. Treatment of the membrane sample with bacterial PI-PLC prior to mixing with Triton X-114 will result in the cleaved GPI-anchored protein being recovered in the aqueous phase. If the first centrifugation at 8800g for 10 min is omitted, the GPI-anchored proteins will also be recovered in the detergent-rich phase (from Hooper and Bashir [1991] with permission of Portland Press, London).

firming the location of the ethanolamine, glucosamine, or mannose labels. Care also has to be taken to ensure that the labeled compounds are not metabolized to another structure prior to incorporation into the protein.

6. Membrane Microdomains and Caveolae

In 1988 we observed that a number of GPI-anchored ecto-enzymes were resistant to solubilization by certain detergents, such as Triton X-100 (Hooper and Turner, 1988a). This detergent insolubility is acquired as the proteins pass through the endoplasmic reticulum and onto the Golgi apparatus (Brown and Rose, 1992), and appears to be an intrinsic property of the association of the lipids in the GPI anchor with sphingolipids and cholesterol in membrane microdomains (Schroeder et al., 1994; Hanada et al., 1995). This sequestration of GPI-anchored proteins in specialized membrane microdomains may be a unifying property that accounts for functions as diverse as intracellular targeting, transmembrane signaling, and potocytosis (*see* Section 3.).

In certain types of mammalian cell, detergent-insoluble microdomains may exist in the plasma membrane as 50–100 nm flask-shaped invaginations called "caveolae" that are characterized by their high content of the 21–24 kDa protein, caveolin/VIP21 (Rothberg et al., 1992). Caveolin is an integral membrane protein of 178 amino acids, which forms an unusual bilayer hairpin loop with both C- and N-terminals oriented on the cytoplasmic face of the membrane (Kurzchalia et al., 1994). The protein has been identified as a major v-Src protein-tyrosine/kinase substrate in Rous sarcoma virus-transformed chick embryo fibroblasts. Both cell transformation and tyrosine phosphorylation of caveolin are dependent on v-Src membrane attachment, which suggests that caveolin may represent a substrate for cellular transformation. Caveolae themselves have been implicated in a number of cellular events, including signal transduction (Schnitzer et al., 1995c), transcytosis (Simionescu, 1983), potocytosis (Anderson, 1993), efflux of cellular free cholesterol (Fielding and

Fielding, 1995), and interaction with the actin-based cytoskeleton (Chang et al., 1994; Lisanti et al., 1994).

6.1. Isolation of Detergent-Insoluble Membrane Microdomains

Since glycosphingolipids are capable of forming more hydrogen bonds than other lipids and, consequently, tend to self-associate, Golgi rafts and plasmalemmal microdomains are characteristically insoluble in cold, nonionic detergents, such as Triton X-100 (Brown and Rose, 1992). Exploitation of this property has led to the development of a simple technique for the isolation of such microdomains in which whole-cell or membrane suspensions are solubilized in cold 1–2% Triton X-100 and then adjusted to 40% (w/v) sucrose by the addition of an equal volume of 80% sucrose. A 5–30% linear sucrose gradient is formed above the sample and the tubes centrifuged at 140,000g for approx 18 h. Detergent-insoluble microdomains form an opaque band in the 10–20% sucrose region (Lisanti et al., 1994; Parkin et al., 1996).

Owing to morphological similarities and the high levels of caveolin in both structures, it has been assumed that detergent-insoluble microdomains isolated by the method described above represent caveolae-enriched fractions (Chang et al., 1994; Lisanti et al., 1994). Although this assumption may be intrinsically correct, it now seems that caveolae cannot be purified on the basis of detergent insolubility alone. Membrane microdomains isolated as such may also contain detergent-resistant domains from organelles other than the plasmalemma (e.g., Golgi apparatus) (Kurzchalia et al., 1995). Such contamination may be avoided to some extent by the preliminary purification of plasma membranes (Chang et al., 1994). However, it has also been demonstrated that, even in the plasma membrane, microdomains exist that lack caveolin and the characteristic flask-shaped morphology of caveolae, but nonetheless, are resistant to detergent solubilization (Schnitzer et al., 1995b).

The problem of contamination by noncaveolar detergent-insoluble plasmalemmal microdomains has been solved by the use of negatively charged colloidal silica to coat epithelial cell apical membranes *in situ* (Schnitzer et al., 1995a,b,c). Following homogenization, these membrane regions can be readily isolated by centrifugation through Nycodenz (5-[*N*-2,3-dihydroxypropylacetamido]-2,4,6-*N*,*N*'-bis[2,3-dihydroxypropyl]isophthalamide; Sigma) density gradients because of their artificially induced high density. Electron microscopy has shown that silica particles are excluded from caveolae, which remain attached to the apical plasma membrane, whereas detergent-insoluble microdomains lacking caveolar morphology are effectively coated. The former membrane microdomains may subsequently be purified to homogeneity by detergent-solubilization followed by buoyant sucrose density gradient centrifugation.

6.2. Protein Components of Detergent-Insoluble Membrane Microdomains

In recent years, a large number of GPI-anchored membrane proteins have been localized to detergent-insoluble membrane microdomains, including the folate receptor (Rothberg et al., 1990), urokinase-type plasminogen activator receptor (Stahl and Mueller, 1995), the differentiation antigen Thy-1 and the scrapie prion protein (Ying et al., 1992), and a transferrin-like iron-binding protein (Danielsen and van Deurs, 1995). However, despite these observations, the inference that such proteins are clustered in caveolae is controversial. In many earlier studies, the localization of GPI-anchored proteins was assessed by indirect immunofluorescence using labeled polyclonal secondary antibodies. However, subsequent studies (Mayor et al., 1994) demonstrated that such proteins only cluster in caveolae under conditions of secondary antibody crosslinking, and that when primary antibodies conjugated to suitable labels are employed, GPI-anchored proteins are shown to be more diffusely distributed over the cell surface.

GPI-anchored proteins in detergent-insoluble membrane microdomains are physically associated with Src-like protein-tyrosine kinases, such that they are able to transduce signals from the extracellular space to the cytoplasm by tyrosine phosphorylation (Brown, 1993). CD36, another known detergent-insoluble microdomain component, has been proposed to function as a bridge between exocytic GPI-anchored proteins and Src-like kinases, which associate with the inner bilayer leaflet. In terms of specific proteins, the Src-like kinases, c-Yes, c-Src, Lyn, Fyn, Lck, c-Fgr, and Jak-2, have all been identified in detergent-insoluble membrane complexes (Lisanti et al., 1994).

Several G-protein-coupled receptors and G-protein-modifying bacterial toxins are also localized or internalized in caveolae, including the β-adrenergic receptor, muscarinic acetylcholine receptor, and cholera toxin. Consequently, it has been suggested that these membrane microdomains may be involved in G-protein-coupled signaling events. Schnitzer et al. (1995a) have shown that caveolae purified to homogeneity by the silica-coating method (Section 6.1.) contain a variety of G-protein subunits and monomeric and trimeric GTPases. Caveolae prepared in this way also contain proteins responsible for the mediation of vesicle formation, docking, and fusion (Schnitzer et al., 1995a). These proteins include vesicle-associated membrane protein-2 (VAMP-2), N-ethylmaleimide-sensitive fusion factor (NSF), soluble NSF attachment protein (SNAP), and annexins II and VI.

In addition, both the plasmalemmal Ca^{2+}-ATPase and the inositol-1,4,5-trisphosphate (IP_3) receptor have been localized to caveolae by *in situ* immunolabeling procedures (Fujimoto, 1993; Fujimoto et al., 1992) and by immunoblotting of homogenous caveolae preparations (Schnitzer et al., 1995c). The colocalization of these proteins may be important in regulating the inositol phospholipid signaling pathway. Here, IP_3 and 1,2-diacyglycerol are formed by the hydrolysis of inositol phospholipids following receptor activation by external ligands. IP_3 and/or its metabolites open the IP_3 receptor channel inducing cellular Ca^{2+} influx (Irvine, 1990). Conversely, increased levels of IP_3 are known to inhibit Ca^{2+}-ATPase activity (Davis et al., 1991). In the light of these observations, it is interesting to note that the annexin family of Ca^{2+}-dependent phospholipid-binding proteins are also located in caveolae and detergent-insoluble membrane fractions (Parkin et al., 1996).

6.3. Lipid Analysis

Investigations into the lipid composition of detergent-insoluble membrane microdomains and caveolae have been few in relation to those regarding the protein components of these structures. As previously discussed, the high glyco-sphingolipid content of such microdomains renders them insoluble in cold, nonionic detergents. Depletion of membrane sphingolipids by inhibition of enzymes involved in their synthesis abolishes the polarized sorting of GPI-anchored proteins and increases their detergent solubility (Futerman, 1995). Similarly, cholesterol is also a preferred lipid component of detergent-insoluble membrane microdomains and is essential for caveolar formation (Schnitzer et al., 1994; Hanada et al., 1995). Caveolin itself has been implicated as a cholesterol-binding protein in synthetic liposomes (Murata et al., 1995), a property that gives rise to the formation of caveolin oligomers.

Several methods are available for the extraction of lipids from membrane suspensions. If sufficient starting material is available, it may be lyophilized and extracted directly with chloroform-methanol (1:1, v/v). Alternatively, lipids may be extracted from wet samples by the addition of hot (70°C) methanol followed by heating for 30 min at the same temperature. Following cooling, 4 mL of chloroform and 1 mL of water are added and the sample is vortexed. After settling, a biphasic system forms and the upper phase is discarded while the lower (lipid-containing) phase is washed three times with 1-mL aliquots of water.

Neutral and acidic lipids are routinely separated on DEAE-Sephadex (A-25) columns following adjustment of samples to a 30:60:8 (v/v) mixture of chloroform-methanol-water. Neutral lipids are eluted in the same solvent mixture and then acidic lipids are eluted in chloroform-methanol-0.8M aqueous sodium acetate (30:60:8, v/v). Neutral glycolipids may be isolated by base hydrolysis of the neutral lipid fraction and subsequent passage through a second Sephadex column (eluted with chloroform-methanol-water, 30:60:8, v/v). The concentrated eluate is then loaded onto a silicic acid column and, following elution of fatty acids and choles- terol with chloroform, neutral glycolipids are eluted in chlo- roform-methanol (80:20, v/v). Gangliosides are isolated from the neutral lipid fraction by elution from a silicic acid col- umn in chloroform-methanol (85:15, v/v) after previously being subjected to base hydrolysis and desalting (e.g., on a Sep-Pak C18-reverse-phase cartridge).

Individual neutral and acidic lipid species may be sepa- rated by thin-layer chromatography on silica gel plates (G- 60; 0.2-mm layer thickness; Merck, Rahway, NJ) in identical solvent systems. Plates are first developed in chloroform- methanol-acetic acid-formic acid-water (35:15:6:2:1, v/v) such that the solvent front rises to a level halfway up the plate. After thorough drying, the plates are then developed fully in a second solvent system consisting of hexane-diisopropyl ether-acetic acid (65:35:2, v/v). Neutral glycolipid species can also be resolved in the latter solvent system, whereas gan- gliosides are most effectively separated in chloroform-metha- nol-0.25% (w/v) KCl (50:45:10, v/v). The reader is referred to the following works for more detailed analytical methods: Macala et al. (1983), Kates (1986), Brown and Rose (1992).

6.4. Glycolipid-Enriched Membrane Microdomains

Axonal extension requires the presence of proteins, such as L1 or N-CAM, which act both as receptors and substrate for growing axons (Doherty et al., 1989; Lemmon et al., 1989). As discussed above (Section 4.), some of these immunoglo- bulin-like glycoproteins possess GPI membrane anchors, including N-CAM, as well as BIG-1 (Yoshihara et al., 1994), rat TAG-1 (Furley et al., 1990), and mouse F3 (Gennarini et al., 1989a,b). It is thought that the process of neurite exten- sion begins with a recognition event in which a ligand expressed on the surface of a neighboring cell or in the extra- cellular matrix is recognized by a receptor protein (probably

GPI-anchored) in the membrane of the growing axon. Subsequent transmembrane signaling events would then be necessary in order to induce a cellular response. In support of this hypothesis, it is known that certain GPI-anchored proteins are physically associated with Src-like kinases (Section 6.2.), although owing to the external membrane orientation of GPI-anchored proteins, such as F3 and the inner bilayer leaflet topology of Src-like kinases, it is apparent that a "transducing" protein would be necessary in order to evoke a cellular response to external ligands. Both GPI-anchored proteins and Src-like kinases (e.g., Lyn, Fyn, Lck, and c-Yes) have been identified in detergent-insoluble membrane microdomains from a number of tissues, such as mouse lung (Lisanti et al., 1994). However, to date, only a single report exists regarding the isolation of such structures from nervous tissue (Olive et al., 1995). In this instance, two populations of F3-containing membrane microdomains were isolated following Triton X-100 solubilization and subsequent sucrose density gradient centrifugation. These fractions were also found to contain the Fyn protein-tyrosine kinase. Consequently, it was proposed that the assembly of these proteins in a complex could be the basis of F3-triggered events occurring in neurons.

7. Conclusions

Over the last decade, the number of membrane proteins recognized as possessing a GPI membrane anchor has increased dramatically. The complete structures of the GPI anchors of several of these proteins have now been solved, and the methods for such structural analyses are becoming increasingly routine and automated. Functional roles for GPI-anchored proteins are less clear-cut. Nevertheless, their very diversity and versatility as cell-interaction and signaling molecules have important ramifications for the control of axonal fasciculation, growth, and guidance. The methodologies outlined in this chapter should be of particular value to those with interests in these aspects of nervous system development.

Acknowledgments

We thank the Wellcome Trust for support of our work on GPI-anchored proteins.

References

Anderson, R. G. W. (1993) Potocytosis of small molecules and ions by caveolae. *Trends Cell Biol.* **3,** 69–72.

Bordier, C. (1981) Phase separation of integral membrane proteins in Triton X-114 solution. *J. Biol. Chem.* **256,** 1604–1607.

Brewis, I. A., Ferguson, M. A. J., Mehlert, A., Turner, A. J., and Hooper, N. M. (1995) Structures of the glycosyl-phosphatidylinositol anchors of porcine and human renal membrane dipeptidase. Comprehensive structural studies on the porcine anchor and interspecies comparison of the glycan core structures. *J. Biol. Chem.* **270,** 22,946–22,956.

Broomfield, S. J. and Hooper, N. M. (1993) Characterization of an antibody to the crossreacting determinant of the glycosyl-phosphatidylinositol anchor of human membrane dipeptidase. *Biochim. Biophys. Acta* **1145,** 212–218.

Brown, D. (1993) The tyrosine kinase connection: how GPI-anchored proteins activate T cells. *Curr. Opinion Immunol.* **5,** 349–354.

Brown, D. A. and Rose, J. K. (1992) Sorting of GPI-anchored proteins to glycolipid-enriched membrane subdomains during transport to the apical cell surface. *Cell* **68,** 533–544.

Carlsson, S. R. (1993) Isolation and characterization of glycoproteins, in *Glycobiology: A Practical Approach* (Fukuda, M. and Kobata, A., eds.), Oxford University Press, Oxford, pp. 1–26.

Chajek-Shaul, T., Halimi, O., Ben-Naim, M., Stein, O., and Stein, Y. (1989) Phosphatidylinositol-specific phospholipase C releases lipoprotein lipase from the heparin releasable pool in rat heart cell cultures. *Biochim. Biophys. Acta* **1014,** 178–183.

Chang, W.-J. Y., Ying, Y., Rothberg, K. G., Hooper, N. M., Turner, A. J., Gambliel, H. A., de Gunzburg, J., Mumby, S. M., Gilman, A. G., and Anderson, R. G. W. (1994) Purification and characterization of smooth-muscle call caveolae. *J. Cell Biol.* **126,** 127–138.

Chaplin, M. F. and Kennedy, J. F. (eds.) (1994) *Carbohydrate Analysis: A Practical Approach.* Oxford University Press, Oxford.

Cross, G. A. M. (1990) Glycolipid anchoring of plasma membrane proteins. *Ann. Rev. Cell Biol.* **6,** 1–39.

Danielsen, E. M. and van Deurs, B. (1995) A transferrin-like GPI-linked iron-binding-protein in detergent-insoluble noncaveolar microdomains at the apical surface of fetal intestinal epithelial cells. *J. Cell Biol.,* **131,** 939–950.

Danielsen, E. M., Cowell, G. M., Norén, O., and Sjöström, H. (1987) Biosynthesis, in *Mammalian Ectoenzymes* (Kenny, A. J. and Turner, A. J., eds.), Elsevier, Amsterdam.

Davis, F. B., Davis, P. J., Lawrence, W. D., and Blas, S. D. (1991) Specific inositol phosphates inhibit basal and calmodulin-stimulated Ca^{2+}-ATPase activity in human erythrocyte membranes in vitro and inhibit binding of calmodulin to membranes. *FASEB J.* **5**, 2992–2995.

Davis, S., Aldrich, T. H., Valenzuela, D. M., Wong, V. V., Furth, M. E., Squinto, S. P., and Yancopoulos, G. D. (1991) The receptor for ciliary neurotrophic receptor. *Science* **253**, 59–63.

Deeg, M. A., Humphrey, D. R., Yang, S. H., Ferguson, T. R., Reinhold, V. N., and Rosenberry, T. L. (1992) Glycan components in the glycoinositol phospholipid anchor of human erythrocyte acetylcholinesterase. Novel fragments produced by trifluoroacetic acid. *J. Biol. Chem.* **267**, 18,573–18,580.

Doherty, P., Barton, C., Dickson, G., Seaton, P., Rowett, L., Moore, S., Gower, H., and Walsh, F. (1989) Neuronal process outgrowth of human sensory neurons on monolayers of cells transfected with cDNAs for five NCAM isoforms. *J. Cell Biol.* **109**, 789–798.

Dotti, C. G., Parton, R. G., and Simons, K. (1991) Polarized sorting of glypiated proteins in hippocampal neurons. *Nature* **349**, 158–161.

Edge, A. S. B., Faltynek, C. R., Hof, L., Reichert, L. E., and Weber, P. (1981) Deglycosylation of glycoproteins by trifluoromethanesulfonic acid. *Anal. Biochem.* **118**, 131–137.

Englund, P. T. (1993) The structure and biosynthesis of glycosyl phosphatidylinositol protein anchors. *Ann. Rev. Biochem.* **62**, 121–138.

Faivre-Sarrailh, C. and Rougon, G. (1993) Are the glypiated adhesion molecules preferentially targeted to the axonal compartment? *Mol. Neurobiol.* **7**, 49–60.

Ferguson, M. A. J. (1991) Lipid anchors on membrane proteins. *Curr. Opinion Struct. Biol.* **1**, 522–529.

Ferguson, M. A. J. (1992a) Chemical and enzymatic analysis of glycosylphosphatidylinositol anchors, in *Lipid Modification of Proteins: A Practical Approach* (Hooper, N. M. and Turner, A. J., eds.), IRL, Oxford, pp. 191–230.

Ferguson, M. A. J. (1992b) Glycosyl-phosphatidylinositol membrane anchors: the tale of a tail. *Biochem. Soc. Trans.* **20**, 243–256.

Ferguson, M. A. J. (1992c) Site of palmitoylation of a phospholipase C-resistant glycosyl-phosphatidylinositol membrane anchor. *Biochem. J.* **284**, 297–300.

Fielding, P. E. and Fielding, C. J. (1995) Plasma membrane caveolae mediate the efflux of cellular free cholesterol. *Biochemistry* **34**, 14,288–14,292.

Fujimoto, T. (1993) Calcium pump of the plasma membrane is localized in caveolae. *J. Cell Biol.* **120**, 1147–1157.

Fujimoto, T., Nakade, S., Miyawaki, A., Mikoshiba, K., and Ogawa, K. (1992) Localization of inositol 1,4,5-trisphosphate receptor-like protein in plasmalemmal caveolae. *J. Cell Biol.* **119**, 1507–1513.

Fukuda, M. and Kobata, A. (eds.) (1993) *Glycobiology: A Practical Approach.* Oxford University Press, Oxford.

Furley, A., Morton, J. S., Manalo, B. D., Karagogeos, D., Dodd, J., and Jessell, T. (1990) The axonal glycoprotein TAG1 is an immunoglobulin superfamily member with neurite outgrowth promoting activity. *Cell* **61**, 157–170.

Futerman, A. H. (1995) Inhibition of sphingolipid synthesis: effects on glycosphingolipid—GPI-anchored protein microdomains. *Trends Cell Biol.* **5**, 377–379.

Gennarini, G., Cibelli, G., Rougon, G., Mattei, M., and Goridis, C. (1989a) The mouse neuronal cell surface protein F3: a phosphatidylinositol-anchored member of the immunoglobulin superfamily related to the chicken contactin. *J. Cell Biol.* **109**, 775–788.

Gennarini, G., Rougon, G., Vitiello, F., Corsi, P., Di Beneditta, C., and Goridis, C. (1989b) Identification and cDNA cloning of a new member of the L2/HNK-1 family of neural surface glycoproteins. *J. Neurosci. Res.* **22**, 1–12.

Guther, M. L. S., Cardoso de Almeida, M. L., Rosenberry, T. L., and Ferguson, M. A. J. (1994) The detection of phospholipase-resistant and -sensitive glycosyl-phosphatidylinositol membrane anchors by western blotting. *Anal. Biochem.* **219**, 249–255.

Hanada, K., Nishijima, M., Akamatsu, Y., and Pagano, R. E. (1995) Both sphingolipids and cholesterol participate in the detergent insolubility of alkaline phosphatase, a glycosylphosphatidylinositol-anchored protein, in mammalian membranes. *J. Biol. Chem.* **270**, 6254–6260.

He, H. T., Finne, J., and Goridis, C. (1987) Biosynthesis, membrane association, and release of N-CAM-120, a phosphatidylinositol-linked form of the neural cell adhesion molecule. *J. Cell Biol.* **105**, 2489–2500.

Homans, S. W., Ferguson, M. A. J., Dwek, R. A., Rademacher, T. W., Anand, R., and Williams, A. F. (1988) Complete structure of the glycosyl phosphatidylinositol membrane anchor of rat brain Thy-1 glycoprotein. *Nature* **333**, 269–272.

Homans, S. W., Edge, C. J., Ferguson, M. A. J., Dwek, R. A., and Rademacher, T. W. (1989) Solution structure of the glycosyl-phosphatidylinositol membrane anchor glycan of *Trypanosoma brucei* variant surface glycoprotein. *Biochemistry* **28**, 2881–2887.

Hooper, N. M. (1992) Identification of a glycosyl-phosphatidylinositol anchor on membrane proteins, in *Lipid Modification of Proteins: A Practical Approach* (Hooper, N. M. and Turner, A. J., eds.), IRL, Oxford, pp. 89–115.

Hooper, N. M. (1993) Determination of mammalian membrane protein anchorage: glycosyl-phosphatidylinositol (G-PI) or transmembrane polypeptide anchor. *Biochem. Ed.* **21**, 212–216.

Hooper, N. M. and Bashir, A. (1991) Glycosyl-phosphatidylinositol-anchored membrane proteins can be distinguished from transmembrane polypeptide-anchored proteins by differential solubilization and temperature-induced phase separation in Triton X-114. *Biochem. J.* **280**, 745–751.

Hooper, N. M. and Turner, A. J. (1987) Isolation of two differentially glycosylated forms of peptidyl-dipeptidase A (angiotensin converting enzyme) from pig brain: a re-evaluation of its role in neuropeptide metabolism. *Biochem. J.* **241**, 625–633.

Hooper, N. M. and Turner, A. J. (1988a) Ectoenzymes of the kidney microvillar membrane. Differential solubilization by detergents can predict a glycosyl-phosphatidylinositol membrane anchor. *Biochem. J.* **250**, 865–869.

Hooper, N. M. and Turner, A. J. (1988b) Ectoenzymes of the kidney microvillar membrane. Aminopeptidase P is anchored by a glycosyl-phosphatidylinositol moiety. *FEBS Lett.* **229**, 340–344.

Hooper, N. M. and Turner, A. J. (eds.) (1992) *Lipid Modification of Proteins: A Practical Approach.* Oxford University Press, Oxford.

Hooper, N. M., Low, M. G., and Turner, A. J. (1987) Renal dipeptidase is one of the membrane proteins released by phosphatidylinositol-specific phospholipase C. *Biochem. J.* **244**, 465–469.

Hooper, N. M., Broomfield, S. J., and Turner, A. J. (1991) Characterization of antibodies to the glycosyl-phosphatidylinositol membrane anchors of mammalian proteins. *Biochem. J.* **273**, 301–306.

Irvine, R. F. (1990) "Quantal" Ca^{2+} release and the control of Ca^{2+} entry by inositol phosphates—a possible mechanism. *FEBS Lett.* **263**, 5–9.

Kates, M. (1986) Techniques of lipidology, in *Laboratory Techniques in Biochemistry and Molecular Biology* (Burdon, R. H. and Knippenberg, P. H., eds.), Elsevier, Amsterdam, pp. 1–464.

Kurzchalia, T. V., Dupree, P., and Monier, S. (1994) VIP21-caveolin, a protein of the trans-Golgi-network and caveolae. *FEBS Lett.* **346**, 88–91.

Kurzchalia, T. V., Hartmann, E., and Dupree, P. (1995) Guilt by insolubility—does a protein's detergent insolubility reflect a caveolar localization? *Trends Cell Biol.* **5**, 187–189.

Lemmon, U., Farr, K., and Lagenaur, T. (1989) L1 mediated axon outgrowth occurs via a homophilic binding mechanism. *Neuron* **2**, 1597–1603.

Lisanti, M. P., Sargiacomo, M., Graeve, L., Saltiel, A. R., and Rodriguez-Boulan, E. (1988) Polarized apical distribution of glycosyl-phosphatidylinositol-anchored proteins in a renal epithelial cell line. *Proc. Natl. Acad. Sci. USA* **85**, 9557–9561.

424 *Turner, Parkin, and Hooper*

Lisanti, M. P., Scherer, P. E., Vidugiriene, J., Tang, Z., Hermanowski-Vosatka, A., Tu, Y., Cook, R. F., and Sargiacomo, M. (1994) Characterization of caveolin-rich membrane domains from an endothelial-rich source—implications for human disease. *J. Cell Biol.* **126,** 111–126.

Low, M. G. (1989) The glycosyl-phosphatidylinositol anchor of membrane proteins. *Biochim. Biophys. Acta* **988,** 427–454.

Macala, L. J., Yu, R. K., and Ando, S. (1983) Analysis of brain lipids by high performance thin-layer chromatography and densiometry. *J. Lipid Res.* **24,** 1243–1250.

Maienschein, V. and Zimmermann, H. (1996) Immunocytochemical localization of ecto-5′-nucleotidase in cultures of cerebellar granule cells. *Neuroscience* **70,** 429–438.

Masterson, W. J. and Magee, A. I. (1992) Lipid modifications involved in protein targeting, in *Protein Targeting: A Practical Approach* (Magee, A. I. and Wileman, T., eds.), IRL, Oxford, pp. 233–259.

Mayor, S., Rothberg, K. G., and Maxfield, F. R. (1994) Sequestration of GPI-anchored proteins in caveolae triggered by crosslinking. *Science* **264,** 1948–1951.

McConville, M. J. and Ferguson, M. A. J. (1993) The structure, biosynthesis and function of glycosylated phosphatidylinositols in the parasitic protozoa and higher eukaryotes. *Biochem. J.* **294,** 305–324.

Montreuil, J., Bouquelet, S., Debray, H., Lemoine, J., Michalski, J.-C., Spik, G., and Strecker, G. (1994) Glycoproteins, in *Carbohydrate Analysis: A Practical Approach* (Chaplin, M. F. and Kennedy, J. F., eds.), Oxford University Press, Oxford, pp. 181–293.

Morris, R. (1985) Thy-1 in developing nervous tissue. *Dev. Neurosci.* **7,** 133–160.

Morris, R. (1992) Thy-1, the enigmatic extrovert on the neuronal surface. *BioEssays* **14,** 715–722.

Mukasa, R., Umeda, M., Endo, T., Kobata, A., and Inoue, K. (1995) Characterization of glycosylphosphatidylinositol (GPI)-anchored NCAM on mouse skeletal cell line C2C12: the structure of the GPI glycan and release during myogenesis. *Arch. Biochem. Biophys.* **318,** 182–190.

Murata, M., Peränen, J., Schreiner, R., Wieland, F., Kurzchalia, T. V., and Simons, K. (1995) VIP21/caveolin is a cholesterol-binding-protein. *Proc. Natl. Acad. Sci. USA* **92,** 10,339–10,343.

Olive, S., Dubois, C., Schachner, M., and Rougon, G. (1995) The F3 neuronal glycosylphosphatidylinositol-linked molecule is localized to glycolipid-enriched membrane subdomains and interacts with L1 and Fyn kinase in cerebellum. *J. Neurochem.* **65,** 2307–2317.

Parkin, E. T., Turner, A. J., and Hooper, N. M. (1996) Calcium resolves two distinct low-density, Triton-insoluble membrane domains from porcine lung. *Biochem. J.* **319,** in press.

Prusiner, S. (1991) Molecular biology of prion diseases. *Science* **252,** 1515–1522.

Relton, J. M., Gee, N. S., Matsas, R., Turner, A. J., and Kenny, A. J. (1983) Purification of endopeptidase-24.11 ("enkephalinase") from pig brain by immunoadsorbent chromatography. *Biochem. J.* **215,** 519–523.

Rodriguez-Boulan, E. and Powell, S. K. (1992) Polarity of epithelial and neuronal cells. *Ann. Rev. Cell Biol.* **8,** 395–427.

Romero, G. and Larner, J. (1993) Insulin mediators and the mechanism of insulin action. *Adv. Pharmacol.* **24,** 21–50.

Rosenberry, T., Roberts, W. L., Haas, R., and Toutant, J.-P. (1990) The glycoinositol phospholipid anchor of human erythrocyte acetyl-cholinesterase, in *Molecular and Cell Biology of Membrane Proteins: Glycolipid Anchors of Cell Surface Proteins* (Turner, A. J., ed.), Ellis Horwood, Chichester, UK, pp. 150–165.

Rothberg, K. G., Ying, Y., Kolhouse, J. F., Kamen, B. A., and Anderson, R. G. W. (1990) The glycophospholipid-linked folate receptor intern-alizes folate without entering the clathrin-coated pit endocytic path-way. *J. Cell Biol.* **110,** 637–649.

Rothberg, K. G., Heuser, J. E., Donzell, W. G., Ying, Y. S., Glenney, J., and Anderson, R. G. W. (1992) Caveolin, a protein component of caveolae membrane coats. *Cell* **68,** 673–682.

Saltiel, A. R. (1994) The paradoxical regulation of protein phosphoryla-tion in insulin action. *FASEB J.* **8,** 1034–1040.

Schnitzer, J. E., Oh, P., Pinney, E., and Allord, J. (1994) Filipin-sensitive caveolae-mediated transport in endothelium: reduced transcytosis, scavenger endocytosis, and capillary permeability of select macro-molecules. *J. Cell Biol.* **127,** 1217–1232.

Schnitzer, J. E., Liu, J., and Oh, P. (1995a) Endothelial caveolae have the molecular transport machinery for vesicle budding docking and fusion, including VAMP, NSF, SNAP, annexins, and GTPases. *J. Biol. Chem.* **274,** 14,399–14,404.

Schnitzer, J. E., McIntosh, D. P., Dvork, A. M., Liu, J., and Oh, P. (1995b) Separation of caveolae from associated microdomains of GPI-anchored proteins. *Science* **269,** 1435–1438.

Schnitzer, J. E., Oh, P., Jacobson, B. S., and Dvork, A. M. (1995c) Caveolae from luminal plasmalemma of rat lung endothelium: microdomains enriched in caveolin, Ca^{2+}-ATPase, and inositol trisphosphate recep-tor. *Proc. Natl. Acad. Sci. USA* **92,** 1759–1763.

Schroeder, R., London, E., and Brown, D. (1994) Interactions between saturated acyl chains confer detergent resistance on lipids and glycosylphosphatidylinositol (GPI)-anchored proteins: GPI-anchored proteins in liposomes and cells show similar behaviour. *Proc. Natl. Acad. Sci. USA* **91,** 12,130–12,134.

Simionescu, N. (1983) Cellular aspects of transcapillary exchange. *Physiol. Rev.* **63,** 1536–1560.

Stahl, A. and Mueller, B. M. (1995) The urokinase-type plasminogen activator receptor, a GPI-linked protein, is localized in caveolae. *J. Cell Biol.* **129,** 335–344.

Stahl, N., Baldwin, M. A., Hecker, R., Pan, K.-M., Burlingame, A. L., and Prusiner, S. B. (1992) Glycosylinositol phospholipid anchors of the scrapie and cellular prion proteins contain sialic acid. *Biochemistry* **31,** 5043–5053.

Stahl, N., Boulton, T. G., Ip, N., Davis, S., and Yancopoulos, G. D. (1994) The tails of two proteins: the scrapie prion protein and the ciliary neurotrophic factor receptor. *Brazil. J. Med. Biol. Res.* **27,** 297–301.

Stevens, V. L. (1995) Biosynthesis of glycosylphosphatidylinositol anchors. *Biochem. J.* **310,** 361–370.

Taguchi, R., Hamakawa, N., Harada-Nishida, M., Fukui, T., Nojima, K., and Ikezawa, H. (1994) Microheterogeneity in glycosylphosphatidylinositol anchor structures of bovine liver 5'-nucleotidase. *Biochemistry* **33,** 1017–1022.

Tiveron, M. C., Barboni, E., Pliego Rivero, B., Gormley, A. M., Seeley, P. J., Grosveld, F., and Morris, R. (1992) Selective inhibition of neurite outgrowth on mature astrocytes by Thy-1 glycoprotein. *Nature* **355,** 745–748.

Turner, A. J. (1990) (ed.) *Molecular and Cell Biology of Membrane Proteins: Glycolipid Anchors of Cell Surface Proteins.* Ellis Horwood, Chichester, UK.

Turner, A. J. (1994) PIG-tailed membrane proteins. *Essays Biochem.* **28,** 113–127.

Udenfriend, S. and Kodukula, K. (1995) How glycosyl-phosphatidylinositol-anchored membrane proteins are made. *Ann. Rev. Biochem.* **64,** 563–591.

Vogel, M., Kowalewski, H., Zimmermann, H., Hooper, N. M., and Turner, A. J. (1992) Soluble low-K_m 5'-nucleotidase from electric ray *(Torpedo marmorata)* electric organ and bovine cerebral cortex is derived from the glycosyl-phosphatidylinositol-anchored ectoenzyme by phospholipase C cleavage. *Biochem. J.* **284,** 621–624.

Walsh, F. S. and Doherty, P. (1991) Glycosylphosphatidylinositol anchored recognition molecules that function in axonal fasciculation, growth and guidance in the nervous system. *Cell Biol. Int. Rep.* **15,** 1151–1166.

Williams, T. A., Hooper, N. M., and Turner, A. J. (1991) Characterization of neuronal and endothelial forms of angiotensin converting enzyme in pig brain. *J. Neurochem.* **57,** 193–199.

Wong, Y. W. and Low, M. G. (1992) Phospholipase resistance of the glycosyl-phosphatidylinositol membrane anchor on human alkaline phoshatase. *Clin. Chem.* **38,** 2517–2525.

Ying, Y.-S., Anderson, R. G. W., and Rothberg, K. G. (1992) Each caveola contains multiple glycosylphosphatidylinositol anchored membrane proteins. *Cold Spring Harbour Symp.* **57,** 593–604.

Yoshihara, Y., Kawasaki, M., Tani, A., Tamada, A., Nagata, S., Kagamiyama, H., and Mori, K. (1994) BIG-1: a new Tag-1/F3 related member of the Ig superfamily with neurite outgrowth promoting activity. *Neuron* **13,** 415–426.

Zamze, S. E., Ferguson, M. A. J., Collins, R., Dwek, R. A., and Rademacher, T. W. (1988) Characterization of the crossreacting determinant (CRD) of the glycosyl-phosphatidylinositol membrane anchor of *Trypanosoma brucei* variant surface glycoprotein. *Eur. J. Biochem.* **176,** 527–534.

Zimmermann, H. (1992) 5'-nucleotidase—molecular structure and functional aspects. *Biochem. J.* **285,** 345–365.

Index